D0151645

The Viscosity of the Earth's Mantle

The Viscosity
of the Earth's Mantle

LAWRENCE M. CATHLES, III

PRINCETON UNIVERSITY PRESS
PRINCETON · NEW JERSEY

Copyright © 1975 by Princeton University Press
Published by Princeton University Press
Princeton and London

All Rights Reserved

Library of Congress Cataloging in Publication Data
will be found on the last printed page of this book

Composed in Linofilm Times Roman by Science Press
and printed in the UNITED STATES OF AMERICA by
Princeton University Press at Princeton, New Jersey

to MARY HELEN

ACKNOWLEDGMENTS

I should like to thank many who made this work possible. Dr. R. H. Dicke introduced me to the subject of glacial uplift and encouraged my pursuit of graduate study. Conversations with him since, although few in number, have been extremely useful to the development of many of the ideas in this book. Dr. Jason Morgan advised in the preparation of the Ph.D. thesis that forms the basis of this work. I am indebted to Dr. Morgan, Dr. A. Dahlen, and Dr. R. A. Phinney for many useful discussions. I owe much to an excellent course in continuum mechanics given by Dr. C. Eringen. A set of lecture notes prepared by Dr. F. Gilbert provided a valuable introduction to Runge-Kutta techniques. The geological analysis owes much to conversations with Dr. Dave Fullerton, Dr. R. Walcott, Dr. A. Bloom, Dr. J. T. Andrews, Dr. E. Dorf, and Dr. S. Judson. Without timely encouragement by Dr. K. Deffeyes, this book would never have been published.

The Geophysical Fluid Dynamics Laboratory provided about seven hours of free computing time. Jim Welch of that Laboratory gave valuable programming assistance in computer graphics. The Aetna Life Insurance Co. provided help in typing. Kennecott Copper Corporation gave assistance and support, as my present employer, and financed the drafting of the numerous illustrations at their Ledgemont Laboratory.

CONTENTS

ILLUSTRATIONS

Appendix Figures

TABLES

The Viscosity of the Earth's Mantle

ONE

Introduction

AT THE CLOSE of the last ice age, climatic changes caused vast loads to shift on the earth's surface. Starting roughly 12,000 years ago, the three kilometers of ice that had covered Canada, the last European glaciers in Fennoscandia and Siberia, as well as other minor glaciers, melted quickly and dumped their meltwaters into the oceans, increasing the average depth of the world's oceans by about 110 meters. The earth's response to this load redistribution was one of fluid flow. By studying the way in which that flow occurred, we can learn much about the long term ($\sim$ 1000 yrs) rheological properties of the earth's mantle and how they vary with depth.

The purpose of this book is to lay the theoretical foundations necessary to model the isostatic (fluid) adjustment of a self-gravitating viscoelastic sphere such as the earth, and to use these foundations, together with geological evidence of the way the earth responded to the Pleistocene load redistributions, to learn about the viscosity of the mantle. The theory is developed in Chapters II and III. In Chapter IV the earth is analyzed in sections, beginning with the lithosphere and ending with flow in the deep mantle. Relevant geologic data are reviewed for each section, and the theory of Chapters II and III is drawn on as needed. An attempt is made to review past literature that bears on the discussion. The last section in Chapter IV summarizes the conclusions reached, and the reasoning by which they were reached, and can be used as a key to the selective reading of Chapter IV. Chapters II and III may be skipped by readers more interested in the conclusions than in the detailed theory that justifies them.

The method of geological analysis is to use data most relevant to deducing the rheological character at a given mantle depth from any geographic locality in which it may be found. Larger scale phenomena potentially probe deeper. Thus, for example, isostatic adjustment data from Utah, Greenland, and the Arctic Archipelago are primarily used to infer the viscosity structure of the uppermost few hundred kilometers of the mantle. Data from Scandinavia are primarily used to determine the viscous properties of the mantle above about 1000 km depth. The

Canadian uplift data and the adjustment of the ocean basins are primarily used to probe the viscous character of the lower mantle. It is assumed that the results from one geographic locality apply to the earth as a whole. In some cases (such as the asthenosphere, which we call here an uppermost mantle low viscosity channel) this is almost certainly not the case. I have tried to warn the reader where geographic variations can be expected. This method should be distinguished from the alternative method of attempting to analyze the viscosity structure of large sections of the mantle from one particular locality (such as Fennoscandia).

An attempt is also made to focus on particular geologic phenomena (i.e., uplift phenomena that can be observed through study of the geomorphology) that can distinguish one mantle structure from another (i.e., channel versus deep flow). This method could be contrasted to more general methods of numerical inversion that have been used in some cases with results similar to those found here (Parsons, 1972).

The method of geophysical analysis is to *assume* that the earth can be modeled as a self-gravitating viscoelastic (Maxwell) solid and to develop the mathematical techniques necessary to predict the isostatic uplift that would occur following a load redistribution similar to the post-Wisconsin one. In this modeling, explicit account is taken of the gradual nature of the load redistribution and of its global character—i.e., that the ocean basins were loaded as the continents were unloaded. The elastic as well as viscous response is accounted for, and gravitational effects are included. Elastic properties and density profiles are taken to be those of Haddon and Bullen (1961). The consequences of non-adiabatic density gradients in the mantle (i.e., density gradients that affect fluid flow) are also allowed for and studied. The model uplift for various models is compared to the uplift observed to determine the extent to which a particular earth model is similar to the earth.

Assumption of a Newtonian mantle viscosity is useful. The load cycle behavior of earth models that have high viscosity lower mantles, non-Newtonian mantle viscosity, or substantial non-adiabatic density gradients are similar to the load cycle behavior of a channel flow model (layered Newtonian viscosity). In the Newtonian channel flow case, peripheral bulges are produced upon loading and troughs upon unloading. This is also true for non-Newtonian models, since strain rate is anomalously large in high stress areas near the edges of the load. Therefore peripheral bulges develop upon loading and troughs upon unloading. Non-adiabatic density gradients restrict flow to a channel after a short time, so channel bulges develop. Thus, computing the single case of a layered Newtonian viscosity permits identification of channel phenomena that could be an indication of a non-linear flow law, non-adiabatic density gradients, or channel flow. The unique characteristics of deep flow in a uniform viscosity Newtonian mantle can also be identified.

The results of the modeling indicate that the mantle is Newtonian and has a quite uniform 10^{22} poise viscosity throughout. The viscosity of the lower mantle may be slightly lower ($.9 \pm .2 \times 10^{22}$ poise) than that of the upper mantle ($1.0 \pm .1 \times 10^{22}$ poise). The viscosity of the mantle is determined remarkably accurately by available uplift data.

The most diagnostic single bit of geological evidence is the uplift history of the east coast of North America. It shows that late glacial and early post-glacial uplift was followed by more recent subsidence. This behavior is permitted only by deep flow in a Newtonian mantle of quite uniform viscosity. With this start, analysis of the Canadian uplift as a whole then permits the viscosity of the lower mantle to be accurately determined.

A 10^{22} poise lower mantle viscosity has fundamental consequences for the material properties of the mantle, mantle convection, polar wandering, and perhaps the chemical evolution of the earth. Some of the consequences for mantle convection are touched on briefly. The implications of solid state theory on mantle viscosity have been discussed elsewhere (see Gordon, 1965, 1967, 1971; Weertman, 1970; Stocker and Ashby, 1973). This book concentrates entirely on deducing the flow laws suggested by isostatic adjustment. We hope that what is observed can be squared with solid state theory.

We are not the first to suggest a lower mantle viscosity of $\sim 10^{22}$ poise. Daly (1934) suggested deep flow was induced by the Pleistocene load redistribution. He reasoned that, if flow were confined to an upper mantle channel of any sort, glaciation should have produced large peripheral bulges. Upon deglaciation, these bulges would collapse. Because he could find no geological evidence of such bulges, Daly proposed a "down punching" hypothesis. An elastic crust forced deep flow and eliminated the bulges. (Daly, 1934, p. 120ff). Soon afterwards Haskell (1937) commented that an elastic crust (or lithosphere) was not necessary to induce deep flow; only a reasonably Newtonian mantle of reasonably uniform viscosity was required.

Haskell's analysis showed only about 20 m of uplift should remain in Fennoscandia at present. Large negative gravity anomalies in that area (~ -30 mgal) led to the belief that a larger amount of uplift remained (~ 210 m). Channel flow models then came into favor because they seemed to provide for a large amount of remaining uplift

(Van Bemmelen and Berlage, 1935; Vening-Meinesz, 1937; Lliboutry, 1971; Walcott, 1972b). Here we show that, although a channel flow model will fit the Fennoscandian uplift data about as well as a deep flow model, channel flow models do not suggest substantial uplift remains at present. We now know the ice load was not applied for longer than ~ 20,000 years. If the load cycle is taken into account, substantial uplift in Fennoscandia is not indicated. Furthermore, detailed studies of gravity in Fennoscandia show that the anomalies correlate with lithological variations and not the present uplift (Honkasalo, 1964, 1966). We show that most anomalies in Fennoscandia are of short enough wavelength to be supported elastically by the lithosphere. The longer wavelength components of the anomaly suggest about 25 m of remaining uplift.

Although we now know better, it seemed for a time that gravity anomalies in Fennoscandia suggested a rigid mantle and only a thin fluid asthenosphere. The suggestion (also incorrect as it turned out) from gravity studies that the earth as a whole possessed an excess equatorial bulge that was markedly larger than its other deviations (see Table I-1) from fluid equilibrium, reinforced the notion of a highly viscous lower mantle.

Munk and MacDonald (1960a, 1960b, p. 281; MacDonald, 1966) suggested that the excess equatorial bulge, C_{20}, might be a fossil rotational bulge left over from past times when the earth was rotating more rapidly. MacDonald suggested a lower mantle viscosity of 7.9×10^{25} poise (MacDonald, 1966, p. 227). Wang (1966) suggested that the excess equatorial bulge might be a result of the Wisconsin glaciation, but McKenzie (1966b, App. 3) and O'Connell (1969, p. 229; 1971) showed that the magnitude of anomaly that could have been caused by the last glaciation was not large enough to account for the magnitude of the observed non-hydrostatic equatorial bulge. McKenzie (1966b, 1967b, 1968) showed that the bulge could not be produced by an inter-

C_{20}	-30 m
C_{22}	15.5 m
S_{22}	-8.9 m
C_{31}	12.4 m
S_{32}	8.2 m

Table I-1 Coefficients of the normalized spherical harmonic description of the earth's geoid (surface of constant gravitational potential). The data are from Kaula (1966). Legendre polynomials are normalized so the square of any harmonic integrated over the unit sphere equals 4π. The unlisted coefficients are all less than 5.7 m. A map of the earth's geoid shows non-hydrostatic bulges as large as 90 m. (See Kaula, 1966.)

action between convection and rotation (as Dicke had suggested might be possible) and concluded that a lower mantle with 2×10^{27} poise viscosity must underlie an upper mantle (above 1000 km depth) of 10^{22} poise to account for the excess rotational bulge. McConnell (1968b) restricted flow to above 800 km depth in modeling the Fennoscandia uplift in order to preserve the observed fossil bulge.

Then Goldreich and Toomre (1969) showed there, in fact, was not a fossil rotational bulge after all. The apparent dominance of the C_{20} coefficient was due to a mathematical bias of the spherical harmonic description. Spherical harmonies overemphasize the amplitude of equatorial anomalies by a factor of $\sqrt{3}$. If the earth's non-hydrostatic gravity field is expanded in spherical harmonies about a pole on the equator, the C_{20} coefficient is again just as large with respect to the other coefficients, but this time it describes anomalies wrapped around the rotational poles along some longitude circle. The earth's non-hydrostatic shape is in fact triaxial with no preference for the present pole of rotation (Goldreich and Toomre, 1969; O'Connell, 1969).

Goldreich and Toomre thus undermined the explicatory significance of the fossile bulge hypothesis, and it no longer provides any particular reason to believe that the viscosity of the lower mantle is high. More than that, however, they suggested that their analysis of polar wandering implied that the viscosity of the lower mantle is less than 6×10^{24} poise and certainly less than 6×10^{25} poise.

Dicke (1966, 1969) pointed out that the very precise data that eclipses provide of changes in the rate of rotation of the earth could be used to determine how rapidly the ocean basins adjusted to the influx of Pleistocene meltwater. Dicke found the eclipse data required a lower mantle viscosity of around 10^{22} poise and a P_2 harmonic decay time of between 860 and 1610 years. The main question in Dicke's analysis, as he points out, is the validity of the eclipse data, not its accuracy if valid. Dicke's determination of P_2 decay time is very close to our determination of ~ 1500 years (see Figure III-16 or IV-26; note the P_2 decay is not exponential).

O'Connell (1969, 1971) performed calculations on eclipse data similar to Dicke's. He pointed out the length of day data could be accounted for if the P_2 load harmonic had a decay time of 2000 or 100,000 years. The 100,000-year decay time could be discarded because of the lack of correlation between the present non-hydrostatic gravity field and the Pleistocene load redistribution. This required a P_2 relaxation time of less than 10,000 to 20,000 years. Eclipse data then require a $\approx 10^{22}$ poise lower mantle.

Bloom (1967) suggested that the rate of isostatic adjustment of the ocean basins relative to the rate of meltwater influx would reflect itself in differences between the range of sea levels recorded by oceanic islands, which act as dip sticks and record the meltwater influx, and the range of sea levels recorded by continents. If the oceans adjust rapidly compared to the influx, continental sea level range will be smaller than the island one; if the ocean basins adjust slowly, it will be larger. At present the sea level range evidence seems to weigh in favor of rapidly adjusting ocean basins. Both Bloom (1967) and Mörner (1969) have shown geologic evidence for hydro-isostatic adjustment (i.e., the adjustment of the ocean basins under the glacial meltwater load). Bloom (1970b) has correctly pointed out that this requires global consideration of the post-Wisconsin isostatic adjustment in order that mantle mass be preserved.

Wellman (1964), in a deceptively short paper, pointed out that, to avoid higher than present sea levels over the last few thousand years, a relaxation time for the ocean basins less than 2000 years is required. Evidence may suggest there was not a Holocene high sea level (see Section IV.A.7). If Holocene high sea levels are ruled out, a low viscosity ($\sim 10^{22}$ poise) lower mantle is required.

Thus we see that, particularly recently, many have speculated about a $\sim 10^{22}$ poise lower mantle. Many others continue to feel that a restriction of flow to the asthenosphere and a rigid or high viscosity lower mantle is more reasonable. The matter of lower mantle viscosity is actively debated at the present time.

The mathematical model we use follows techniques developed in the study of free oscillation of the earth. The work of Backus (1967) was found particularly useful. We use his Scaler and Tensor Representation Theorums and employ the notation he developed.

Viscous flow in a sphere has been treated by various workers. Darwin solved the viscous decay of a P_2 deformation in 1879. Niskannen (1939–1949) included the effect of an elastic crust on the adjustment of a viscous sphere, although in light of recent work by Brotchie and Silvester (1969) it appears he overemphasized its effect. Takeuchi and Hasegawa (1964) considered the problem again, and Anderson and O'Connell (1967) calculated relaxation times for several earth models employing Hasagawa's technique. McKenzie (1966) developed an approach to layered viscosity. None of the above considered the effects of self-gravitation.

O'Connell (1969, 1971) calculated the viscous decay of a layered sphere, each layer having constant η, ρ, with the densities and viscosities changing from layer to layer. He took account of perturbation of the gravity field by considering the redistribution of mass resulting from the deformation of the boundaries between different density layers. As we shall see, this is not a strictly correct approach, as under viscous flow the density of each "layer" will change as it is deformed to mimic exactly the density of the layer it displaces. This is so *unless* the density gradient is non-adiabatic (i.e., not due to elastic compression but due to compositional changes or phase changes). As a result, O'Connell overemphasized gravity results. O'Connell's results also differ from ours in that he assumed that the decay even of low order load harmonics was exponential. We find the decay of a given harmonic, under properly accounted for core-mantle boundary conditions, cannot be characterized by a single decay time.

Parsons (1972) considered the adjustment of a self-gravitating visco-elastic sphere without non-adiabatic density gradient and with uniform non-adiabatic gradients. He used Laplace transform techniques; we do not. His results are comparable to those presented here. The methods used, although not identical, are equivalent.

No one to my knowledge has considered in detail the effects of phase changes on isostatic adjustment to glacial loading. Several have considered phase changes from the point of view of the filling of sedimentary basins, notably MacDonald (1960a, b) and O'Connell and Wasserburg (1967). Others have considered phase changes from the viewpoint of mantle convection (Knopoff, 1964; Turcotte and Oxburgh, 1969; Turcotte and Schubert, 1971). We are indebted to their work in parts of our treatment.

Throughout this book, emphasis will be placed on physical derivations rather than mathematical ones because of a preference of the author for the former. An attempt is made at the start of each section, particularly in the first two chapters, to tell the reader where the section is going and why.

TWO

Physical and Mathematical Foundations

A. Equations of Motion of Self-gravitating Elastic and Self-gravitating Viscous Spheres

IN THIS SECTION we derive the equations of motion for simple elastic and simple viscous self-gravitating spheres. After a brief introduction to Lagrangian and Eulerian coordinates, we start by writing down the usual Lagrangian equation for conservation of momentum. We then identify a certain material element and its surface in Eulerian co-ordinates and convert the Lagrangian equation to a Eulerian one. This identification is valid only for a very short time interval δt about the time t_0 at which we desire a valid equation. That is, the Eulerian equation holds only in the interval

$$t_0 - \frac{\delta t}{2} < t < t_0 + \frac{\delta t}{2}$$

whereas, the original Lagrangian equation (expressing conservation of momentum) is valid at all t. After converting to the Eulerian description, we remain in that description. Boundary conditions always apply to spatially fixed boundaries, although some corrections are necessary to allow for the spatial deformation of the material surfaces.

We find that the equations of motion describing self-gravitating vis-cous and elastic spheres are different. Our elastic equations of motion agree with the free oscillation equations of Backus (1967) and others. The difference is discussed and seen to be an entirely reasonable and proper one. If gravitation is neglected, the difference disappears. It is shown in Section III.A.1.c. that the difference also disappears for small scale deformations. It is of importance to us only because we consider deformation of the whole earth.

1. Lagrangian and Eulerian Coordinates.

A continuous medium can be viewed in two equivalent ways. In one way it may be viewed as a dense collection of material particles, the identity of each of which is known by knowing its position at some particular time; for example, at $t = 0$. Suppose $\mathbf{X}$ is the initial location of a particle, and hence its "name." The velocity or any other property of a given particle will be a function of its identity and time. Then

$$v_L = v_L\,(\mathbf{X}, t)$$
$$T_L = T_L\,(\mathbf{X}, t)$$

This is known as the Lagrangian or Material description.

Equivalently we may seek the properties of the material at a given point in space at a given time. Clearly the spatial location of a particle $\mathbf{x}$ depends on the identity of the particle $\mathbf{X}$, and the time. In this system

$$v_E = v_E\,(\mathbf{x}(\mathbf{X}, t), t)$$
$$T_E = T_E\,(\mathbf{x}(\mathbf{X}, t), t)$$

This is known as the Eulerian or Spatial description.

The Lagrangian and Eulerian descriptions are equivalent at any initial instant of time. As soon as there is *motion*, they differ. In particular, the time derivative of a function T is:

$$\frac{d}{dt}\,[T_L(\mathbf{X}, t) = T_E(\mathbf{x}(\mathbf{X}, t), t)]$$

$$\frac{dT_L}{dt} = \frac{\partial T_E}{\partial t} + \frac{\partial T_E}{\partial x_i}\frac{\partial x_i}{\partial t}$$

The same is true of any quantity, which suggests the general operator

$$\frac{d}{dt} = \frac{\partial}{\partial t} + v \cdot \nabla \tag{II-1}$$

Physically, suppose we have a procession of particles moving past an observation point $\mathbf{x}$. Suppose each particle remains at a constant temperature, but each particle is hotter than its neighbor.

$$\nabla T \longrightarrow$$

$$\cdot \quad \cdot \quad \cdot \quad \cdot \quad \cdot \quad \boxed{\cdot} \quad \cdot \qquad \frac{dT}{dt} = 0.$$

$$v \quad \longrightarrow \mathbf{x}$$

The temperature at $\mathbf{x}$ will decrease with time. That decrease or advection will be

$$\frac{\partial T}{\partial t} = -\mathbf{v} \cdot \nabla T. \tag{II-2}$$

Any type of *motion* will suffice to give rise to this effect. For example, suppose the chain of particles suffers an instantaneous displacement. The temperature at $\mathbf{x}$ after the displacement $\mathbf{u}$ will equal the temperature at $\mathbf{x}$ before $\mathbf{u}$ minus $\mathbf{u} \cdot \nabla \mathrm{T}$. Mathematically $\frac{dT}{dt} = 0$, displacement $= \mathbf{u}H(t)$, where $H(t)$ is the Heaviside step function. $\mathbf{v}$ then equals $\mathbf{u}\delta(t)$, and we can integrate (II-2):

$$\int_0^{\delta t} \frac{\partial T}{\partial t} dt = \int_0^{\delta t} -\mathbf{u}\delta(t) \cdot \nabla T \, dt,$$

$$T - T_0 = -\mathbf{u} \cdot \nabla T. \tag{II-3}$$

2. Derivation of the Equation of Motion.

Let $\mathcal{U}$ be a material volume—i.e., a volume which may change in spatial magnitude but always contains the same material. Conservation of mass states:

$$\frac{d}{dt} \int_{\mathcal{U}} \rho \, d\mathcal{U} = 0.$$

Since $\frac{d}{dt}$ is the material derivative, $\mathcal{U}$ is a constant as far as it is concerned, and we can move it inside the integral. We can then switch to a spatial volume V instantaneously equal to $\mathcal{U}$

$$\int_V \frac{d}{dt} (\rho dV) = 0$$

$$\frac{d\rho}{dt} = \frac{\partial \rho}{\partial t} + \mathbf{v} \cdot \nabla \rho$$

$$\frac{d(dV)}{dt} = \nabla \cdot \mathbf{v} \, dV$$

(see Eringen, 1967, p. 72),

so

$$\int_V \left[\frac{\partial \rho}{\partial t} + \nabla \cdot (\rho \mathbf{v})\right] dV = 0.$$

Thus, we get the usual continuity equation in the Eulerian or Spatial system

$$\frac{\partial \rho}{\partial t} + \nabla \cdot (\rho \mathbf{v}) = 0. \tag{II-5}$$

Integrating over the spatial volume and using the divergence theorem, we see that this equation simply states that the rate of change of mass in a particular spatial volume must equal the flux of material into that volume. This would be another, perhaps easier, method of derivation.

If inertial forces vanish (they do for our purposes), the conservation of linear momentum requires that the body forces **F** acting on an element of the body be balanced by the stresses acting on the surface of that element. At the *instant* t in question, let the material volume coincide with a spatial volume V and a spatial surface S. Then

$$\int_V \mathbf{F}\, dV + \int_S \mathbf{T} \cdot d\mathbf{a} = 0. \tag{II-6}$$

Using the divergence theorem we get

$$\nabla \cdot \mathbf{T} + \mathbf{F} = 0. \tag{II-7}$$

Equation II-7 holds at any instant of time

$$t_0 - \frac{\delta t}{2} < t < t_0 + \frac{\delta t}{2}$$

where $\delta t \rightarrow 0$. It is an Eulerian Equation; **T** and **F** are Eulerian quantities.

If the material is self-gravitating or in a gravitational field, it will be hydrostatically pre-stressed. This pre-stressed initial condition causes no deformation of interest. It can be conveniently subtracted out. Strain due to new applied stresses (not due to gravitational pre-stress) will be related to a perturbation stress τ:

$$\tau \equiv \mathbf{T} + (p_0)_E \mathbf{I}. \tag{II-8}$$

τ is related to strain by the constitutive equation. "E" emphasizes that p_0, is Eulerian, as it must be to belong in (II-7).

The difference between the viscous and elastic equations of motion results from different expressions for $(p_0)_E$. Love (1911, p. 89) first pointed out that $(p_0)_E$ must be expressed in the elastic equations of motion differently than might have been at first supposed. Below we review his reasoning, put it in mathematical form, and show how it leads to a difference between the elastic and viscous equations of motion.

The initial zero order pressure of a material element does not change in δt (See Love, 1911, p. 89). We may apply stresses to the body and cause perturbation of the pressure of material elements within that body, but these perturbations are superimposed on the initial pressure, which remains unchanged. It would be unreasonable to suppose that the zero order pressure of a material element could change in an instant, spontaneously, for no reason. Thus:

$$\frac{d(p_0)_L}{dt}\bigg|_{\text{over } \delta t} = 0. \tag{II-9}$$

Until there is motion, both Eulerian and Lagrangian descriptions of pressure are equivalent.

$$(p_0)_E = (p_0)_L. \quad \text{(no motion)} \tag{II-10}$$

When movement occurs, (II-9) and (II-1) imply

$$\frac{\partial(p_0)_E}{\partial t} = -\mathbf{v} \cdot \nabla (p_0)_E.$$

Integrating over any small time interval δt, we see in the Eulerian system:

$$p_0\big|_{t_0+\delta t} = p_0\big|_{t_0} - \int_{t_0}^{t_0+\delta t} \mathbf{v} \cdot \nabla p_0 \, dt. \tag{II-11}$$

Unless $\mathbf{v} = (\text{const.}) \cdot \delta(t - t_0)$,

$$p_0\big|_{t_0+\delta t} = p_0\big|_{t_0}.$$

If, however, the material elements of the body suffer an instantaneous (neglecting inertia) elastic displacement of the form

$$\mathbf{u} = \mathbf{u}\,H(t - t_0),$$

then

$$\mathbf{v} = \mathbf{u}\,\delta(t - t_0),$$

and

$$p_0\big|_{t_0+\delta t} - p_0\big|_{t_0} = -\mathbf{u}\cdot\nabla p_0.$$

Thus, the condition for equilibrium *after* the elastic displacement is:

$$\nabla\cdot(\mathbf{T} = \tau - p_0\big|_{t_0+\delta t}\mathbf{I}) + \mathbf{F} = 0.$$

$$\boxed{\nabla\cdot\tau - \nabla p_0\big|_{t_0} + \nabla(\mathbf{u}\cdot\nabla p_0) + \mathbf{F} = 0}$$

(II-12)
Elastic Equation
of Motion

Viscous displacements, on the other hand, are always continuous functions of time. In the viscous case

$$\int_{t_0}^{t_0+\delta t} \mathbf{v}\cdot\nabla p_0\,dt \xrightarrow[\delta t\to 0]{} 0.$$

Consequently, the viscous equation of motion is:

$$\boxed{\nabla\cdot\tau - \nabla p_0 + \mathbf{F} = 0}$$

(II-13)
Viscous Equation
of Motion

Notice there is a distinct difference between the elastic and viscous equation of motion. One contains an advection term as a result of an initial instantaneous elastic displacement; the other (the viscous equation of motion) does not.[1]

---

[1]This derivation may warrant discussion. We calculate the deformation at any instant. Viscously, we calculate the instantaneous viscous velocity at all points in the fluid. Viscously there is never a sudden displacement. Only after some time is there any movement of the material elements. By the time we allow any motion to occur, the

The body force terms in Equations (II-12), (II-13), for a self-gravitating earth may be written:

$$\mathbf{F} = \rho\mathbf{g}, \tag{II-14}$$

$$\nabla\phi \equiv -\mathbf{g}, \tag{II-15}$$

$$\nabla^2\phi \equiv 4\pi G\rho. \tag{II-16}$$

Note that our definition (II-15) differs from some gravitational notation conventions but is consistent with the usual physical meaning of potential.

3. Linearization of the Equations of Motion

Assume a zero order state of hydrostatic equilibrium

$$\nabla p_0 - \mathbf{F}_0 = 0,$$

$$\nabla p_0 + \rho_0 \nabla \phi_0 = 0. \tag{II-17}$$

Further let us assume

$$\rho(\mathbf{x}, t) = \rho_0(x_1) + \rho_1(x_1, x_2, x_3, t), \tag{II-18}$$

$$\phi(\mathbf{x}, t) = \phi_0(x_1) + \phi_1(x_1, x_2, x_3, t), \tag{II-19}$$

$$\nabla^2\phi_0 = 4\pi G\rho_0. \tag{II-20}$$

---

analysis has long since finished. This is not true in the elastic case, where equilibrium must be established *after* an instantaneous displacement, which immediately follows the application of perturbing stresses. Forces must balance at the displaced position. This was first noted by Love (1911, p. 89), who initially left out the added term in (II-12) and later corrected his error. Heuristically, the added term enables material elements to carry their zero order pressure field with them as they undergo any instantaneous (elastic) displacement.

There are ways other than the one I have presented to obtain the conclusions of Love and Equations II-12 and II-13. One such method, shown me by Dr. C. Eringen, is given in Appendix I. It utilizes the full elegance of the formalism of modern continuum mechanics. In doing so, it perhaps obscures to some extent (to the non-continuum mechanicist) the assumptions made. For that reason I have retained the derivation as it originally occurred to me.

Another is presented by Backus (1967, Eq. 5.32, 5.46). He obtains Equations II-12 and II-13 by relating Lagrangian stress to strain in the elastic case but Eulerian stress to strain rate in the viscous case. Although his method leads to precisely the same results as ours, and although it is an approach which is undoubtedly more familiar to most geophysicists, it should be pointed out that continuum mechanicists currently regard Eulerian and Lagrangian stress as indistinguishable in first order (linear) theory. The notion that they were distinguishable is due to a certain confusion in classical elastic theory, Eringen (1967, pp. 108–110).

Plugging (II-17–II-20) into (II-5, II-12–II-16) and discarding terms of second order or higher, we find

$$\frac{\partial \rho_1}{\partial t} + \nabla \cdot (\rho_0 \mathbf{v}) = 0.$$

Integrating

$$\rho_1 + \nabla \cdot (\rho_0 \mathbf{u}) = 0 \tag{II-21}$$

$$\nabla \cdot \tau + \nabla(\mathbf{u} \cdot \nabla p_0) - \rho_0 \nabla \phi_1 - \rho_1 \nabla \phi_0 = 0 \tag{II-22 Elastic}$$

$$\nabla \cdot \tau - \rho_0 \nabla \phi_1 - \rho_1 \nabla \phi_0 = 0 \tag{II-23 Viscous}$$

$$\nabla^2 \phi_0 = 4\pi G \rho_0 \tag{II-24}$$

$$\nabla^2 \phi_1 = 4\pi G \rho_1 \tag{II-25}$$

$$\mathbf{g}_0 = -\nabla \phi_0 = -\partial_r \phi_0 \hat{\mathbf{r}} \equiv -g_0 \hat{\mathbf{r}} \tag{II-26}$$

$$\nabla p_0 + \rho_0 \nabla \phi_0 = 0$$

If we use (II-26) and (II-21), the field equation may be written in a more easily interpretable form:

$$\nabla \cdot \tau - \rho_0 \nabla(g_0 u_{x_1}) + \rho_0 g_0 (\nabla \cdot \mathbf{u})\hat{\mathbf{x}}_1 - \rho_0 \nabla \phi_1 = 0 \tag{II-27a Elastic}$$

$$\nabla \cdot \tau + \left(\rho_0 g_0 \nabla \cdot \mathbf{u} + u_{x_1} \frac{\partial \rho_0}{\partial x_1}\right)\hat{\mathbf{x}}_1 - r_0 \nabla \phi_1 = 0 \tag{II-27b Viscous}$$

where $x_1 = r$ or z.[2]

4. Boundary Conditions

If the material surface coincides with the Eulerian surface initially identified with it, boundary conditions for the Eulerian equations

[2]The elastic equation of motion is identical to the Free Oscillation Equation of Backus (1967) with $\omega = 0$. The derivation was influenced by his but differs in several respects.

$$\partial_r = \frac{\partial}{\partial r}.$$

would follow straight from (II-22), (II-23), (II-25) by the usual Gaussian pillbox technique. For example, integrate (II-25) over a small pillbox volume, apply the divergence theorem to the $\nabla \cdot \nabla \phi_1$ term, and let the volume shrink to zero by reducing the thickness of the pillbox to zero. The terms on the right go to zero as the volume goes to zero, unless a surface mass distribution σ (of zero thickness) has been applied.

In our case σ will be zero everywhere except perhaps at the free surface of the earth. In general

$$\nabla \phi_{1_{\text{Top}}} \cdot \hat{\mathbf{n}} - \nabla \phi_{1_{\text{Bottom}}} \cdot \hat{\mathbf{n}} = 4\pi G\sigma \tag{II-28}$$

must hold over all interfaces in the media.

Similar operation on (II-23) shows that the *viscous* boundary conditions require

$$\tau \cdot \hat{\mathbf{r}} \tag{II-29}$$

to be continuous across all interfaces in the media.

The elastic boundary conditions are complicated by the $\nabla(\mathbf{u} \cdot \nabla p_0)$ term. By (II-17) and (II-26)

$$\nabla(\mathbf{u} \cdot \nabla p_0) = -\nabla(U\rho_0 g_0),$$

where $U = \mathbf{u} \cdot \hat{\mathbf{r}}$. This term can contribute to the boundary conditions anywhere there is a discontinuity in ρ_0. Suppose (i.e., at the surface, $r = 1$) $\rho_0 = \rho_0 H(1 - r)$. Then

$$\int_{\text{pillbox}} - \nabla(U\rho_0 g_0) \cdot \mathbf{A} \, dr =$$

$$\int - U g_0 \rho_0 \delta(1 - r) dr = -\rho_0 g_0 U \quad \text{(per unit area)}.$$

Thus, for the elastic case

$$\tau \cdot \hat{\mathbf{r}} - U g_0 \rho_0 \hat{\mathbf{r}} \qquad\qquad\qquad\qquad \text{(II-30)}$$

must be continuous across all interfaces.

This would be the end of the matter if it were not for the fact our initial assumption is invalid. The material surface of the deformed body will not in general coincide with the Eulerian (spatial) surface initially identified with it. We must correct for any overlap of the material surface with respect to the assumed Eulerian surface if we are to deduce boundary conditions that are physically correct for use with the Eulerian equation of motion.

To first order accuracy it does not matter that the stress is applied to $S + U$ rather than to S. $\mathbf{u} \cdot \nabla \tau$ is a second order term, $\mathbf{u}$ and τ both being first order. We must, however, take into account the weight of the overlap material and its gravitational attraction for the earth—i.e., the interaction of the mass in the overlap of the material surface above or below the assumed geometrical surface of the earth (which is a first order term $\rho_0 \mathbf{u}$/unit area) with the zero order gravitational field of the earth, and the interaction of the first order gravitational field of the overlap material with the zero order mass of the earth. We must make adjustments anywhere boundary conditions are applied. In the case of the earth this means at the core-mantle boundary and at the free surface.

Remembering tension is positive by convention, the weight of material displaced U above the mathematical surface is

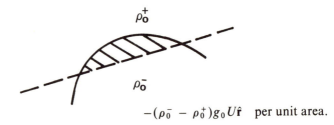

$$-(\rho_0^- - \rho_0^+) g_0 U \hat{\mathbf{r}} \quad \text{per unit area.}$$

This must be balanced by the stress. Thus, for the viscous case

$$\hat{\mathbf{r}} \cdot \tau^+ - \hat{\mathbf{r}} \cdot \tau^- - (\rho_0^- + \rho_0^+) g_0 U \hat{\mathbf{r}} = 0.$$

The viscous boundary conditions (II-29) thus become:

$$\boxed{\begin{array}{c} [\hat{\mathbf{r}} \cdot \tau + \rho_0 g_0 U \hat{\mathbf{r}}]_-^+ = 0 \\[4pt] \text{at } r = a,b \end{array}} \qquad\qquad \text{Viscous (II-31)}$$

where

a = radius of the core-mantle boundary and
b = radius of the surface of the earth.

The elastic boundary conditions become:

$$[\boldsymbol{\tau} \cdot \hat{\mathbf{r}} - Ug_0\rho_0\hat{\mathbf{r}}]^+_- + [Ug_0\rho_0]^+_- = 0$$

$$\boxed{\begin{array}{l} [\boldsymbol{\tau} \cdot \hat{\mathbf{r}}]^+_- = 0 \\[6pt] \text{at } r = a, b \end{array}}$$

Elastic (II-32)

The ϕ_1 boundary conditions adjust:

$$\nabla^2\phi_1 = 4\pi G(\rho_0^- - \rho_0^+)U = \partial_r\phi_1^+ - \partial_r\phi_1^-$$

Thus (II-28) becomes

$$\boxed{\begin{array}{l} [\partial_r\phi_1 + 4\pi G\rho_0 U]^+_- = 4\pi G\sigma \\[6pt] \text{at } r = a, b. \end{array}}$$

Viscous or Elastic (II-33)

a. Discussion

The difference between the viscous and elastic boundary conditions should cause no surprise if it is remembered that material elements carry their zero pressure field with them when they undergo elastic displacement. Thus, local displacement of the surface of an elastic half-space from its equilibrium position involves *no* buoyant forces.

NO BUOYANT FORCES

ELASTIC ½-SPACE

We may argue in a reductio ad absurdum vein: If elastic deformations did involve buoyant forces, elastic deformations greater than Archimedes limit would be ruled out. But a long bar of low rigidity, standing in a gravitational field could easily be deformed in excess of

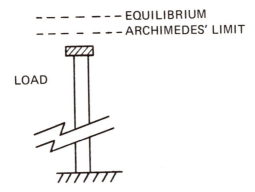

Archimedes limit by a given load. It only need be made long enough. Thus, elastic deformation cannot invoke buoyant forces.[3]

The situation would be quite different in a fluid half-space. A boat floats because it pierces the zero order hydrostatic pressure field of the ocean. At equilibrium the zero order pressure field is in no way disturbed or altered. The ocean is in a state of zero perturbation stress: $\tau = 0$.

b. Boundary Conditions for an Elastic Solid Overlying an Inviscid Fluid

The elastic upper strata will carry its zero order pressure field with it as it deforms. A fluid (inviscid) substratum will adjust immediately to any deformation of its boundary with the elastic upper stratum. It will oppose that deformation only with buoyant force. Those buoyant forces will be

$$\rho_{\text{fluid}} \, g_0 \, U,$$

where U = vertical deformation of the elastic-fluid boundary. Mathematically from (II-31), (II-32):

$$[\tau_e \cdot \hat{\mathbf{r}}]^+ = [\tau_\nu \cdot \hat{\mathbf{r}} + \rho_f g_0 U \hat{\mathbf{r}}]_- .$$

$\tau_\nu = 0$ in an inviscid fluid (always in equilibrium), so

$$\boxed{[\tau_e \cdot \hat{\mathbf{r}}]^+ = \rho_f g_0 U \hat{\mathbf{r}}}$$

(II-34)

Elastic Layer over Inviscid Fluid

[3]See Appendix I for other differences between solid and fluids.

c. Simplification of the Potential Boundary Conditions.

The potential boundary conditions (II-33)

$$[\partial_r \phi_1 + 4\pi G \rho_0 U]_-^+ \Big|_{\text{at } r=a,b} = 4\pi G \sigma$$

may be simplified at both $r = a$ and $r = b$, since above $r = a$ and below $r = b$, ϕ_1 satisfies Laplace's equation:

$$\nabla^2 \phi_1 = 0 \qquad\qquad\qquad\qquad\qquad (\text{II-35})$$

Clearly this is true above $r = a$ where there is no mass at all. That it is true below $r = b$ rests on the observation that flow in a (inviscid) fluid core caused by slow (inertia neglected) harmonic deformation of order greater than zero will be *isoporic*, that is, will involve no dilitation (even though the fluid may be compressible). Further, being inviscid, even if the core is density stratified, only the surface layer of the core will be redistributed to accommodate the deformation of the core-mantle boundary. Thus $(\rho_1)_{\text{Core}} = 0$ and (II-35) follows, even in the density stratified case.

(II-35) enables us to expand ϕ_1 in $\begin{array}{c} r \geq b \\ r \leq a \end{array}$ in a harmonic series:

$$\phi_1(r \geq b, x_2, x_3) = \sum_{l=0}^{\infty} \sum_{m=-l}^{l} \Phi_l^m(b)\left(\frac{b}{r}\right)^{l+1} Y_l^m(x_2, x_3), \qquad (\text{II-36})$$

$$\phi_1(r \leq a, x_2, x_3) = \sum_{l=0}^{\infty} \sum_{m=-l}^{l} \Phi_l^m(a)\left(\frac{r}{a}\right)^{l} Y_l^m(x_2, x_3). \qquad (\text{II-37})$$

The expansions are chosen so they are finite in the region of their validity. Φ_0^0 is assumed $= 0$. Differentiating, we see

$$\partial_r \phi_1^+(b) = -\frac{(l+1)}{b} \phi_1^+(b),$$

$$\partial_r \phi_1^-(a) = \frac{l}{a} \phi_1^-(a).$$

Remembering $\sigma = 0$ except possibly at $r = a$, *if* we define g_1:

$$g_1 \equiv 4\pi G \rho_0 U + \partial_r \phi_1 \tag{II-38}$$

(II-33) becomes

$$g_1^- = -\frac{(n+1)}{b}\phi_1 - 4\pi G\sigma \quad \text{at } r = b$$

$$g_1^+ = \frac{l}{a}\phi_1 + 4\pi G\rho_c U \quad \text{at } r = a \tag{II-39}$$

Alternatively *if* we make the physically more meaningful definition:

$$\hat{g}_1 \equiv \partial_r \phi_1 \tag{II-40}$$

(II-33) becomes

$$\hat{g}_1^- = -\frac{(n+1)}{b}\phi_1 - 4\pi G(\rho_0^- U + \sigma) \quad \text{at } r = b$$

$$\hat{g}_1^+ = 4\pi G(\rho_c - \rho_m)U + \frac{l}{a}\phi_1 \quad \text{at } r = a \tag{II-41}$$

It should be noted here that the pressure developed at the core-mantle boundary by the anomalous gravitational field of the deformed mantle and load may be conveniently expressed:

$$P_a = \rho_c \phi_1 \tag{II-42}$$

We might note also that

$$g_1 \equiv -\nabla\phi_1$$
$$\therefore [g_1]_r = -\partial_r\phi_1 \hat{r} = -\hat{g}_1 \hat{r} \tag{II-43}$$

Thus $\hat{g}_1$ is just the r component of the attractive $(-\hat{r})$ first order gravitational field intensity.

1) Discussion. The above derivation of the boundary conditions at the core-mantle boundary is a physical derivation. As such, it facilitates an understanding of what effects may be expected and which are reasonable. It is not the usual method for obtaining core-mantle boundary conditions. The usual method is to solve for the deformation of a homogeneous fluid sphere by direct power series methods. The homogeneous sphere one chooses may either be small (in which case the solution is propagated to the core-mantle boundary) or of a size equal to the core itself. Longman (1966) solves for elastic deformation coefficients the first way; Kaula (1963) and Takeuchi (1962) use the second approximation. We have verified that our boundary conditions are equivalent to Kaula's and Takeuchi's. They should also be equivalent to Longman's (See Appendix IV).

B. Constitutive Equations to Model the Earth

The physical properties of a medium are embodied solely in the constitutive relations. Here we derive the constitutive equations of a non-heat-conducting Maxwell body. This derivation could be made from several theoretical considerations (following Eringen) as was done in the derivation of the stress-strain relations for a Stokesian fluid in Appendix I. It is more instructive, however, to construct the constitutive relation of a viscoelastic solid from those of a Stokesian fluid and elastic solid.

1. Constitutive Relation of a Maxwell Body

We start with the constitutive relations for a simple elastic solid and a simple viscous fluid:

$$_e t_{kl} = \lambda_e \, _e e_{rr} \delta_{kl} + 2\mu_e \, _e e_{kl} \tag{II-44}$$
elastic solid

$$_v t_{kl} = (-\Pi + \lambda_v \, _v \dot{e}_{rr})\delta_{kl} + 2\mu_v \, _v \dot{e}_{kl} \tag{II-45}$$
linear Stokesian fluid

(e means elastic, v viscous, Π = thermodynamic pressure).[4] In a Maxwell body, displacements add tensorally, but "viscous" and "elastic" stresses are equal. The simple spring and dashpot model looks:

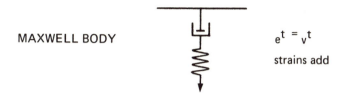

MAXWELL BODY $_e t = _v t$

                                               strains add

---

[4]Note that elastic and viscous constitutive equations are not identical. See Appendix I for discussion of viscous equation.

Contracting the above equation

$$_et_{kk} = (3\lambda_e + 2\mu_e)_ee_{kk},\tag{II-46}$$

$$_vt_{kk} = -3\Pi + 3(\lambda_v + 2\mu_v)\dot{e}_{kk}.\tag{II-47}$$

Introducing deviatoric stress and strain

$$\mathbf{t}^D + \frac{1}{3}t_{kk}\mathbf{I} = \mathbf{t},$$

$$_vt_{kl}^D = 2\mu_v\,_v\dot{e}_{kl}^D,\tag{II-48}$$

$$_et_{kl}^D = 2\mu_e\,_ee_{kl}^D.\tag{II-49}$$

Since displacements add but stresses are identical

$$_e\mathbf{t} = _v\mathbf{t} = \mathbf{t},\tag{II-50}$$

$$\dot{\mathbf{e}} = _v\dot{\mathbf{e}} + _e\dot{\mathbf{e}}.\tag{II-51}$$

Thus for a Maxwell body

$$\dot{e}_{kl}^D = \frac{t_{kl}^D}{2\mu_v} + \frac{\dot{t}_{kl}^D}{2\mu_e},\tag{II-52}$$

$$\dot{e}_{kk} = \frac{\dot{t}_{kk}}{3\lambda_e + 2\mu_e} + \frac{t_{kk} + 3\Pi}{3\lambda_v + 2\mu_v},\tag{II-53}$$

or

$$\dot{t}_{kl}^D + \frac{\mu_e}{\mu_v}t_{kl}^D = 2\mu_e\,\dot{e}_{kl}^D,\tag{II-54}$$

$$\dot{t}_{kk} + \frac{3\lambda_e + 2\mu_e}{3\lambda_v + 2\mu_v}(t_{kk} - 3\Pi) = (3\lambda_e + 2\mu_e)\dot{e}_{kk}.\tag{II-55}$$

Adding (II-54) and $\frac{1}{3}$ (II-55) δ_{kl} we see

$$\dot{t}_{kl} + \frac{\mu_e}{\mu_v}t_{kl} + \left(\frac{1}{3}\frac{3\lambda_e + 2\mu_e}{3\lambda_v + 2\mu_v}(t_{kk} + 3\Pi) - \frac{\mu_e}{\mu_v}t_{kk}\right)\delta_{kl}\tag{II-56}$$

$$= 2\mu_e\dot{e}_{kl} + \lambda_e\dot{e}_{rr}\delta_{kl}.$$

Equation (II-56) is the stress-strain relation of a Maxwell body. For quasi-static situations, pressure may be argued to be equal to thermo-dynamic pressure;[5] i.e., $\frac{1}{3} t_{kk} = -\Pi$. In this case λ_v type viscous dissi-pation is unimportant. (II-56) reduces to:

$$\dot{t}_{kl} + \frac{\mu_e}{\mu_v}\left(t_{kl} - \frac{1}{3}t_{rr}\delta_{kl}\right)$$
$$= 2\mu_e \dot{e}_{kl} + \lambda_e \dot{e}_{rr}\delta_{kl}. \tag{II-57}$$

2. The Theorem of Correspondence.

Let a bar (—) denote the Laplace transform and s be the Laplace transform variable. The Laplace transform of (II-57) is:

$$s\bar{t}_{kl} + \frac{\mu_e}{\mu_v}\bar{t}_{kl} - \left[\frac{1}{3}\frac{\mu_e}{\mu_v}\bar{t}_{rr}\delta_{kl}\right] = 2\mu_e s\bar{e}_{kl} + \lambda_e s\bar{e}_{rr}\delta_{kl}. \tag{II-58}$$

Contracting (II-58), we see

$$\bar{t}_{kk} = (3\lambda_e + 2\mu_e)\bar{e}_{kk}.$$

Substituting in (II-58) for $\bar{t}_{kk}$

$$\bar{t}_{kl} = \left(\lambda_e + \frac{2}{3}\mu_e \frac{\mu_e/\mu_v}{s + \mu_e/\mu_v}\right)\bar{e}_{rr}\delta_{kl} + 2\mu_e\left(\frac{s}{s + (\mu_e/\mu_v)}\right)\bar{e}_{kl}.$$

Thus

$$\bar{t}_{kl} = \lambda(s)\bar{e}_{rr}\delta_{kl} + 2\mu(s)\bar{e}_{kl}$$

$$\lambda(s) = \frac{\lambda_e s + \frac{\mu_e}{\mu_v}K_e}{s + \frac{\mu_e}{\mu_v}} \tag{II-59}$$

$$\mu(s) = \frac{\mu_e s}{s + \frac{\mu_e}{\mu_v}}.$$

[5] See Appendix I.

This demonstrates the *Theorem of Correspondence:* We can obtain the Laplace transformed solution to the linear homogeneous viscoelastic Maxwell body problem from the ordinary solution to the corresponding elastic problem by replacing λ_e and μ_e by $\lambda(s)$, $\mu(s)$.

3. Discussion and Examples.

Interpretation is easy in the case of the spring and dashpot model. When a stress is applied, there is an immediate elastic displacement, followed by a steady viscous extension which continues at a rate proportional to the stress for as long as the stress is applied. When the stress is released, the steady extension ceases and there is an immediate recovery of the initial elastic displacement.

In three dimensions, the Maxwell fluid may be thought of as a slurry of bubbles (Bragg's Bubble Model). Each individual bubble is free to deform elastically. At the same time it is free to slide over its neighbor in viscous flow.

Suppose for example we shear the fluid between two infinite plates:

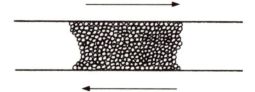

As in the case of the spring and dashpot, following the application of the shearing stress there will be an instantaneous elastic displacement. This will be followed by a steady viscous shear. When the shearing stress is released, the initial elastic displacement will be recovered.

An interesting illustrative case to consider is the uniaxial compression of a confined Maxwell body. We would expect the immediate response to be that of an elastic solid. The ultimate response must be that of a fluid.

Consider the Maxwell body to be encased in a rigid container and subject to a pressure $-p$ at $z = a$

MAXWELL BODY

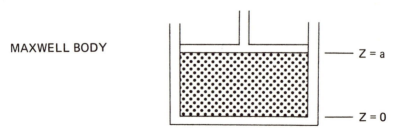

Boundary Conditions:

$t_{zz} = -p$
all e_{ij} except $e_{zz} = 0$
free slip at walls implies **t** diagonal
and $t_{xx} = t_{yy} \equiv \sigma$

$$t_{ij} = \lambda e_{kk}\delta_{ij} + 2\mu e_{ij}.$$

Elastic Solution

$$e_{zz} = -\frac{p}{\lambda + 2\mu}, \tag{II-60}$$

$$\sigma = t_{xx} = t_{yy} = -\frac{\lambda p}{\lambda + 2\mu}. \tag{II-61}$$

Invoking the Theorm of Correspondence, we may write the Laplace transform of the viscoelastic problem:

$$\bar{\sigma} = \frac{-\bar{p}\left(\lambda_e s + \frac{\mu_e}{\mu_v} K_e\right)}{(\lambda_e + 2\mu_e)s + \frac{\mu_e}{\mu_v} K_e} \tag{II-62}$$

$$\bar{e}_{zz} = \frac{-\bar{p}\left(s + \frac{\mu_e}{\mu_v}\right)}{(\lambda_e + 2\mu_e)s + \frac{\mu_e}{\mu_v} K_e} \tag{II-63}$$

$$(\lambda_e + 2\mu_e)s\bar{e}_{zz} + \frac{\mu_e}{\mu_v} K_e\bar{e}_{zz} = -\bar{p}s - \bar{p}\frac{\mu_e}{\mu_v}$$

$$(\lambda_e + 2\mu_e)\frac{\partial e_{zz}}{\partial t} + \frac{\mu_e}{\mu_v} K_e e_{zz} = -\frac{\partial p}{\partial t} - p\frac{\mu_e}{\mu_v}$$

$$\boxed{e_{zz} = \frac{-p\left(1 + \frac{\mu_e}{\mu_v} t\right)}{(\lambda_e + 2\mu_e) + \frac{\mu_e}{\mu_v} K_e t}} \tag{II-64}$$

$$(\lambda_e + 2\mu_e)s\bar{\sigma} + \frac{\mu_e}{\mu_v} K_e\bar{\sigma} = -\bar{p}\lambda_e s - \frac{\mu_e}{\mu_v} K_e\bar{p}$$

$$(\lambda_e + 2\mu_e) \frac{\partial \sigma}{\partial t} + \frac{\mu_e}{\mu_v} K_e \sigma = - \frac{\partial p}{\partial t} \lambda_e - \frac{\mu_e}{\mu_v} K_e p$$

$$\sigma = \frac{-p \left(\lambda_e + \frac{\mu_e}{\mu_v} K_e t\right)}{(\lambda_e + 2\mu_v) + \frac{\mu_e}{\mu_v} K_e t} \qquad \text{(II-65)}$$

Equations (II-64) and (II-65) are solutions to the viscoelastic problem.

1) If $t = 0$ or $\mu_v = \infty$

$$\sigma = \frac{-p\lambda_e}{\lambda_e + 2\mu_e} \qquad \text{(II-66)}$$

$$e_{zz} = \frac{-p}{\lambda_e + 2\mu_e} \qquad \text{(II-67)}$$

which is the solution to the simple elastic problem.

2) If t is large

$$\sigma = -p$$
$$e_{zz} = -p/K_e$$

which is the response to be expected of a linear Stokesian fluid.

Thus, indeed, the immediate response is elastic. As time goes on, the initial strain increases until it equals the fluid response.

We may interpret what has happened:

1) Initially $t_{zz} = 0$. The body is in a state of hydrostatic equilibrium in which perturbation stresses in all directions are equal to zero. Molecules in the body are on the average equally spaced (neglecting gravitational compression).

2) Immediately after loading, the body is elastically deformed. Stresses on the frictionless walls of the confining chamber are not equal to the applied stress in the $\hat{z}$ direction. Planes of molecules in the body are more compressed perpendicular to $\hat{z}$ than parallel to $\hat{z}$.

3) After sufficient time has elapsed, viscous flow has returned the body to a state of hydrostatic equilibrium. All the molecules are

again on the average equally spaced. Stress in all directions is equal and equal to $-t_{zz}$. This rearranging of the molecules allows the piston to sink a slight additional amount.

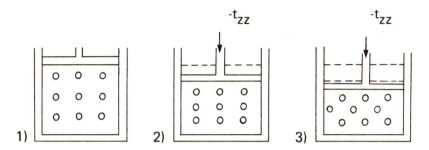

4. Phase Changes, Cracks, and Voids.

Suppose a phase change from one mineral assemblage to another occurs at a given depth, at a certain temperature and pressure. If a load is applied to the surface, the pressure at that level will be increased. Oversimplifying for a moment, we can distinguish two extreme cases:

1) If the phase change boundary migrates very rapidly as compared to the isostatic adjustment of the earth to surface load, the effect of the phase change at the surface will appear like an *elastic* response.

2) If on the other hand the phase boundary migrates very slowly compared to the fluid adjustment of the whole body, the phase change region will have *no effect* other than the effect it naturally has on the density and perhaps viscosity of fluid.

In either case the effects at the phase change layer are easy to incorporate into a Maxwell body model. One just alters the viscosity, elastic properties, and non-adiabatic (fluid) density gradient at the appropriate depth in the appropriate fashion.

Which of these two cases is more appropriate for the earth has implications for the temperature structure of the mantle and for the existence and nature of mantle convection. We discuss this in greater detail in Section III.A.3 and in Chapter IV.

If voids and cracks are present in the surface material, the application of pressure will tend to cause their closure, resulting in a certain consolidation which would not be retrieved after unloading. This effect is of course unobservable and hence unimportant in Pleistocene glacial phenomena. It should be and generally is remembered in observations attending the filling of reservoirs, however.

C. Matrix Methods of Solution

Suppose we start with a system of linear, first order differential equations of the form:

$$\partial_z \mathbf{u} \;=\; \mathbf{A}(z)\mathbf{u} \tag{II-68}$$

or

$$z\partial_z \mathbf{u} \;=\; \mathbf{A}(z)\mathbf{u} \tag{II-69}$$

These equations give the rate of change of $\mathbf{u}$ with respect to z at a particular depth in terms of the value of $\mathbf{u}$ there. Given a value of $\mathbf{u}$ at a particular z, say z_0, it is clear that we can bootstrap our way up or down to obtain $\mathbf{u}$ at any other level.

Exact, closed form methods of solving (II-68) and (II-69) for certain cases are also available. These methods will be useful to us in weighing the relative importance of various terms in our equations and in checking the accuracy of the Runge-Kutta solutions. The exact solutions we shall discuss here fall under the general heading of Propagator Matrix Techniques. Full discussion is made in Gantmacher (1960) and Gilbert and Backus (1966). We discuss below, as simply as possible, only those parts of the subject of direct interest to us. We follow closely the above sources, especially Gantmacher.

1. Matrix Techniques.

Symbolically (II-68) and (II-69) may be solved exactly as they would be if $\mathbf{u}$ were not a vector and $\mathbf{A}$ not a matrix:

$$\mathbf{u}(z) \;=\; e^{\int_{z_0}^{z} \mathbf{A}(\xi)d\xi}\,\mathbf{u}(z_0) \tag{II-70}$$

$$\mathbf{u}(z) \;=\; x^{\int_{z_0}^{z} \mathbf{A}(\xi)d\xi}\,\mathbf{u}(z_0) \tag{II-71}$$

It can be shown the constant vector $\mathbf{u}(z_0)$ must multiply the matrix function from the right. It can also be demonstrated that the solution $\mathbf{u}(z)$, given $\mathbf{u}(z_0)$, is unique (Gantmacher, 1960, Vol. II, pp. 114, 125–131).

A material whose properties vary continuously as a function of z may be approximated by one whose properties are constant in thin layers. Assume each layer has a thickness D_n. (II-70) and (II-71) then become:

$$\mathbf{u}(z) = e^{A\left(z_{n-1}^{n}\right)D_{n}} e^{A\left(z_{n-2}^{n-1}\right)D_{n-1}} \ldots e^{A\left(z_{0}^{1}\right)D_{1}} e^{A\left(z_{00}^{0}\right)z_{00}} \mathbf{u}(z_{00}), \quad \text{(II-72)}$$

$$\mathbf{u}(z) = x^{A\left(z_{n-1}^{n}\right)D_{n}} x^{A\left(z_{n-2}^{n-1}\right)D_{n-1}} \ldots x^{A\left(z_{0}^{1}\right)D_{1}} x^{A\left(z_{00}^{0}\right)z_{00}} \mathbf{u}(z_{00}). \quad \text{(II-73)}$$

In this case there are assumed to be n layers between z_0 and z. The various elements in the product chain of (II-72) and (II-73) are matrices and have the effect of propagating the solution at z_0, $\mathbf{u}$ (z_0), through the layer each represents. That is in

$$\mathbf{u}(z_0 + D_1) = e^{A\left(z_0^{1}\right)D_1} \mathbf{u}(z_0).$$

$\mathbf{u}$ (z_0) is propagated to the top of the first layer. These product elements are called *propagator matrices*.

The general case may be represented:

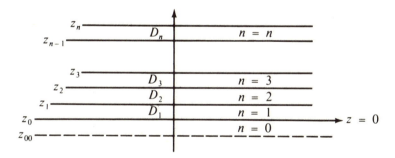

We start the solution at $z = z_{00}$, i.e., some arbitrary depth into the homogeneous substratum. Then the solution vector at z_0, $(z = 0)$ is:

$$\mathbf{u}(z_0) = e^{A\left(z_{00}^{0}\right)z_{00}} \mathbf{u}(z_{00}).$$

Propagating through the various layers:

$$\mathbf{u}(z_1) = \mathbf{P}(z_1, z_0) \mathbf{u}(z_0)$$

.
.
.

$$\mathbf{u}(z_n) = \mathbf{P}(z_n, z_{n-1}) \mathbf{P}(z_{n-1}, z_{n-2}) \ldots \mathbf{P}(z_1, z_0) \mathbf{u}(z_0).$$

So altogether

$$\mathbf{u}(z_n) = \mathbf{P}_n\mathbf{P}_{n-1}\mathbf{P}_{n-2} \ldots \mathbf{P}_3\mathbf{P}_2\mathbf{P}_1 \, e^{\mathbf{A}\left(\overset{0}{z_{00}}\right)z_{00}}\mathbf{u}(z_{00}), \qquad\qquad \text{(II-74)}$$

$$\mathbf{u}(z_n) = \mathbf{P}\, e^{\mathbf{A}\left(\overset{0}{z_{00}}\right)z_{00}}\mathbf{u}(z_{00}). \qquad\qquad\qquad\qquad \text{(II-75)}$$

A common requirement is that the solution vector remain finite in the substratum. The condition

$$e^{\mathbf{A}\left(\overset{0}{z_{00}}\right)z_{00}} \, \mathbf{u}(z_{00}) \xrightarrow[z_{00} \to -\infty]{} \quad \text{finite} \qquad\qquad \text{(II-76)}$$

places two conditions on $\mathbf{u}$. The remaining two arbitrary components of $\mathbf{u}$ are determined by the particular boundary conditions applied to the top of the stack of layers.

Functions of matrices, such as $e^{\mathbf{A}D}$ may be evaluated by simple expansion, i.e.:

$$e^{\mathbf{A}D} = 1 + \mathbf{A}D + \frac{\mathbf{A}D^2}{2!} + \frac{\mathbf{A}D^3}{3!} + \ldots \qquad\qquad \text{(II-77)}$$

This still does not obtain the solution in closed form—at least not directly.

A better manner in which to proceed is to note (Gantmacher, 1964, *I*, pp. 95ff) that the behavior of a function $f(\mathbf{A})$ is completely determined by its behavior at the eigenvalues of $\mathbf{A}$, λ_i. If there are four eigenvalues, all of multiplicity one

$$\begin{aligned}
f(\mathbf{A}) = &\frac{(\mathbf{A} - \lambda_2\mathbf{I})(\mathbf{A} - \lambda_3\mathbf{I})(\mathbf{A} - \lambda_4\mathbf{I})}{(\lambda_1 - \lambda_2)(\lambda_1 - \lambda_3)(\lambda_1 - \lambda_4)} f(\lambda_1) \\[2mm]
&+ \frac{(\mathbf{A} - \lambda_1\mathbf{I})(\mathbf{A} - \lambda_3\mathbf{I})(\mathbf{A} - \lambda_4\mathbf{I})}{(\lambda_2 - \lambda_1)(\lambda_2 - \lambda_3)(\lambda_2 - \lambda_4)} f(\lambda_2) \\[2mm]
&+ \frac{(\mathbf{A} - \lambda_1\mathbf{I})(\mathbf{A} - \lambda_2\mathbf{I})(\mathbf{A} - \lambda_4\mathbf{I})}{(\lambda_3 - \lambda_1)(\lambda_3 - \lambda_2)(\lambda_3 - \lambda_4)} f(\lambda_3) \\[2mm]
&+ \frac{(\mathbf{A} - \lambda_1\mathbf{I})(\mathbf{A} - \lambda_2\mathbf{I})(\mathbf{A} - \lambda_3\mathbf{I})}{(\lambda_4 - \lambda_1)(\lambda_4 - \lambda_2)(\lambda_4 - \lambda_3)} f(\lambda_4).
\end{aligned} \qquad \text{(II-78)}$$

If there are two distinct eigenvalues, each of multiplicity two

$$f(\mathbf{A}) = \left[\frac{f(\lambda_1)}{(\lambda_1 - \lambda_2)^2} + \left(\frac{-2}{(\lambda_1 - \lambda_2)^3} f(\lambda_1)\right.\right.$$

$$\left.\left. + \frac{f'(\lambda_1)}{(\lambda_1 - \lambda_2)^2}\right)(\mathbf{A} - \lambda_1\mathbf{I})\right](\mathbf{A} - \lambda_2\mathbf{I})^2$$

$$\text{(II-79)}$$

$$+ \left[\frac{f(\lambda_2)}{(\lambda_2 - \lambda_1)^2} + \left(\frac{-2}{(\lambda_2 - \lambda_1)^3} f(\lambda_2)\right.\right.$$

$$\left.\left. + \frac{f'(\lambda_2)}{(\lambda_2 - \lambda_1)^2}\right)(\mathbf{A} - \lambda_2\mathbf{I})\right](\mathbf{A} - \lambda_1\mathbf{I})^2.$$

(Gantmacher, 1960, Vol. I, pp. 103ff)

The above equations express $f(\mathbf{A})$ in closed form. Only a few matrix multiplications need be performed.

2. Runge-Kutta Techniques

Runge-Kutta Techniques are generally applicable to systems of the form

$$\partial_z\mathbf{u} = \mathbf{F}(z, \mathbf{u}).$$

Our case is in the form

$$\partial_z\mathbf{u} = \mathbf{A}(z)\mathbf{u}. \tag{II-80}$$

If the step size of integration is h,

$$\mathbf{u}_{n+1} = \mathbf{u}_n + \frac{1}{6}(\mathbf{k}_1 + 2\mathbf{k}_2 + 2\mathbf{k}_3 + \mathbf{k}_4) \tag{II-81}$$

$$\mathbf{k}_1 = h\mathbf{A}(z_n)\mathbf{u}_n$$

$$\mathbf{k}_2 = h\mathbf{A}\left(z_n + \frac{h}{2}\right)\left(\mathbf{u}_n + \frac{\mathbf{k}_1}{2}\right)$$

$$\tag{II-82}$$

$$\mathbf{k}_3 = h\mathbf{A}\left(z_n + \frac{h}{2}\right)\left(\mathbf{u}_n + \frac{\mathbf{k}_2}{2}\right)$$

$$\mathbf{k}_4 = h\mathbf{A}(z_n + h)(\mathbf{u}_n + \mathbf{k}_3)$$

(Hildebrand, 1962, p. 103).

The four **k** vectors are easily obtained by matrix multiplication of **A** at the depth indicated $z_n + \dfrac{h}{2}$, etc. and the solution vector **u** at z_n and the other **k**'s. If $\mathbf{k}_1$ is evaluated first, $\mathbf{k}_2$ second, etc., all **k**'s are easily evaluated. (II-81) then yields the solution vector at z_{n+1}, $\mathbf{u}_{n+1}$. The process can then be repeated with $n = n + 1$, and this cycle repeated till the solution at the surface is obtained.

3. Inhomogeneous Systems of Equations

Gilbert and Backus (1966) have shown the solution to

$$\partial_z \mathbf{u} = \mathbf{A}(z)\mathbf{u} + \mathbf{b}(z) \tag{II-83}$$

may be expressed

$$\mathbf{u}(z) = \int_{z_0}^{z} \mathbf{P}(z, \zeta)\mathbf{b}(\zeta)d\zeta + \mathbf{P}(z, z_0)\mathbf{u}(z_0). \tag{II-84}$$

If $\mathbf{b}(z)$ varies in small steps, remaining constant over shells of thickness H, (II-84) may be rewritten:

$$\mathbf{u}(z) = \sum_{k=1}^{NF} \int_{\zeta = z_k - H/2}^{\zeta = z_k + H/2} \mathbf{P}(z, \zeta)\mathbf{b}(\zeta)d\zeta + \mathbf{P}(z, z_0)\mathbf{u}(z_0),$$

$$\mathbf{u}(z) = \sum_{k=1}^{NF} \mathbf{P}(z_{NF}, z_k)\mathbf{b}(z_k)H + \mathbf{P}(z_{NF}, z_0)\mathbf{u}(z_0), \tag{II-85}$$

where $z_1 = z_0, z_{NF} = z$.

   P is just the usual propagator. Runge-Kutta integration may be used instead of **P** if desired. Thus we see $\mathbf{u}(z)$ is just the sum of $\mathbf{u}(z_0)$ propagated from z_0 to z and the contributions of the various $\mathbf{b}(z_k)$ propagated from their points of origin, z_k, to z.

THREE

Formulation of the Theory For Application to the Earth

IN THE FIRST PART of this section we solve several simple cases involving a flat earth with g = constant and $\partial_z \rho_0 = 0$.[1] The results of these calculations will be of use to us in developing the full self-gravitating viscoelastic solutions which we shall do in Section III.B.

A. Flat Earth, Constant Gravitation

1. Interpretation of Equation of Motion and Reduction to Runge-Kutta Form

a. Meaning of Hydrostatic Pre-Stress.

Let $\mathbf{g}$ = constant = $-\nabla \phi_0 = -g_0 \hat{\mathbf{z}}$. Equations (II-22) and (II-23) then become:

$$\nabla \cdot \tau - \rho_0 g_0 \nabla u_z + g_0 \rho_0 \nabla \cdot \mathbf{u} \hat{\mathbf{z}}, \qquad \text{Elastic (III-1)}$$

$$\nabla \cdot \tau + (g_0 \rho_0 \nabla \cdot \mathbf{u} + g_0 u_z \partial_z \rho_0) \hat{\mathbf{z}} = 0. \qquad \text{Viscous (III-2)}$$

If the material is assumed to be incompressible, $\nabla \cdot \mathbf{u} = 0$. Only the hydrostatic pre-stress term remains in (III-1), and its meaning is easily determined.

$$\nabla \cdot \tau - \rho_0 g_0 \nabla u_z = 0. \qquad \text{Elastic Incompressible (III-3)}$$

Suppose, for example, $\partial_x u_z$ = neg. const. = $-C$. Then since elastic deformations carry their initial pressure fields with them, we see that at a given level z in space, the pressure to the left will be greater than the pressure to the right. The decrease in pressure from left to right is just $\rho_0 g_0 \nabla u_z$, $\nabla_x u_z$ being negative. Since tensional forces are by con-

---

[1] $\partial_z \rho_0 = \dfrac{\partial \rho_0}{\partial z}$

$$p_\ell > p_r$$

vention positive, and for equilibrium

$$\nabla \cdot \tau + \mathbf{F} = 0,$$
$$\nabla \cdot \tau - \rho_0 g_0 \nabla_x u_z = 0,$$

which is just (III-3). New stresses are thus induced under non-uniform deformation of a pre-stressed medium. They are induced *because* material elements carry their pressure fields with them under elastic deformation.

The viscous incompressible equation is:

$$\nabla \cdot \tau + g_0 u_z \partial_z \rho_0 \hat{\mathbf{z}} = 0.$$

The added term here is simply interpreted as a buoyancy term arising from a disturbance u_z in the zero order density gradient. If the density is getting greater with depth, $\partial_z \rho_0$ is negative. If u_z is also to be negative, $\nabla \cdot \tau$ = negative, and a medium will be in equilibrium only when there exists a net stress holding the buoyant, lighter, depressed material down (compressive stress).

In the viscous case, the buoyancy terms arise because the material elements (in an incompressible viscous fluid) *do not* carry their pressure field with them under deformation. We might note $\partial_z \rho_0$ is the density gradient as it exists in the earth with the adiabatic density gradient subtracted out. If we call this density gradient the non-adiabatic density gradient and denote it $(\partial_z \rho_0)_{NA}$.

$$(\partial_z \rho_0)_{NA} \equiv \partial_z \rho_0 - (\partial_z \rho_0)_{\text{Adiabatic}}.$$

The adiabatic density gradient is the density gradient which would arise solely due to compression by hydrostatic pressure. We shall discuss this in more detail later.

b. Reduction of (III-1) and (III-2) to Runge-Kutta Form.

The details of incorporating the constitutive equations, Fourier transforming, and reducing (III-1) and (III-2) to Runge-Kutta form, are contained in Appendix II. In that appendix it is also shown that the resulting 6-equation system separates into an uncoupled 4-equation and

2-equation system which describe motions which are divergent and divergentless on surfaces $\perp$ to $\hat{z}$, respectively. It is shown that, if the divergentless part of stress or strain is zero on any surface $\perp$ $\hat{z}$, it must be zero everywhere in the media. This means the type of motion produced in the media (whether described by the 4-equation or 2-equation system) depends entirely upon the boundary conditions applied. In the cases we consider, the deformation is entirely covered by the 4-equation system. We comment in Appendix II that the incompressible elastic equation may be simply obtained from the compressible equations by letting $\lambda_e \to \infty$. The incompressible elastic equations may be interpreted as viscous equations, if we are careful about the body force terms which involve total displacement and not velocities or rate of displacement.

We assume the viscous flow to be incompressible. All dilatation (i.e. change in volume or density) is elastic.

Simplifying the expressions of Appendix II by making them non-dimensional, we see (App. II-10) that the compressible elastic system corresponding to (III-1) is:

$$
\partial_z
\begin{bmatrix}
2\mu^* i k \bar{u}_x \\
2\mu^* k \bar{u}_z \\
i \bar{\tau}_{xz} \\
\bar{\tau}_{zz}
\end{bmatrix}
= k
\begin{bmatrix}
0 & 1 & 2\tilde{\mu}^{-1} & 0 \\
-\tilde{\sigma}^{-1}\tilde{\lambda} & 0 & 0 & 2\tilde{\sigma}^{-1} \\
2\tilde{\mu}\tilde{\sigma}^{-1}(\tilde{\lambda} + \tilde{\mu}) & \dfrac{-\rho_0 g_0}{2k\mu^*} & 0 & \tilde{\lambda}\tilde{\sigma}^{-1} \\
\dfrac{\rho_0 g_0}{2k\mu^*} & 0 & -1 & 0
\end{bmatrix}
\begin{bmatrix}
2\mu^* i k \bar{u}_x \\
2\mu^* k \bar{u}_z \\
i \bar{\tau}_{xz} \\
\bar{\tau}_{zz}
\end{bmatrix}
$$

$$\text{Elastic (III-4)}$$

$$\tilde{\mu} \equiv \frac{\mu}{\mu^*} \qquad \tilde{\lambda} \equiv \frac{\lambda}{\mu^*}$$

$$\tilde{\sigma} \equiv \tilde{\lambda} + 2\tilde{\mu}$$

$\mu^* \equiv$ rigidity of the lowest layer

The incompressible viscous system corresponding to (III-2) is:

$$\partial_z \begin{bmatrix} 2\eta^* i k \bar{v}_x \\ 2\eta^* k \bar{v}_z \\ i\bar{\tau}_{xz} \\ \bar{\tau}_{zz} \end{bmatrix}$$

$$= k \begin{bmatrix} 0 & 1 & 2\tilde{\eta}^{-1} & 0 \\ -1 & 0 & 0 & 0 \\ 2\tilde{\eta} & 0 & 0 & 1 \\ 0 & 0 & -1 & 0 \end{bmatrix} \begin{bmatrix} 2\eta^* i k \bar{v}_x \\ 2\eta^* k \bar{v}_z \\ i\bar{\tau}_{xz} \\ \bar{\tau}_{zz} \end{bmatrix} + \begin{bmatrix} 0 \\ 0 \\ 0 \\ g_0 \partial_z \rho_0 \bar{u}_z \end{bmatrix} \qquad \text{Viscous (III-5)}$$

$$\bar{u}_z = \int_0^t \bar{v}_z \, dt$$

$\eta^* \equiv$ Viscosity of lowest layer

$$\tilde{\eta} \equiv \frac{\eta}{\eta^*}$$

c. The Wavenumber Dependence of Hydrostatic Pre-Stress.

All elements in the matrix part of (III-4) are of order 1 except

$$\frac{\rho_0 g_0}{2 k \mu_e^*}.$$

These are the hydrostatic pre-stress terms. When

$$\frac{\rho_0 g_0}{2 k \mu_e^*} \ll 1,$$

or, for $\mu_e^* = 10^{12}$ dyne-cm^{-2}, $\rho_0 g_0 = 4 \times 10^3$ gm-cm^{-2}-sec^{-2}

$$\lambda \ll 30{,}000 \text{ km.}$$

the hydrostatic pre-stress will be unimportant. This condition will be met for deformations small enough to be treated in the flat space approximation.[2]

The buoyancy term in the viscous case is not so easily discarded since

[2]See Figure III-2

it involves $\bar{u}_z$, which can be quite large compared to $\bar{v}_z$. However, it can be discarded if we assume for the moment that $(\partial_z \rho_0)_{NA} = 0$.

Assuming $\lambda \ll 30,000$ km and $(\partial_z \rho_0)_{NA} = 0$, we investigate: (a) the harmonic response of homogenous incompressible elastic or viscous half-space, (b) the response of a layered elastic or viscous half-space, (c) flow restricted to a channel, (d) the effect of an elastic crust, (e) a compressible elastic half-space, (f) a viscoelastic half-space.

2. Calculation of Examples

a. Background Theory.

$$\partial_z \begin{bmatrix} 2\mu^* ik\bar{u}_x \\ 2\mu^* k\bar{u}_z \\ i\bar{\tau}_{xz} \\ \bar{\tau}_{zz} \end{bmatrix} = k \begin{bmatrix} 0 & 1 & 2\tilde{\mu}^{-1} & 0 \\ -1 & 0 & 0 & 0 \\ 2\tilde{\mu} & 0 & 0 & 1 \\ 0 & 0 & -1 & 0 \end{bmatrix} \begin{bmatrix} 2\mu^* ik\bar{u}_x \\ 2\mu^* k\bar{u}_z \\ i\bar{\tau}_{xz} \\ \bar{\tau}_{zz} \end{bmatrix} \qquad \text{(III-6)}$$

$$\partial_z \bar{\mathbf{u}} = \mathbf{A}\bar{\mathbf{u}} \qquad \text{(III-7)}$$

Setting det $(\mathbf{A} - \alpha\mathbf{I}) = 0$ we find that $\mathbf{A}$ has two distinct eigenvalues $\pm k$, each of multiplicity two.

The general solution to (III-7) may be written

$$\mathbf{u}(z) = e^{\int_{z_0}^{z} \mathbf{A}(\xi) d\xi} \bar{\mathbf{u}}(z_0) \qquad \text{(III-8)}$$

If the half-space consists of n layers above a homogeneous half-space

from (II-72)

$$\bar{\mathbf{u}}(z) = e^{\mathbf{A}\left(z_{n-1}^{n}\right)D_n} e^{\mathbf{A}\left(z_{n-2}^{n-1}\right)D_{n-1}} \cdots e^{\mathbf{A}\left(z_0^1\right)D_1} e^{\mathbf{A}\left(z_{00}^0\right)z_{00}} \bar{\mathbf{u}}(z_{00}).$$

$A(z_{n-1}^n)$ is simply matrix A with the values of μ pertinent to the layer between z_{n-1} and z_n. The solution, in this case, is propagated by a series of propagator matrices from z_{00}, an arbitrary level in the substratum, to the surface z_n.

Define a *propagator* matrix P_n:

$$\mathbf{P}_n \equiv e^{\mathbf{A}\left(z_{n-1}^n\right)D_n}$$

$\mathbf{P}_n$ propagates a solution at the bottom of a layer to the top of that layer.

$$\bar{\mathbf{u}}(z_n) = \mathbf{P}_n \mathbf{P}_{n-1} \cdots \mathbf{P}_2 \mathbf{P}_1 e^{\mathbf{A}\left(z_{00}^0\right)z_{00}} \bar{\mathbf{u}}(z_{00}), \tag{III-9}$$

$$e^{\mathbf{A}z} = \mathbf{G}^+ + \mathbf{G}^- = \mathbf{P}. \tag{III-10}$$

From equation (II-79)

$$\mathbf{G}^+ = \frac{e^{kz}}{4k^2}\left[\left((\mathbf{A}^2 + k\mathbf{I})^2 - \frac{1}{k}(\mathbf{A} - k\mathbf{I})(\mathbf{A} + k\mathbf{I})^2\right)\right. \tag{III-11}$$
$$\left. + z(\mathbf{A} - k\mathbf{I})(\mathbf{A} + k\mathbf{I})^2\right]$$

$$\mathbf{G}^- = \frac{e^{-kz}}{4k^2}\left[\left((\mathbf{A}^2 - k\mathbf{I})^2 + \frac{1}{k}(\mathbf{A} + k\mathbf{I})(\mathbf{A} - k\mathbf{I})^2\right)\right.$$
$$\left. + z(\mathbf{A} + k\mathbf{I})(\mathbf{A} - k\mathbf{I})^2\right]$$

$$\mathbf{G}^+ = \frac{e^{kz}}{2}\left[\begin{pmatrix} 1 & 0 & \tilde{\mu}^{-1} & 0 \\ 0 & 1 & 0 & \tilde{\mu}^{-1} \\ \tilde{\mu} & 0 & 1 & 0 \\ 0 & \tilde{\mu} & 0 & 1 \end{pmatrix} + kz \begin{pmatrix} 1 & 1 & \tilde{\mu}^{-1} & \tilde{\mu}^{-1} \\ -1 & -1 & -\tilde{\mu}^{-1} & -\tilde{\mu}^{-1} \\ \tilde{\mu} & \tilde{\mu} & 1 & 1 \\ -\tilde{\mu} & -\tilde{\mu} & -1 & -1 \end{pmatrix}\right]$$

$$\mathbf{G}^- = \frac{e^{-kz}}{2}\left[\begin{pmatrix} 1 & 0 & -\tilde{\mu}^{-1} & 0 \\ 0 & 1 & 0 & -\tilde{\mu}^{-1} \\ -\tilde{\mu} & 0 & 1 & 0 \\ 0 & -\tilde{\mu} & 0 & 1 \end{pmatrix} + kz \begin{pmatrix} -1 & 1 & \tilde{\mu}^{-1} & -\tilde{\mu}^{-1} \\ -1 & 1 & \tilde{\mu}^{-1} & -\tilde{\mu}^{-1} \\ \tilde{\mu} & -\tilde{\mu} & -1 & 1 \\ \tilde{\mu} & -\tilde{\mu} & -1 & 1 \end{pmatrix}\right]$$

Let

$$S = kD \sinh kD$$
$$C = kD \cosh kD$$

$$CP = \cosh kD + kD \sinh kD$$
$$CM = \cosh kD - kD \sinh kD$$
$$SP = \sinh kD + kD \cosh kD$$
$$SM = \sinh kD - \cosh kD$$

Then the propagator for a layer of thickness D is

$$\mathbf{P} = \begin{pmatrix} CP & C & \tilde{\mu}^{-1}SP & \tilde{\mu}^{-1}S \\ -C & CM & -\tilde{\mu}^{-1}S & \tilde{\mu}^{-1}SM \\ \tilde{\mu}SP & \tilde{\mu}S & CP & C \\ -\tilde{\mu}S & \tilde{\mu}SM & -C & CM \end{pmatrix} \quad \text{(III-12)}$$

b. The Harmonic Load Response of a Homogeneous Incompressible Elastic or Viscous Half-Space.

The solution must remain *finite* in the substratum. Thus:

$$\mathbf{G}_0^- \bar{\mathbf{u}}(z_{00}) = 0.$$

Suppose

$$\bar{\mathbf{u}}(z_{00}) = \begin{pmatrix} A \\ B \\ C \\ D \end{pmatrix}.$$

Since $\tilde{\mu} = 1$ in the substratum, applying the boundary conditions at $z = 0$,

$$\mathbf{G}_0^- \begin{pmatrix} A \\ B \\ C \\ D \end{pmatrix} = 0 \quad \rightarrow \quad \begin{array}{l} A = C \\ B = D. \end{array}$$

So *in the substratum:*

$$\bar{\mathbf{u}}(z) = \mathbf{G}^+ \begin{pmatrix} A \\ B \\ A \\ B \end{pmatrix}.$$

At the top of the substratum ($z = 0$)

$$\bar{\mathbf{u}}(0) = \begin{pmatrix} A \\ B \\ A \\ B \end{pmatrix} = \mathbf{G}_0^+ \, \bar{\mathbf{u}}(0).$$

Thus if we consider a unit harmonic load *applied to z = 0;* i.e.

$$\begin{array}{ll} \bar{\tau}_{xz} = 0 \\ \bar{\tau}_{zz} = -1 \end{array} \quad \longrightarrow \quad \begin{array}{l} A = 0 \\ B = -1, \end{array}$$

we see in the unlayered case:

$$\bar{\mathbf{u}}(z) = \mathbf{G}^+ \bar{\mathbf{u}}(0) \quad \bar{\mathbf{u}}(0) = \begin{pmatrix} 0 \\ -1 \\ 0 \\ -1 \end{pmatrix}.$$

$$\begin{bmatrix} 2\mu^* ik\bar{u}_x \\ 2\mu^* k\bar{u}_z \\ i\bar{\tau}_{xz} \\ \bar{\tau}_{zz} \end{bmatrix} = e^{kz} \left[\begin{pmatrix} 0 \\ -1 \\ 0 \\ -1 \end{pmatrix} + kz \begin{pmatrix} -1 \\ 1 \\ -1 \\ 1 \end{pmatrix} \right]$$

(III-13)
Homogeneous
Half-space
Solution

In an elastic media the vertical deflection of a one dyne harmonic load is thus

$$\bar{u}_z(z = 0) = \frac{-1}{2\mu^* k}.$$

If the surface were loaded with a harmonic load of $\rho g h$ dynes,

$$\bar{u}_z(z = 0) = \frac{-\rho g h}{2\mu^* k}. \qquad\qquad (\text{III-14})$$

In a viscous media $\bar{u}_z |_{z=0}$ is the velocity of uplift at the surface, rather than the surface displacement. μ^* becomes η^*, the viscosity of the substrate. For a harmonic load of amplitude $\rho g h$ applied directly to a uniform viscosity half-space (i.e., the substrate) from (III-13):

$$\bar{v}_z\big|_{z=0} = \frac{-\rho g h}{2\eta^* k}.$$ (III-15)

Since $\bar{v}_z\big|_{z=0} = \dfrac{\partial h}{\partial t}$, if the initial ($t = 0$) amplitude is h_0,

$$h(t) = h_0 e^{-t/\tau^*}$$ (III-16)

and

$$\tau^* = \frac{2\eta^* k}{\rho g}$$ (III-17)

τ^* is the decay time of the substrate to the application of a harmonic load of wave number k.

Figure III-2 shows the dependence of stress or strain rate with depth for a homogeneous elastic or viscous half-space. It can be seen that by the time or depth of $\frac{1}{2}\lambda$ is reached, both stress and elastic displacement (or fluid velocity) are strongly attenuated.

c. The Harmonic Load Response of a Layered Elastic or Viscous Half-Space

If, instead, we apply a vertical load on top of the n-th layer,

$$\bar{\tau}_{xz} = 0$$
$$\bar{\tau}_{zz} = -1.$$

We propagate the solution at $z = 0$,

$$\bar{u}_0 = \bar{u}(0) = \begin{pmatrix} A \\ B \\ A \\ B \end{pmatrix} = A \begin{pmatrix} 1 \\ 0 \\ 1 \\ 0 \end{pmatrix} + B \begin{pmatrix} 0 \\ 1 \\ 0 \\ 1 \end{pmatrix}$$

to the surface and apply the boundary conditions:

$$\begin{pmatrix} \tilde{} \\ \tilde{} \\ 0 \\ -1 \end{pmatrix} = \mathbf{P}_n \mathbf{P}_{n-1} \cdots \mathbf{P}_3 \mathbf{P}_2 \mathbf{P}_1 \mathbf{G}_0^+ \bar{\mathbf{u}}_0. \tag{III-18}$$

This gives us two conditions which uniquely determine A and B. $\mathbf{P}_1 \ldots$ $\mathbf{P}_n$ came directly from (III-12) with $\tilde{\mu}$ appropriate for each layer $1 \ldots n$. $\mathbf{P}_n \mathbf{P}_{n-1} \ldots \mathbf{P}_1 \mathbf{G}_0^+$ is one way to "propagate" $\bar{\mathbf{u}}_0$ to the free surface. An equally good way is to use a Runge Kutta techniques.

$$\partial_z \bar{\mathbf{u}} = \mathbf{A}(z) \bar{\mathbf{u}}.$$

Knowing $\mathbf{u}_0 = A \begin{pmatrix} 1 \\ 0 \\ 1 \\ 0 \end{pmatrix} + B \begin{pmatrix} 0 \\ 1 \\ 0 \\ 1 \end{pmatrix}$, we can use the above equation to ob-

tain $\mathbf{u}$ at the free surface. The boundary conditions then uniquely specify A and B.

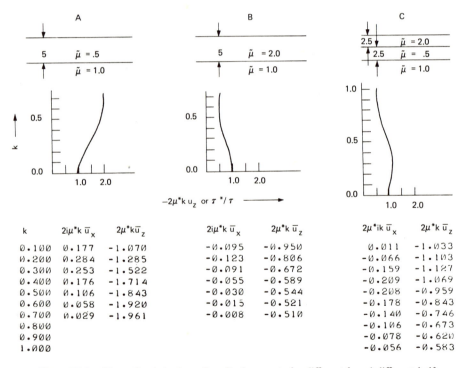

Figure III-1. Normalized elastic surface displacements for different k and different half-space earth models. Calculated by propagator matrix technique and checked by Runge-Kutta techniques.

Simple examples were calculated using both the propagator and Runge-Kutta techniques. Agreement between the two methods of calculation and of both with the theoretical solution for the homogeneous case (III-13) served to verify both against theoretical errors. Both methods of calculation agreed to at least three places, and the execution time was nearly identical for 100 Runge-Kutta steps. Program listings are given in Cathles (1971). The results of several simple calculations are shown in Figures III-1 and III-2.

It can be seen from Figure III-1 that if the half-space is layered, but the frequency high enough, the deformation will be as if the first layer were infinite. As the frequency diminishes, the deformation "looks" deeper and deeper into the media. Finally at very low frequencies the response will be as if only the substratum were present. The number of layers can be directly inferred from the number of inflections in a plot of k versus $2\mu^*k\bar{u}_z$. Note that since only kD appears in $\mathbf{P}$, we can scale

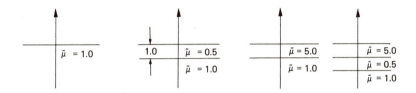

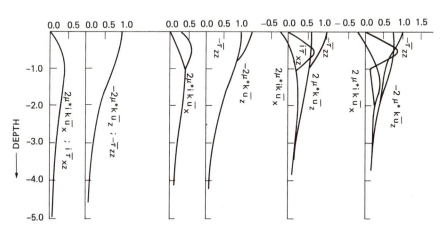

Figure III-2. "Normalized" stress and elastic strain as a function of depth for various models subject to $\bar{\tau}_{xz} = 0$, $\bar{\tau}_{zz} = -1$ at surface. $k = 1.0$, $\lambda = 6.28$. (Propagator matrix technique used for solution.)

up the layer thicknesses provided we scale down the frequency so as to keep kD constant.

The horizontal displacement at the surface of a layered structure is not zero, even though the applied horizontal traction at the surface is. Figure III-2 gives "normalized" displacements as a function of depth for several types of layered structures. It can be seen from Figure III-2:

(1) Variation in the elastic (or viscous) parameters with depth alters the variation of stress with depth.

(2) Both stress and elastic displacement are strongly diminished by the time a depth of $\frac{1}{2}\lambda$ is attained.

(3) The surface horizontal displacement is not zero even though the surface horizontal traction is. The phase of the horizontal displacement on top of a more rigid layer overlying a less rigid layer is opposite that of a less rigid layer overlying a more rigid layer.

(4) Figure III-2 can easily be converted to displacement cross-section by retransforming and looking at a single frequency.

$$2\mu^* k u_x = \text{Re} \int_{-\infty}^{\infty} 2\mu^* ik\bar{u}_x e^{ikx} dk$$

$$\underset{k=\text{const}}{=} 2\mu^* k \text{Re}[ie^{ikx}]\bar{u}_x$$

$$= -2\mu^* k\bar{u}_x \sin kx$$

$$2\mu^* k u_z = 2\mu^* k\bar{u}_z \cos kx$$

Thus flowlines can be roughly constructed:

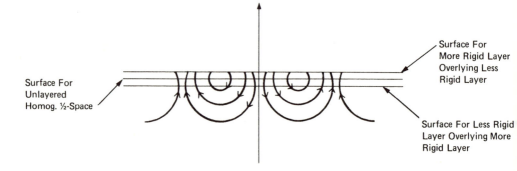

(5) Again, all diagrams and calculations are equally valid for an instantaneous picture of the flow of an incompressible fluid. Displace-

ment $\bar{u}_x$ or $\bar{u}_z$ is simply taken as rate of flow, μ^* as the Newtonian viscosity, η^*.

Very simple cases can be solved algebraically. Consider for example the case of a single viscous layer overlying a viscous half space of different viscosity. Since the solution must remain finite at large depths, we need only solve:

$$
\begin{pmatrix} \tilde{} \\ \tilde{} \\ 0 \\ -1 \end{pmatrix} = \mathbf{P} \begin{pmatrix} A \\ B \\ A \\ B \end{pmatrix}
$$

Here $\mathbf{P}$ is given by (III-12). The solution of the above equation is:

$$
2\mu^* k\bar{u}_z = \frac{(\tilde{\mu} + \tilde{\mu}^{-1})S'C' + kD(\tilde{\mu} - \tilde{\mu}^{-1}) + (S'^2 + C'^2).}{2C'S'\tilde{\mu} + (1 - \tilde{\mu}^2)k^2D^2 + (\tilde{\mu}^2 S'^2 + C'^2)}.
$$

(III-19)

Where $S' = \sinh kD$, $C' = \cosh kD$. The right side has limits of -1 for $kD \to 0$, and $\dfrac{-1}{\tilde{\mu}}$ for kD large. These are the appropriate uniform (non-layered) half-space limits for large and small harmonic wavelengths, respectively.

For the viscous case, the displacement of the surface can be described analogously to (III-16):

$$
h(t) = h_0 e^{-t/\tau},
$$

where

$$
\boxed{\tau = \tau^* \mathfrak{R}}
$$

(III-20)

$$
\boxed{\mathfrak{R} = \frac{2C'S'\tilde{\eta} + (1 - \tilde{\eta})k^2D^2 + (\tilde{\eta}S'^2 + C'^2)}{(\tilde{\eta} + \tilde{\eta}^{-1})S'C' + kD(\tilde{\eta} - \tilde{\eta}^{-1}) + (S'^2 + C'^2)}}
$$

(III-21)

τ^* is given by (III-17). $\tilde{\eta} \equiv \dfrac{\eta_{\text{channel}}}{\eta^*}$; η^* is the viscosity of the substrate beneath the channel.

d. Fluid Flow Confined to a Layer.

Suppose $z = 0$ is a perfectly rigid boundary above which is either a fluid or deformable elastic layer of thickness D.

$z = D$

$z = 0$

The boundary conditions at $z = 0$ are $\bar{u}_z = \bar{u}_x = \bar{u}_y = 0$. Let us apply only normal stresses to the free surface. So at $z = D$

$$\bar{\tau}_{xz} = 0$$
$$\bar{\tau}_{zz} = -1$$

Then the general solution in the layer is

$$\bar{u}(z) = P(z, 0)\bar{u}_0,$$

where

$$\bar{u}_0 = \begin{pmatrix} 0 \\ 0 \\ C' \\ D' \end{pmatrix},$$

and

$$\bar{u}(D) = \begin{pmatrix} \sim \\ \sim \\ 0 \\ -1 \end{pmatrix} = P(D, 0) \begin{pmatrix} 0 \\ 0 \\ C' \\ D' \end{pmatrix}.$$

Using P as in (III-12) it is easy to solve the above two simultaneous equations. The resulting $\bar{u}(D)$ is:

$$\bar{u}(D) = \begin{bmatrix} 2\mu^* ik\bar{u}_x \\[6pt] 2\mu^* k\bar{u}_z \\[6pt] i\bar{\tau}_{xz} \\[6pt] \bar{\tau}_{zz} \end{bmatrix}_{z=D} = \begin{bmatrix} \dfrac{\tilde{\mu}^{-1} k^2 D^2}{(C^2 + k^2 D^2)} \\[10pt] \dfrac{\tilde{\mu}^{-1}(kD - CS)}{(C^2 + k^2 D^2)} \\[10pt] 0 \\[6pt] -1 \end{bmatrix}. \qquad \text{(III-22)}$$

Here $C \equiv \cosh kD$
       $S \equiv \sinh kD$

$$\tilde{\mu} \equiv \frac{\mu}{\mu^*}; \; \mu^* = \text{reference rigidity or viscosity.}$$

As kD becomes large, (III-22) goes into the homogeneous half-space limit and becomes identical to (III-13) at $z = 0(\tilde{\mu} = 1)$.

As kD becomes small, i.e., as the wavelength of the harmonic load applied to the surface becomes large with regard to the thickness of the fluid layer,

$$\bar{\mathbf{u}}(D) \underset{kD \text{ small}}{=} \begin{bmatrix} 2\mu^* ik\bar{u}_x \\ 2\mu^* k\bar{u}_z \\ i\bar{\tau}_{xz} \\ \bar{\tau}_{zz} \end{bmatrix} = \begin{bmatrix} \tilde{\mu}^{-1}((kD)^2 + 0(kD)^4) \\ \tilde{\mu}^{-1}\left(\frac{-2}{3}(kD)^3 + 0(kD)^5\right) \\ 0 \\ -1 \end{bmatrix} . \quad \text{(III-23)}$$

Thus with errors of the order of $(kD)^5$, for large λ,

$$2\mu^* k\bar{u}_z = \frac{2}{3}(kd)^3, \quad\quad\quad\quad\quad\quad\quad\quad\quad\quad \text{(III-24)}$$

for a fluid channel of viscosity η^*, τ becomes:

$$\tau = \frac{3\eta^*}{\rho g k^2 D^3} = \frac{\tau^*}{\frac{2}{3}(kD)^3} \quad\quad\quad\quad\quad\quad\quad \text{(III-25)}$$

The decay time of a load harmonic is proportional to $\frac{1}{k^2}$ not k as in the viscous half-space case (III-17). This result is well known and due originally to Takeuchi. (See also Jeffreys, 1959, §11.092 p. 342). As shown by equation III-56, flow in a thin channel is directly analogous to heat conduction.

A decay time that does not assume kD small (i.e., load harmonic wavelengths large with respect to the channel thickness) follows directly from (III-22):

$$\tau = \tau^* \mathcal{L} \qu\quad\quad\quad\quad\quad\quad\quad\quad\quad\quad\quad\quad \text{(III-26)}$$

Where

$$\mathcal{L} = \frac{C^2 + k^2 D^2}{\tilde{\eta}^{-1}(CS - kD)} \tag{III-27}$$

(III-26 and 27) describe the adjustment of a harmonic load of any wavelength applied to a viscous channel of viscosity $\tilde{\eta}\eta^*$ and thickness D.

e. Harmonic Loading of an Elastic "Lithosphere" over a Fluid Substratum.

If a harmonic load is applied to a fluid half-space, the surface of the fluid will deform until the weight of the fluid displaced from the equilibrium level balances the applied load (Archimedes' principle). If an elastic crust (or lithosphere) covers the fluid, part of the applied load will be supported by the crust, part by the buoyant forces of the fluid beneath acting through the crust.

The buoyant forces arise from displacement of the bottom surface of the crust. Consequently, we can distinguish two polar situations: (1) The wavelength of the harmonic load is so small that the crust appears effectively infinite. The bottom of the crust is not deformed at all and the entire load is supported by the crust. (2) The wavelength of the harmonic load is so great that the deformation of the crust requires essentially no force. The crust supports none of the load. The deformation at the bottom of the crust is the full Archimedean deformation, and the fluid supports the whole load. The crust thus acts like a low-pass filter.

To solve more exactly, let us assume that no horizontal stresses are applied to the top surface of the crust. Let us specify that the bottom of the crust is deformed an amount $\bar{u}_z^B$, and that this gives rise to a vertical stress of magnitude $\bar{\tau}_{zz}^B = \rho g \bar{u}_z^B$, where ρ is the density of the fluid beneath the crust. We assume in addition that $\bar{\tau}_{zx}^B = 0$, i.e., that the fluid substratum has zero viscosity. So

$$\bar{\mathbf{u}}\,(\text{Top}) = \begin{bmatrix} A \\ B \\ 0 \\ C' \end{bmatrix}$$

$$\bar{\mathbf{u}} \text{ (Bottom)} = \begin{bmatrix} \widetilde{2\mu^* k\bar{u}_z^B} \\ 0 \\ \rho g\bar{u}_z^B \end{bmatrix}$$

$$\bar{\mathbf{u}} \text{ (Bottom)} = \mathbf{P}(-H, 0)\,\bar{\mathbf{u}} \text{ (Top)}.$$

We have three equations and three constants to determine. We solve

$$\begin{bmatrix} 2\mu^* k\bar{u}_z^B \\ 0 \\ \rho g\bar{u}_z^B \end{bmatrix} = \begin{bmatrix} -kDC & C - kDS & S - kDC \\ S + kDC & kDS & kDC \\ -kDS & S - kDC & C - kDS \end{bmatrix} \begin{bmatrix} A \\ B \\ C' \end{bmatrix}, \tag{III-28}$$

where

$$D \equiv -H$$
$$C = \cosh kD$$
$$S = \sinh kD.$$

$$A = 2\mu^* ik\bar{u}_x^T = \frac{2\mu^* k(k^2 H) - \rho g kH}{S + kHC}\,\bar{u}_z^B \tag{III-29}$$

$$\bar{u}_x^T \underset{kH \ll 1}{=} -i\bar{u}_z^B \frac{\rho g}{4k\mu^*} \doteq 0$$

$$B = 2\mu^* k\bar{u}_z^T = \frac{2\mu^* k\bar{u}_z^B[CS + kH] + \rho g\bar{u}_z^B S^2}{S + kHC} \tag{III-30}$$

$$\bar{u}_z^T \underset{kH \ll 1}{=} \left[1 + \frac{\rho g}{4\mu^*}\right]\bar{u}_z^B \doteq \bar{u}_z^B,$$

$$C' = \bar{\tau}_{zz}^T = \frac{2\mu^* k[S^2 - k^2 H^2] + \rho g[CS + kH]}{S + kHC}\,\bar{u}_z^B \tag{III-31}$$

$$\bar{\tau}_{zz}^T \underset{kh \ll 1}{=} \rho g\bar{u}_z^B.$$

We see that $\bar{u}_x^T$, $\bar{u}_z^T$, $\bar{\tau}_{zz}^T$ behave as expected as λ becomes large. From (III-31) we can compute

$$\alpha \equiv \frac{\bar{\tau}_{zz}^T}{\bar{u}_z^B \rho g}, \tag{III-32}$$

$$\alpha = \frac{\dfrac{2\mu^*k}{\rho g}\,[S^2 - k^2H^2] + [CS + kH]}{S + kH}.$$

(III-33)

If $H = 30$ km, $\mu^* = 10^{12}$ dyne-cm^{-2}, $\rho g = 3.3 \times 10^3$ gm-cm^{-2}-sec^{-2}, $k = \dfrac{2\pi}{\lambda}$ and λ is measured in units of 30 km, then

$$\alpha = \frac{200k[S^2 - k^2H^2] + [CS + kH]}{S + kHC}.$$

(III-34)

(III-34) is evaluated and plotted in Figure III-3.[3]

Roughly, the dimensions of the Baltic ice lake stage of the Fennoscandian glacial load were 1800×800 km. The dimensions of ancient Lake Bonneville were roughly 140×290 km. The Fourier transform of loads of these dimensions will give spectral distributions with considerable energy at wavelengths about twice these dimensions. Most

---

[3]We could have made the same calculation assuming a compressible crust. These calculations merely introduce a factor multiplying the first term of equation III-33, so the equation would read:

$$\alpha = \left\{ \frac{2\mu^*k}{\rho g}\,\frac{\lambda + \mu}{\lambda + 2\mu}\,[S^2 - k^2H^2] + [CS + kH] \right\} \Big/ [S + kHC]$$

The compressibility of the crust has little effect, as can be seen in Figure III-3.

Historically this subject was first approached by considering the flexure of the crust to bending moments Jeffreys (1959, §6.06). This approach involves a parameter D, known as the flexural rigidity, which is useful in characterizing the elastic strength of the crust (Walcott 1970). The flexure approach may be generalized to cover distributed loads. In doing so, it may be seen that such an approach is entirely equivalent to the one we have developed above. In fact, our α may be alternatively defined:

$$\alpha = 1 + \frac{k^4D}{\rho g}$$

where k = wave number, ρ = density of fluid substratum, g = gravitational field strength, and D = flexural rigidity.

$$D = \frac{EH^3}{12(1 - v^2)}$$

where H = thickness of elastic layer, E = Young's modulus, v = Poisson's ratio. (See Nadai, 1950, Vol. II, p. 302 Eq. 10-83.)

Brotchie and Silvester have considered the effects of the elastic crust in the case of the Canadian glaciation (1969).

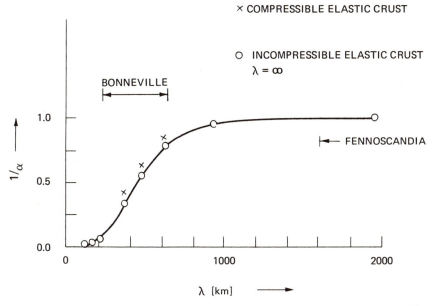

Figure III-3. Displacement of base of elastic crust with flexural rigidity 6×10^{23} N-m overlying an inviscid mantle as a function of isostatic displacement for various wavelength harmonic loads. $D = 6000 \times 10^{20}$ N-m could correspond to a 30 km crust with $\lambda = \mu = 1.0 \times 10^{12}$ dyne-cm^{-2} or to a 43 km crust with $\lambda = \mu = 3.34 \times 10^{11}$ dyne-cm^{-2}. α^{-1} is tabulated for the incompressible case:

k	α^{-1}	$\lambda[km]$
.10	1.00	1,880
.20	.95	942
.30	.79	626
.40	.55	470
.50	.33	376
.60	.20	314
.70	.12	270
.80	.076	235
.90	.050	210
1.00	.034	188

significant spectral energy for Fennoscandia will thus lie between 1600 km and 3200 km and for Lake Bonneville between 240 km and 580 km. More exact analysis depends on more precise Fourier decomposition of the particular load involved, but we can see from Figure III-3 that the crust is likely to be of importance for loads of the dimension of Lake Bonneville, but of little or no real importance for loads of the dimension of Fennoscandia or larger (save for "minor" details). These matters are discussed in far greater detail in Section IV.B.

From (III-13) we see that the surface of a fluid of viscosity η subject to harmonic load $\bar{\tau}_{zz}$ will deform at a rate

$$\bar{v}_z = \frac{\bar{\tau}_{zz}}{2\eta k}.$$

Thus

$$\bar{v}_z^B = \frac{\bar{\tau}_{zz}^B}{2\eta k} = \frac{\partial \bar{u}_z^B}{\partial t},$$

$$\bar{\tau}_{zz}^B = {}^{applied}\bar{\tau}_{zz}^T - {}^{supported}\bar{\tau}_{zz}^T,$$

$${}^{supported}\bar{\tau}_{zz}^{T} \overset{(\text{III-32})}{=} \rho g \alpha \bar{u}_z^B.$$

Thus

$$\frac{\partial \bar{u}_z^B}{\partial t} = \frac{{}^{app}\bar{\tau}_{zz}^T - \rho g \alpha}{2\eta k} \bar{u}_z^B,$$

$$\frac{\rho g \alpha \bar{u}_z^B - {}^{app}\bar{\tau}_{zz}^T}{\rho g \alpha \bar{u}_{z0}^B - {}^{app}\bar{\tau}_{zz}^T} = e^{-\rho g t \alpha / 2\eta k}. \tag{III-35}$$

Suppose there is no load applied to the top of the crust, but the bottom of the crust is initially deformed an amount $\bar{u}_{z0}^B$. Then the crust and fluid will adjust to equilibrium:

$$\bar{u}_z^B(t) = \bar{u}_{z0}^B e^{-\rho g t \alpha / 2\eta k}. \tag{III-36a}$$

If $\bar{u}_{z0}^B = 0$ and top of crust is loaded ${}^{app}\bar{\tau}_{zz}^T$, the crust will deform:

$$\bar{u}_z^B(t) = \frac{{}^{app}\bar{\tau}_{zz}^T}{\rho g \alpha} (1 - e^{-\rho g t \alpha / 2\eta k}). \tag{III-36b}$$

(III-36 a,b) are exactly the same as the corresponding equations for the viscous half-space without crust except ρg is replaced by $\rho g \alpha(k)$. $\alpha(k)$ runs from 1 for large λ to large values for short λ (see Figure III-3) We thus see that the elastic crust has two effects:

(1) For loading, full isostatic adjustment is achieved at[4]

$$\frac{{}^{app}\bar{\tau}_{zz}^T}{\rho g \alpha} \tag{III-37}$$

rather than

$$\frac{{}^{app}\bar{\tau}_{zz}^T}{\rho g}$$

[4] See Brotchie and Silvester (1969) for examples of this.

(2) The rate of isostatic adjustment toward equilibrium is augmented[5] to

$$e^{-\rho g \alpha t / 2 \eta k} \quad \text{from} \quad e^{-\rho g t / 2 \eta k} \tag{III-38}$$

where

$$\alpha = \alpha(k).$$

f. Harmonic Load Response of a Compressible Elastic Half-Space.

We proceed exactly as in the incompressible case (Section III.A.2.b.)

$$G^{+} = \frac{e^{kz}}{2}$$

$$\cdot \left[\begin{pmatrix} 1 & \tilde{\mu}\tilde{\sigma}^{-1} & (\tilde{\lambda} + 3\tilde{\mu})\tilde{\sigma}^{-1}\tilde{\mu}^{-1} & 0 \\ \tilde{\mu}\tilde{\sigma}^{-1} & 1 & 0 & (\tilde{\lambda} + 3\tilde{\mu})\tilde{\sigma}^{-1}\tilde{\mu}^{-1} \\ \tilde{\mu}\tilde{\sigma}^{-1}(\tilde{\lambda} + \tilde{\mu}) & 0 & 1 & -\tilde{\mu}\tilde{\sigma}^{-1} \\ 0 & \tilde{\mu}\tilde{\sigma}^{-1}(\tilde{\lambda} + \tilde{\mu}) & -\tilde{\mu}\tilde{\sigma}^{-1} & 1 \end{pmatrix} \right.$$

$$\left. + kz\tilde{\sigma}^{-1}(\tilde{\lambda} + \tilde{\mu}) \begin{pmatrix} 1 & 1 & \tilde{\mu}^{-1} & \tilde{\mu}^{-1} \\ -1 & -1 & -\tilde{\mu}^{-1} & -\tilde{\mu}^{-1} \\ \tilde{\mu} & \tilde{\mu} & 1 & 1 \\ -\tilde{\mu} & -\tilde{\mu} & -1 & -1 \end{pmatrix} \right] \tag{III-39}$$

$$G^{-} = \frac{e^{-kz}}{2}$$

$$\cdot \left[\begin{pmatrix} 1 & -\tilde{\mu}\tilde{\sigma}^{-1} & -(\tilde{\lambda} + 3\tilde{\mu})\tilde{\sigma}^{-1}\tilde{\mu}^{-1} & 0 \\ -\tilde{\mu}\tilde{\sigma}^{-1} & 1 & 0 & -(\tilde{\lambda} + 3\tilde{\mu})\tilde{\sigma}^{-1}\tilde{\mu}^{-1} \\ -\tilde{\mu}\tilde{\sigma}^{-1}(\tilde{\lambda} + \tilde{\mu}) & 0 & 1 & \tilde{\mu}\tilde{\sigma}^{-1} \\ 0 & -\tilde{\mu}\tilde{\sigma}^{-1}(\tilde{\lambda} + \tilde{\mu}) & \tilde{\mu}\tilde{\sigma}^{-1} & 1 \end{pmatrix} \right.$$

$$\left. + kz\tilde{\sigma}^{-1}(\tilde{\lambda} + \tilde{\mu}) \begin{pmatrix} -1 & 1 & \tilde{\mu}^{-1} & -\tilde{\mu}^{-1} \\ -1 & 1 & \tilde{\mu}^{-1} & -\tilde{\mu}^{-1} \\ \tilde{\mu} & -\tilde{\mu} & -1 & 1 \\ \tilde{\mu} & -\tilde{\mu} & -1 & 1 \end{pmatrix} \right] \tag{III-40}$$

[5]This result has been noted previously by McConnell (1965). His method of derivation is different from the method presented here.

As before

$$P = G^+ + G^-. \tag{III-41}$$

The homogeneous half-space solution may be obtained

$$G^-\bar{u}_0 = 0 \tag{III-42}$$

$$(G^+\bar{u}_0)_{\text{rows } 3/4} \underset{z=0}{=} \begin{pmatrix} 0 \\ -1 \end{pmatrix}. \tag{III-43}$$

The rows of the first matrix in either (III-39) or (III-40) are doubly degenerate. The space of the matrix is spanned by either the first and second rows or the third and fourth rows, and this space includes the single vector represented by the second matrix. Thus (III-42) and (III-43) represent four linearly independent equations. Their solution is:

$$\bar{u}_0 = \begin{pmatrix} \dfrac{-1}{\bar{\lambda} + \tilde{\mu}} \\ \dfrac{-\tilde{\sigma}}{\tilde{\mu}(\bar{\lambda} + \tilde{\mu})} \\ 0 \\ -1 \end{pmatrix}. \tag{III-44}$$

From (III-41), (III-42),

$$\bar{u}(z) = P(z)\bar{u}_0 = G^+\bar{u}_0, \tag{III-45}$$

$$\begin{bmatrix} 2\mu^* ik\bar{u}_x \\ 2\mu^* k\bar{u}_z \\ i\bar{\tau}_{xz} \\ \bar{\tau}_{zz} \end{bmatrix} = e^{kz} \left[\begin{pmatrix} \dfrac{\tilde{\mu}}{\bar{\lambda} + \tilde{\mu}} \\ \dfrac{-\tilde{\sigma}}{\bar{\lambda} + \tilde{\mu}} \\ 0 \\ -1 \end{pmatrix} + kz \begin{pmatrix} -1 \\ 1 \\ -1 \\ 1 \end{pmatrix} \right]. \tag{III-46}$$

(III-46) has been derived in other ways by the author and by Durney (1964). It also goes into the incompressible case (III-13) as $\lambda \to \infty$. (Remember $\tilde{\sigma} = \bar{\lambda} + 2\tilde{\mu}$.)

g. Harmonic Load Response of a Viscoelastic Half-Space.

In this section we investigate the extent to which the elastic and viscous equation of motion may be considered decoupled. This is an important lemma to the particular method of solution we shall use. Insight is provided by the elastic response of a viscoelastic half-space which can be solved analytically.

From (III-46) the vertical elastic response at $z = 0$ of a compressible half-space to a harmonic load $\rho g h$ is:[6]

$$\bar{u}_z \big|_{z=0} = \frac{-\rho g h}{2k} \left(\frac{1}{\mu} + \frac{1}{\lambda + \mu} \right).$$

The theorem of correspondence may be applied directly. Substituting from (II-59), $\lambda(s)$, $\mu(s)$ for λ, μ, the transformed viscoelastic solution is:

$$\bar{\bar{u}}_z \big|_{z=0} = \frac{-\rho g}{2k} \left\{ \frac{\bar{h}}{\mu_e} + \frac{\bar{h}}{s\mu_v} + \frac{sh}{(\lambda_e + \mu_e)s + \frac{\mu_e}{\mu_v} K_e} \right.$$

$$\left. + \frac{\frac{\mu_e}{\mu_v} h}{(\lambda_e + \mu_e)s + \frac{\mu_e}{\mu_v} K_e} \right\}.$$

The inverse is:

$$\bar{u}_z(t) = \frac{-\rho g}{2k} \cdot$$

$$\left\{ \frac{h}{\mu_e} + \frac{h}{\lambda_e + \mu_e} + \int_0^t \frac{h(\tau)}{\mu_v} d\tau + h \left[\frac{1}{K_e} - \frac{1}{\lambda_e + \mu_e} \right] (1 - e^{-at}) \right\},$$

$$\text{(III-47)}$$

where $a \equiv (\mu_e/\mu_v) K_e / (\lambda_e + \mu_e)$.

1) Interpretation for the Case of Glacial Loading and Unloading. Equation (III-47) is valid for a constant applied load equal to $-\rho g h$. This constant load causes an immediate elastic response $\dfrac{-\rho g h}{2k}$ $\cdot \left(\dfrac{1}{\mu_e} + \dfrac{1}{\mu_e + \lambda_e} \right)$ and a steady linear viscous displacement $\dfrac{-\rho g h t}{2k\mu_v}$. The last term represents the transition from the immediate solid-elastic

---

[6]Note that we let $\bar{\tau}_{zz} = -\rho g h$ instead of -1. In the uniform half-space we have also let $\mu^* = \mu$.

response to the ultimate fluid-elastic response we have previously discussed in Section II.B.3. At $t = 0$, the sum of the first, second, and fourth terms (the initial elastic response of the half-space) is:

$$-\frac{\rho g h}{2k}\left(\frac{1}{\lambda_e + \mu_e} + \frac{1}{\mu_e}\right).$$

At $t = \infty$ the sum of the first, second, and fourth terms (the ultimate elastic response of the half-space) is:

$$\frac{-\rho g h}{2k}\left(\frac{1}{K_e} + \frac{1}{\mu_e}\right).$$

Since $K_e = \lambda_e + \frac{2}{3}\mu_e$, if $\lambda_e = \mu_e$ the magnitude of the solid-elastic fluid-elastic transition is 6.6% of the initial elastic response. From (III-47) the solid-elastic fluid-elastic response will be 90% complete when $at = 2.3$ or

$$t_{90\%} = \frac{2.3\,\mu_v}{\mu_e}\left(\frac{\lambda_e + \mu_e}{\lambda_e + \frac{2}{3}\mu_e}\right). \tag{III-48}$$

For $\lambda_e = \mu_e = 10^{12}$ dyne-cm^2 and μ_v between 3×10^{21} and 3×10^{22} poise

$$t_{90\%} = 200 \text{ to } 2000 \text{ years.}$$

We see that the solid-elastic fluid-elastic transition is of small magnitude (being only about 7% of the total elastic response) and relatively rapid (being 90% complete in 666 years for a 10^{22} poise mantle). As we might expect the rate of transition is independent of k. The transition occurs in each volume element with no relation to boundary conditions or surface effects. In most cases we shall consider that the solid-elastic fluid-elastic response is much more rapid than the viscous isostatic adjustment. Errors are negligible for small scale loads because in these cases the magnitude of the total elastic response is generally much less than the magnitude of isostatic adjustment. Thus with little error (III-47) can be written:

$$\bar{u}_z\big|_{z=0} = \frac{-\rho g h}{2k}\left\{\frac{1}{\mu_e} + \frac{1}{K_e} + \int_0^t \frac{dt}{\mu_v}\right\}. \tag{III-49}$$

In other words, the viscoelastic response to a given harmonic load at any moment is the sum of 1) an elastic response appropriate to the fluid, and 2) a steady viscous flow.

If the load changes with time, the same conclusions pertain. In particular an initial harmonic deformation of a viscoelastic solid will decay exponentially (as for a simple viscous solid). The elastic response need only be taken into account. The position of the surface of a viscoelastic solid that is depressed at $t = 0$ an amount h_0 (harmonic amplitude) can be expressed:

$$h = h_0 e^{-\rho g t / 2\mu_v k} \left[1 - \frac{\rho g}{2k} \left(\frac{1}{\mu_e} + \frac{1}{K_e} \right) \right] \qquad \text{(III-50)}$$

Notice that there is an immediate elastic uplift following (at $t = 0$) the removal of whatever forces caused the surface deformation.

2) Decoupling of the Elastic and Viscous Equations of Motion. The solid-elastic fluid-elastic transition due to initial unloading is small in magnitude and rapid enough to be lost geologically in the much greater immediate elastic uplift attendant to unloading and in the uncertainties regarding the history of load removal. The transition effects that will occur as isostatic adjustment proceeds are even smaller (7% of a 30% elastic response in Canada gives a maximum value of about 2% of the fluid adjustment) and thus negligible.

If these two transition effects were the only phenomena connecting the viscous and elastic equations of motion, having argued that they are both insignificant to our investigation, the elastic and viscous equations would decouple. The total displacement at the surface would simply be the sum of the viscous and elastic displacements. This would remain equally true at large scale and when hydrostatic pre-stress and self-gravitation are taken into account. There is an additional source of coupling, however, and that is the alteration of the stress distribution by the variation of the elastic or viscous parameters with depth (see Figure III-2).

It may be possible to argue in some general way that variation of elastic parameters with depth should only produce second order variation of the fluid adjustment. This seems reasonable, but the only way I could convince myself that it was necessarily so was to compute the response of a non-self-gravitating viscoelastic medium whose elastic properties varied with depth using numerical techniques (see Cathles, 1971, p. 85 and Appendix VIII).

For vertical gradients in the elastic parameters, greater than those found in the mantle, viscous flow was found to be unaffected within the accuracy of the method. In the incompressible flat space case this was within 1.5% of deformation or stress at all depths. In the compressible case, viscous rate of flow was altered by about 2%. Flow was also essentially unaffected by elastic gradients in a spherical geometry (±4%). There is some question whether the indication of slight alterations is due to errors in calculation. For reasons discussed in the reference cited above, I suspect this is true at least to some extent. Although the answer might be of some interest to students of continuum mechanics, a precise answer is of no direct interest here.

We conclude, with at most a small error, that the elastic and viscous equations of motion may be assumed to decouple. This is a substantial conceptual simplification. It permits easy handling of the special boundary conditions associated with hydrostatically pre-stressed self-gravitating bodies, where elastic and viscous displacements must be distinguished from the start. It permits calculations to be made in a more physically direct fashion than through Laplace transform techniques.

3. Phase Changes and Non-Adiabatic Density Gradients

As commented earlier, only non-adiabatic density gradients count in the buoyant terms of (III-2) or (III-5). If the density gradient is adiabatic, a material element will simply change its density as it moves through the zero order hydrostatic pressure field, always adapting, chameleon-like, to the density of its environment. Thus, if $\frac{\partial \rho}{\partial z}$ is purely adiabatic, no buoyant forces operate.

For our purposes, there are two ways the density gradient may depart from adiabatic: compositional or phase change density gradients.[7] A compositional change is the case at the core-mantle boundary, a situation we have already discussed in considering boundary conditions (Section II.A.4.). A region of phase change is also a region of non-

---

[7]Non-adiabatic density gradients exist in the absence of phase changes, but their magnitudes for the case of the earth are probably small enough to be neglected. At atmospheric pressure, the coefficient of volume expansion of olivine is 30×10^{-6} °C^{-1} (Clark, 1966, Table 6-8). Thus, if the bottom of the mantle were 1000°C cooler than its adiabatic temperature, the non-adiabatic density gradient integrated over the entire mantle would be only about .12 gm/cc (i.e., 4 gm/cc $\times$ 30 $\times$ 10^{-6} $\times$ 10^3). This is a maximum estimate since the coefficient of volume expansion decreases with increasing pressure. In this study, we assume that non-adiabatic density gradients in the mantle that are unrelated to phase transitions or compositional changes are small enough to be neglected.

adiabatic density gradients if the temperature profile through it is non-adiabatic. An adiabatic temperature profile can be maintained only if there is constant (slow) fluid motion through the phase change region (i.e., some form of mantle convection). Seismological evidence indicates there are phase changes in the earth's upper mantle between 350 and 650 km. Thus a determination of the degree to which there are non-adiabatic density gradients in this region can tell us something of the existence and nature of mantle convection.

There may also be phase changes at the base of the earth's crust. Glacial uplift phenomena place restrictions on the nature of this phase transition.

It is convenient to consider the effect of a phase transition zone on fluid flow and elastic response separately, although this is of course somewhat artificial. In the section immediately following, phase boundary migration in an adiabatic mantle, we *assume* that an adiabatic mantle is physically achievable. Section III.A.3.b. discusses the conditions under which such a mantle can exist.

a. Pseudo-Elastic Responses of Phase Boundaries to Sudden Load Redistributions

1) In an Adiabatic Mantle. A reasonably sudden change or redistribution of surface loads can produce only adiabatic changes in the mantle. (There is not enough time for the process to be isothermal.) An already adiabatic mantle should remain so through an adiabatic load redistribution. Phase transition zones should be found at the same pressure horizons as previously. The adiabatic temperature curve of Figure III-4 will just shift to the left an amount appropriate for the pressure perturbation. The phase transition zone will immediately migrate and always be at equilibrium position.

This may appear strange; a gedanken experiment is perhaps useful. Enclose a bit of mantle material at 1000 km depth in a thermally insulated sack and pull it to the surface. The material in the sack will cool as elastic expansion follows the pressure release, and it will cool more as it goes through the phase transition at 430 km depth. But if the mantle temperature profile is adiabatic, the sack material will everywhere be indistinguishable (in temperature, density and composition) from the mantle material just outside.

Now rerun the experiment. Start again at 1000 km depth but this time apply a slight pressure perturbation to the sack so that the material in the sack is always at a pressure $\Delta_A P$ greater than the outside mantle material. At 1000 km depth the pressure perturbation will com-

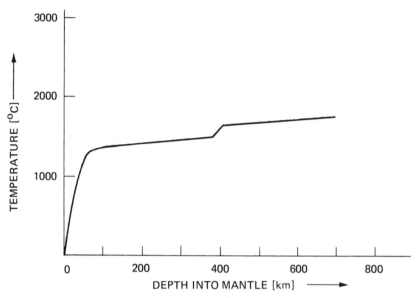

Figure III-4. The adiabatic temperature profile of Turcotte and Oxburgh. The profile shows step at 400 km depth associated with the olivine-spinel phase transition. The temperature profile becomes non-adiabatic near the surface (mostly within the lithosphere). As Turcotte and Oxburgh point out, the adiabatic temperature profile has some support from temperature profiles inferred from electrical conductivity measurements. The figure, reproduced with the author's permission, is from D. L. Turcotte and E. R. Oxburgh, *Journal of Geophysical Research, 74,* p1466, © American Geophysical Union.

press the sack material, and it will heat slightly. The sack material will look precisely like mantle material at a slightly greater depth $\left(1000 \text{ km} + \dfrac{\Delta_A P}{\rho g}\right)$.

As the sack is raised, the similarity with mantle material a distance $\dfrac{\Delta_A P}{\rho(z)g}$ deeper will be preserved. The phase transition will begin and finish sooner. If mantle properties (temperature, density, phase composition) are plotted versus pressure, the sack will trace out a curve displaced an amount $\Delta_A P$ to the right of the mantle curve.

In principle, application of a pressure perturbation to the whole mantle is no different from application of such a perturbation to the test material in our sack. We thus see that increasing the pressure on an adiabatic mantle will cause an immediate migration of any phase boundaries to the identical pressure horizon at which they previously existed.

Under a surface load, this immediate migration will appear like an augmented elastic response. Immediately after loading (of a half-space) the adiabatic phase boundary will be elevated an amount $(\Delta z)_{max}$.

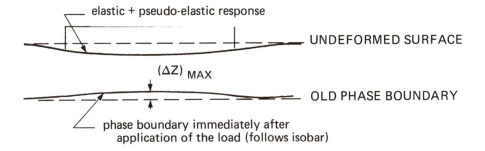

phase boundary immediately after
application of the load (follows isobar)

We shall see in a few pages that flow will occur through the phase boundary as if it weren't there. After full isostatic adjustment is achieved, $\Delta_A P = 0$, the phase boundary will be at its original position and Δz will again be everywhere zero.

We can estimate $(\Delta z)_{max}$: The total change in density due to phase changes between 350 and 700 km depth is probably between 0.5 and 1.0 gm/cc (see Figure III-9). The density just above the phase change zone is probably about 3.5 gm/cc. Phase boundary migration to application of a 1 km thick ice load will be about $1000\left(\dfrac{1.0}{3.5}\right) = 286$ m. This migration will cause a decrease in total mantle thickness of $\dfrac{\Delta\rho}{\rho + \Delta\rho}$ · (286 m) or of between 36 m and 64 m depending on the value of $\Delta\rho$. (0.5 or 1.0 gm/cc). The total isostatic adjustment expected to a 1 km ice load is about 300 m so the pseudo-elastic response associated with the (adiabatic) mantle phase boundaries is between 12 and 21% of the isostatic response.

The immediate elastic uplift following glacial unloading will probably be difficult to observe. The post-glacial isostatic uplift will appear as if the load were 10—20% smaller than it was. Of course for smaller loads the pressure perturbation at 350–650 km depth may not be as large as at the surface.

2) Phase Boundary Migration in a Mantle or Crust Where the Temperature Profile Is not Adiabatic. O'Connell and Wasserburg (1967) have studied the rate of phase boundary migration when the temperature gradient in the vicinity of the phase transition is not adiabatic but has some non-zero value (say α). We summarize relevant parts of their analysis below.

Suppose the temperature at which the phase change takes place is linearly related to the pressure at which it occurs:

$$T_c = GP_c - F \qquad \text{(Clapeyron equation)(III-51)}$$

Given the temperature of the phase boundary, we immediately know its pressure and vice versa. Consider the uniform loading of a half-space which undergoes a phase change at some depth z_p. The increase over hydrostatic pressure due to the load is $\Delta_A P$ and is the same at all depths.

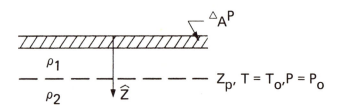

Suppose the temperature gradient at z_p is α, that the phase transition liberates L calories/gram of material transformed and that the heat capacity of the second phase is c cal/gm $-$ °C, the location of the phase boundary immediately after application of the load, $\Delta_A P$, will then be determined by:

$$T_0 + \frac{L}{c} + \alpha \Delta_m z = G(P_0 + \Delta_A P + \rho g \Delta_m z) - F,$$

or since (III-51) still holds

$$\Delta_m z = - \frac{G \Delta_A P - \dfrac{L}{c}}{G \rho g - \alpha}. \qquad \text{(III-52)}$$

Note ρ is the density of the lighter (upper) phase.

If $L/c \leq G \Delta_A P$ the phase boundary is what O'Connell and Wasserburg call "overdriven" and the boundary can migrate a little way, $\Delta_m z$, immediately. Migration to the ultimate equilibrium position must await the conductive dissipation of latent heat (L/c). When this is accomplished (III-52) becomes

$$(\Delta_m z)_{\substack{\text{ultimate} \\ \text{equilibrium}}} = -\frac{\Delta_A P}{\rho g - \dfrac{\alpha}{G}}. \qquad \text{(III-53)}$$

Notice that the pre-existing temperature gradient, α, adjacent to the phase change zone has the effect of magnifying or reducing the magnitude of ultimate migration depending on the slope of the clapyron curve, G. The magnification may be expressed

$$M \equiv \frac{\text{Magnification of}}{\text{phase zone migration}} = \frac{1}{1 - \dfrac{\alpha}{\rho g G}}. \qquad \text{(III-54)}$$

If M is greater than one the phase boundary may migrate to lower pressure horizons than it originally occupied. Note that at ultimate equilibrium the total thickness of mantle material will have been reduced by the conversion of lighter phases to denser phases an amount ΔH:

$$\Delta H = \frac{\Delta \rho}{\rho + \Delta \rho} \left(\frac{\Delta_A P}{\rho g} \right) M. \qquad \text{(III-55)}$$

Here we assume no fluid flow takes place above the phase transition.

Table III-1 gives a feeling for the magnitude of the quantitites mentioned for the olivine-spinel phase transition and for the phase transition that O'Connell and Wasserburg proposed for the base of the crust. It can be seen M is probably only significant for the crustal phase transition and that for glacial size loads the phase boundaries are not "overdriven." This means all phase boundary migrations must await conductive dissipation of latent heat.

It should be pointed out that O'Connell's and Wasserburg's analysis is valid only for a non-adiabatic mantle. In an adiabatic (isentropic) mantle

$$\frac{\partial T}{\partial P} = G = \frac{\partial T}{\partial z} \frac{\partial z}{\partial P} = \frac{\alpha}{\rho g}$$

Thus the denominator of (III-52) is zero and the expression blows up.

For a non-adiabatic mantle, since from Table III-1 none of the phase boundaries is overdriven, the significance of their response to glacial

Location of Phase Change and Reference	L Latent heat of transition [cal/gm]	c Heat Capacity of denser phase [cal/gm°K]	α Temperature gradient adjacent to phase transition zone [°K/cm]	G Slope of Clapyron Curve [°K/kbar]	ρ Density of Lighter Phase [gm/cc]	$\Delta\rho$ Change in density due to phase transition	$(L/c)/G\Delta_A P$ for $\Delta_A P = 1$ kbar (10 Km water)	M Magnification of phase zone migration by temperature gradient	ΔH Ultimate change in mantle and/or crustal thickness above phase transition due to application of $\Delta_A P = .1$ kbar (1 km water load), assuming no flow occurs above the phase transition
Crust-mantle Boundary (O'Connell and Wasserburg, 1968)	15–50	.2	12.4×10^{-5}	75	2.8	.6	1 – 3.3	2.44	−155 m
~430 Km Olivine-Spinel (Turcotte and Oxburgh, 1969) (Turcotte and Schubert, 1971)	40	.3	$.45 \times 10^{-5}$	25	3.5	.28	5.4	~1.0	−21 m
~650 Km Spinel-Oxide (Turcotte and Schubert, 1971)			$.4 \times 10^{-5}$	−20 to −100				1.0	+

Table III-1 Parameters describing phase transitions in a non-adiabatic mantle. Notice that only if $\frac{(L/c)}{G\Delta_A P}$ is less than one is the phase boundary overdriven and an immediate migration can occur without awaiting conductive dissipation of latent heat. None of the phase boundaries appears overdriven.

load redistributions hinges on how fast the latent heat of transition can be conducted away to permit phase boundary migration. The dissipation of heat will not be assisted by the harmonic nature of the load. Any load harmonics large enough to be of interest to us decay too slowly to be of interest in the calculation of the rate of migration of the phase boundary. For example, consider a harmonic load of wavelength 30 km producing a thermal fluctuation of the same wavelength. If we assume an elastic solid (no convection) with constant conductivity and no heat sources, the general heat flow equation

$$\rho c \frac{dT}{dt} = \nabla \cdot (K \nabla T) + A,$$

becomes

$$\rho c \frac{\partial T}{\partial t} = K \nabla^2 T.$$

Assuming $T = T(t) \cos kx$,

$$\frac{\partial T}{\partial t} = \frac{-k^2 K}{\rho c} T.$$

$$\boxed{T = T_0 e^{-Kk^2 t/\rho c}}$$

(III-56)

The relaxation time for $\lambda = 30$ km is 825,000 years. ($K = .005$ cal/cm-sec-°C, $\rho = 2.8$, $c = .2$ cal/gm °C). Thus, without serious error we may neglect the harmonic nature of the load and consider heat conduction to take place perpendicular to a planar phase boundary.

O'Connell and Wasserburg have studied how heat is transported from a moving planar phase boundary and hence how the phase boundary migrates. At early times the solution is reasonably straightforward and may be solved quite adequately by what O'Connell and Wasserburg call the Stephan approximation. From O'Connell's and Wasserburg's work we may construct a table for suitable earth parameters. Table III-2 lists the percentage of the full expected migration of the crustal phase boundary at various times.

From Table III-2 we can conclude that, if the glacial load cycle is less than about 50,000 years, a phase boundary at the base of the crust will not distort isostatic adjustment substantially. Under these circum-

Time Load is Applied t (yrs)	Fraction of distance phase boundary has migrated toward equilibrium position	
	$L = 50$	$L = 15$
10^6	28%	80%
10^5	12%	48%
50,000	8.7%	33%
30,000	6.7%	22%
10,000	3.9%	12%

Values of Parameters used:

$$K = .005 \text{ cal/cm-sec-}°C$$
$$\kappa = .0089 \text{ cm}^2/\text{sec}$$
$$c = .2 \text{ cal/gm-}°C$$
$$\rho = 2.8 \text{ gm}$$
$$\alpha = .124 \times 10^{-3}°C/\text{cm}$$
$$G = 75°C/\text{kbar}$$
$$F = 320°C$$
$$\Delta_A P = 10 \text{ kbar}$$

Total migration of phase boundary = 9.15 km

Table III-2 Percentage of full crustal phase boundary migration as a function of time. Data deduced from O'Connell and Wasserburg (1967, Fig. 11, 12)

stances only a 33% phase boundary migration is expected for $L = 15$ cal/gm. Since the crust above the phase transition is not fluid this indicates a $(.33)(155 \text{ m}) = 52 \text{ m}$ depression[8] of the surface will result from the application of a 1 km thick ice load due to phase boundary migration. The isostatic adjustment expected for such a load is 1000 m/3.4 = 294 m. Thus the depression due to phase boundary migration is 18% of the isostatic adjustment. Of course not all of this will be recovered 12,000 years after unloading.

If an ice load is applied for a million years or so and $L = 15$ cal/gm, the surface depression associated with the migration of a phase boundary at the bottom of the crust grows to about 42% of the depression expected of isostatic adjustment.[9] This is large enough to be observable, and isostatic adjustment in areas of long duration glacial load

[8]See Table III-1.
[9]The phase migration surface depression is $(.8)(155)/294$ or 42% of the isostatic depression.

such as Greenland and Antarctica perhaps offers a test of O'Connell and Wasserburg's hypothesis. If L is greater than 15 cal/gm the test becomes progressively less definitive.

If no flow takes place above the phase transition zone (350 to 650 km depth) the situation is similar to the case of the crustal phase transition. The magnitude of surface depression that could be caused by the mantle phase transition is rather small however. In the previous section we estimated that the maximum surface depression that could be caused by full migration of the mantle phase boundaries was between 10 and 20% of the expected isostatic response. The load cycle is fast compared to the time required to reach thermal equilibrium.   From Table III-2 we might estimate the migration surface depressions would be at most a third of these values. This is insignificant.

The possibility of flow above the mantle phase transition zones distinguish the pseudo-elastic response of this zone from that at the crust-mantle boundary. Flow in the channel above the mantle phase zone will eventually erase the load, and the phase boundary will migrate back to its former position. Such erasure is not possible in the elastic crust and so the pseudo-elastic response associated with conversion of "crust" to "mantle" will be permanent to the extent the load is permanent.

b. The Effect of Phase Transition Zones on Fluid Flow in the Mantle

At ultimate equilibrium, an applied load will attain complete isostatic equilibrium. Any mantle phase boundaries will be undeformed and in equilibrium also.

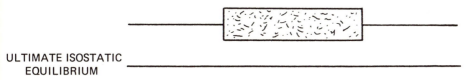

ULTIMATE ISOSTATIC
   EQUILIBRIUM

                                         PHASE BOUNDARY

As the above situation is approached, if the phase boundary cannot migrate fast enough to keep up with the viscous flow deforming it, the phase boundary will be deformed. This deformation will induce buoyant forces. These buoyant forces, operating at the phase boundary, will tend to impede the viscous adjustment.

BUOYANT FORCES TEND
TO ENCOURAGE CHANNEL
FLOW

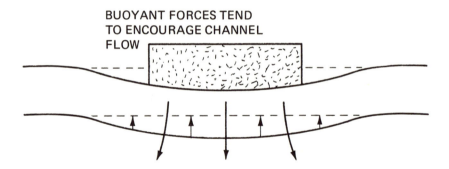

After a certain point, the buoyant forces will balance the non-compensated surface load and will restrict flow entirely to the channel above the phase boundary. This channel flow is in fact an alternate way to achieve phase boundary "migration" and equilibrium.

If the phase boundary does not migrate, the pressure perturbation caused by the load redistribution is relieved only by fluid flow in the channel above the phase boundary.

For the same reasons that a phase boundary will migrate immediately to a pressure perturbation, fluid flow can occur through an adiabatic phase transition zone unaltered and unaffected. In an adiabatic mantle, a fluid element migrating (adiabatically)[10] downward through a region of phase transition will remain everywhere at the same temperature, phase composition, and density as the mantle adjacent to it. Only if the preexisting temperature profile of the mantle is non-adiabatic will the migrating fluid element become hotter, richer in light phase composition, and (for both reasons) lighter than the material adjacent to it and subject to impeding buoyant forces. Of course attempts to flow through the phase change region will tend to establish an adiabatic temperature profile through the mantle.

The degree of adiabaticity of the mantle has been discussed in the literature and has some interesting implications. If latent heat is liberated (by hypothesis) as the composition of the mantle becomes progressively enriched in the denser phase, an adiabatic temperature gradient will be steeper in the region of phase change than elsewhere in the mantle (where the increase in temperature is just associated with increased elastic compression). This is shown in Figure III-4 which reproduces Turcotte's and Oxburgh's (1969) adiabatic mantle temperature profile. The step in the temperature profile through the phase change region will tend to be ironed out by thermal conduction. It can

[10]Glacial uplift flow is rapid enough to be considered adiabatic.

be maintained over long periods *only* if there is a continual flow of material through it (i.e., if heat is continually brought to the phase change region by fluid advection).

Clearly the rate at which steady flow must take place through the region of phase change to maintain an adiabatic temperature gradient depends strongly on the thickness of the phase change zone. If the thickness is zero, since a discontinuity in temperature in a substance of non-zero thermal conductivity is not physically reasonable, an adiabatic temperature gradient through the "zone" is not possible and a phase change will always act to stop fluid transit (convection). Knopoff pointed this out in an early paper on mantle convection (Knopoff, 1964, p. 106). Turcotte and Schubert (1971, p. 7983) show that it is possible to have a zone of transition of non-zero thickness for a univariant phase transition provided there is flow through the boundary at greater than a critical rate. For a reasonable estimate of mantle parameters at the olivine-spinel phase transition they find a critical velocity of 2 mm/yr. Verhoogen (1965) pointed out that a multivariant phase transition could produce a zone of phase transition of non-zero thickness even without any flow through the boundary. Turcotte and Schubert (1971) feel the olivine-spinel phase transition is univariant, and that this is perhaps supported by seismic evidence indicating a narrow transition zone for the olivine-spinel phase transition.

c. Our Approach to Phase Boundaries in the Crust and Mantle

If the mantle is adiabatic, loading will cause an immediate migration of the phase transition zones, but flow through those zones will not be affected. The migration will look like a reduced elastic compressibility at the transition depths and can be modeled in this way. For the larger scale ice loads this will cause the post-glacial uplift to appear as if it were caused by a slightly (10–20%) lighter load redistribution than was actually the case.

In a non-adiabatic mantle, migration of phase boundaries must await the conductive dissipation of latent heat. In this case, the magnitude of the pseudo-elastic effects in the mantle are small, but significant buoyant forces build up at the deformed phase transition zones, and the flow pattern is altered. Channel flow phenomena should be promoted.

Our approach to the mantle phase transitions will be to develop a model in which a non-adiabatic density gradient of appropriate magnitude may be inserted in the upper mantle. The consequences of a phase transition in a non-adiabatic model may thus be elucidated and compared to the adiabatic case where $\Delta\rho_{NA} = 0$, everywhere.

We shall see later that Greenland and Antarctica might answer the question of whether the boundary between the crust and mantle is a phase transition with a small enough latent heat to be of any interest to us. The effect is probably small. The boundary between crust and mantle may not be a phase boundary under continents, but rather a compositional boundary (that may be gradational). Our approach will be to assume a crust-mantle phase boundary does not exist.

B. Spherical Earth

1. Elastic Deformation

With the discussion of the isostatic adjustment of a flat earth and the effects of phase boundaries behind us, we continue to increase the complexity of our model. In this section we consider the effects of a spherical geometry (like the earth), of the non-adiabatic density jump at the core-mantle boundary, of non-adiabatic density gradients in the upper mantle, and of the effects of self-gravitation. We first consider elastic deformation, then viscous deformation, and finally viscoelastic deformation. We pay special attention to the P_1 load harmonic. Finally we compute decay spectra for five earth models that we shall use extensively in the last chapter.

a. Simplest Case: No Gravity

$$\nabla \cdot \tau = 0. \tag{III-57}$$

In case $g_0 = g_1 = 0$ (II-22) reduces to: $\nabla \cdot \tau = 0$. As shown in Appendix III, the elastic displacement $\mathbf{u}$ and stress tractions $\hat{\mathbf{r}} \cdot \tau$ may be expressed:

$$\mathbf{u} = \hat{\mathbf{r}}U + \nabla_1 V,$$
$$\hat{\mathbf{r}} \cdot \tau = \hat{\mathbf{r}}P + \nabla_1 Q,$$

where $\nabla_1 V$ is the surface gradient of V, V having been projected onto the unit sphere. U, V, P, Q may implicitly be considered surface harmonic coefficients of order n (see end of Appendix III). This is not strictly necessary, however and we therefor omit the bar signifying spatial transform (i.e. harmonic component) over U, V, τ, etc.

The Runge-Kutta form of (III-57) is:

$$
r\partial_r
\begin{bmatrix}
\mu^* U \\
\mu^* V \\
P \\
Q
\end{bmatrix}
$$

$$
=
\begin{bmatrix}
-2\tilde{\lambda}\tilde{\beta}^{-1} & -\tilde{\lambda}\tilde{\beta}^{-1}\nabla_1^2 & \beta^{-1}r & 0 \\
-1 & 1 & 0 & \tilde{\mu}^{-1}r \\
4r^{-1}\tilde{\gamma} & 2\tilde{\gamma}r^{-1}\nabla_1^2 & -4\tilde{\mu}\tilde{\beta}^{-1} & \nabla_1^2 \\
-2r^{-1}\tilde{\gamma} & -r^{-1}[(\tilde{\gamma}+\tilde{\mu})\nabla_1^2 + 2\tilde{\mu}] & -\tilde{\lambda}\tilde{\beta}^{-1} & -3
\end{bmatrix}
\begin{bmatrix}
\mu^* U \\
\mu^* V \\
P \\
Q
\end{bmatrix},
$$

$$
\text{(III-57RK)}
$$

where

$$
\begin{aligned}
\tilde{\gamma} &\equiv \tilde{\mu}(3\tilde{\lambda} + 2\tilde{\mu})/\tilde{\beta} & \tilde{\mu} &\equiv \mu/\mu^* \\
\tilde{\beta} &\equiv \tilde{\lambda} + 2\tilde{\mu} & \tilde{\lambda} &\equiv \lambda/\mu^* \\
\mu^* &= \text{rigidity at some particular location.}
\end{aligned}
$$

The derivation of (III-57) is sketched in Appendix III. Note $\nabla_1^2 P_n^m(\theta,\phi)$ equals $-n(n+1)P_n^m(\theta,\phi)$. n is the order of the surface harmonic.

It is convenient to express the elastic displacements in "normalized form." Normalization has the advantage that the solution may be readily compared to the flat-space case where the normalized displacement is invariant with respect to k. For example, at the surface of a uniform earth, (III-46) indicates the solution at large k of (III-57RK) becomes:

$$
\begin{bmatrix}
2\mu^* ik\bar{u}_x \\
2\mu^* k\bar{u}_z \\
i\bar{\tau}_{xz} \\
\bar{\tau}_{zz}
\end{bmatrix}_{z=0}
=
\begin{bmatrix}
\dfrac{-\tilde{\mu}}{\tilde{\lambda}+\tilde{\mu}} \\[2mm]
\dfrac{-\tilde{\beta}}{\tilde{\lambda}+\tilde{\mu}} \\[2mm]
0 \\[1mm]
-1
\end{bmatrix}_{\tilde{\lambda}=\tilde{\mu}=1}
=
\begin{bmatrix}
-0.5 \\
-1.5 \\
0.0 \\
-1.0
\end{bmatrix}.
\qquad \text{(III-58)}
$$

k in the case of the sphere is *defined* in terms of asymptotic limit of large k

$$k = \frac{n + \frac{1}{2}}{b} \text{ (see Appendix V).}$$

In the case of an earth with an inviscid fluid core starting vectors are uniquely determined

$$Q = 0$$
$$P = \rho_c g_0(a)U$$
$$U = U$$
$$V = V$$

so at the core-mantle boundary the starting vectors are

$$\mathbf{u}_0 = \begin{bmatrix} \mu^* \\ 0 \\ \rho_c g_0(a) \\ 0 \end{bmatrix} A + \begin{bmatrix} 0 \\ \mu^* \\ 0 \\ 0 \end{bmatrix} B = \begin{bmatrix} \mu^* U \\ \mu^* V \\ P \\ Q \end{bmatrix}_{\text{Core-Mantle Boundary}}$$

In the case of the homogeneous sphere, we can adopt the incompressible flat-space starting vectors as likely candidates. The actual value of the starting vectors is not important as after a very short distance the results do not depend on the initial value. Essentially, a marble at the center of the earth may have *any* reasonable stress distribution on its surface; the earth's response to surface loading will be unaffected.

The starting vectors are simply propagated from either a very small radius ($\tilde{r} = .001$) or the core-mantle boundary ($\tilde{r} = .5$) to the surface ($\tilde{r} = 1$) where the surface boundary conditions are applied. If a normal stress of 1 dyne and no shear is applied at the surface, the surface boundary conditions are

$$P = -1$$
$$Q = \quad 0$$

These two equations are sufficient uniquely to determine A and B.

Figure III-5 shows the normalized deformation of a homogeneous elastic sphere with and without an inviscid fluid core to $\tilde{r} = 0.5$ for an incompressible and compressible mantle. As can be seen from the graph, the presence of a fluid core, which opposes deformation of the core-mantle boundary with only buoyant forces, permits greater

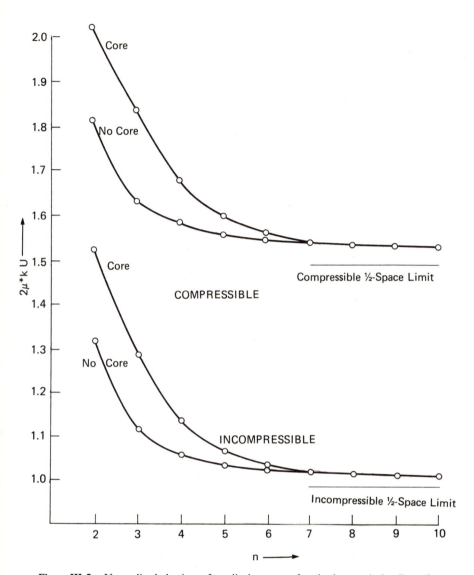

Figure III-5. Normalized elastic surface displacements for the harmonic loading of a sphere with and without a fluid core for both $\lambda = \infty$ and $\lambda = \mu$. Other parameter values were: $\tilde{\lambda} = \tilde{\mu} = 1.0$, $g_0 = 1000$ cm/sec^2, $\rho_c = 10.52$ gm/cc, $\tilde{r}_c = .5$, $\tilde{r}_s = 1.0$, $r^* = 6371$ km, $k = \dfrac{n + \frac{1}{2}}{r_s}$, $\tilde{\mu} = \dfrac{\mu}{\mu^*}$, $\mu^* = 10^{12}$ dyne-cm^{-2}. Boundary conditions at the surface were $P_s = -1$, $Q_s = 0$; at the core mantle boundary, the boundary conditions were $P_c = \rho_c g_0 U_c$, $Q_c = 0$. Curves were computed using (III.57) and starting vectors as described in the text. The results were checked by analytical calculation and computation using propagator methods.

deformation at low order numbers for the same applied stress. However, past $n = 7$ the behavior of the sphere is essentially the same as that of the half-space. The core is too deep to be felt at the surface, and the fact the body is spherical and not flat has lost significance.

In the case of the homogeneous sphere, the Runge-Kutta calculated deformation may be checked against an exact analytical Frobenius solution, and may also be compared to a propagator matrix solution (II-78). This was done, and agreement was excellent.

b. More Complicated Cases of Elastic Deformation

Next in order of complexity is the non-gravitating equation with the hydrostatic pre-stress term added (g_0 = constant).

$$\nabla \cdot \tau - \rho_0 g_0 \nabla U = 0 \qquad\qquad \text{(III-59)}$$

(See p. 77 for (III-59RK).)

If we add to the hydrostatic pre-stress the interaction between the change of density due to compression and some given, fixed, radial gravitational field $-g_0\hat{\mathbf{r}}$, i.e., the $\rho_1 - g_0$ interaction, the system becomes

$$\nabla \cdot \tau - \rho_0 g_0 \nabla U + \rho_0 g_0 \nabla \cdot \mathbf{u}\hat{\mathbf{r}} = 0 \qquad\qquad \text{(III-60)}$$

(See p. 77 for (III-60RK).)

Last, we can relax the condition that g_0 is constant and allow our elastic sphere to become self-gravitating. The self-gravitating differential equation

$$\nabla^2 \phi_1 = 4\pi G \rho_1$$

introduces two additional scalars (i.e., ϕ_1, and some form of its derivative g_1 or $\hat{g}_1$), making our propagator 6 × 6 rather than 4 × 4. Self-gravitation enters the equation of motion via the body force interaction. The equation of motion becomes

$$\nabla \cdot \tau - \rho_0 \nabla(g_0 U) + g_0\rho_0 \nabla \cdot \mathbf{u}\hat{\mathbf{r}} - \rho_0 \nabla \phi_1 = 0 \qquad\qquad \text{(III-61)}$$

The Runge-Kutta form of this equation is given on page 78 in (III-61RK) (see also Appendix III).

Starting vectors for the 6 × 6 system follow from (II-39), (II-42), and (II-34).

$$r\partial_r \begin{bmatrix} \mu^*U \\ \mu^*V \\ P \\ Q \end{bmatrix} = \begin{bmatrix} -2\tilde{\lambda}\tilde{\beta}^{-1} & -\tilde{\lambda}\tilde{\beta}^{-1}\nabla_1^2 & \tilde{\beta}^{-1}r & 0 \\[4pt] -1 & 1 & 0 & \tilde{\mu}^{-1}r \\[4pt] 4r^{-1}\gamma^2 - \dfrac{2\rho_0 g_0 \tilde{\lambda}\tilde{\beta}^{-1}}{\mu^*} & \left(2\tilde{\gamma}r^{-1} - \dfrac{\rho_0 g_0 \tilde{\lambda}\tilde{\beta}^{-1}}{\mu^*}\right)\nabla_1^2 & -4\tilde{\mu}\tilde{\beta}^{-1} + \dfrac{\rho_0 g_0}{\mu^*}r\tilde{\beta}^{-1} & -\nabla_1^2 \\[4pt] \dfrac{\rho_0 g_0}{\mu^*} - 2r^{-1}\tilde{\gamma} & -r^{-1}[(\tilde{\gamma}+\tilde{\mu})\nabla_1^2 + 2\tilde{\mu}] & -\tilde{\lambda}\tilde{\beta}^{-1} & -3 \end{bmatrix} \begin{bmatrix} \mu^*U \\ \mu^*V \\ P \\ Q \end{bmatrix}$$

$$\text{(III-59RK)}$$

$$r\partial_r \begin{bmatrix} \mu^*U \\ \mu^*V \\ P \\ Q \end{bmatrix} = \begin{bmatrix} -2\tilde{\lambda}\tilde{\beta}^{-1} & -\tilde{\lambda}\tilde{\beta}^{-1}\nabla_1^2 & \tilde{\beta}^{-1}r & 0 \\[4pt] -1 & 1 & 4\tilde{\mu}\tilde{\beta}^{-1} & \tilde{\mu}^{-1}r \\[4pt] 4r\tilde{\gamma} - \dfrac{2\rho_0 g_0 \tilde{\lambda}\tilde{\beta}^{-1}}{\mu^*} & \left(2\gamma r^{-1} - \dfrac{\rho_0 g_0}{\mu^*}\right)\nabla_1^2 & 4\tilde{\mu}\tilde{\beta}^{-1} & -\nabla_1^2 \\[4pt] \rho_0 g_0 - 2r^{-1}\tilde{\gamma} & -r^{-1}[(\tilde{\gamma}+\tilde{\mu})\nabla_1^2 + \tilde{\mu}] & -\tilde{\lambda}\tilde{\beta}^{-1} & -3 \end{bmatrix} \begin{bmatrix} \mu^*U \\ \mu^*V \\ P \\ Q \end{bmatrix}.$$

$$\text{(III-60RK)}$$

$$\partial_r \begin{bmatrix} \mu^*U \\ \mu^*V \\ P \\ Q \\ \phi_1 \\ g_1 \end{bmatrix} = r^{-1} \begin{bmatrix} -2\tilde{\lambda}\tilde{\beta}^{-1} & -\tilde{\lambda}\tilde{\beta}^{-1}\nabla_1^2 & r\tilde{\beta}^{-1} & 0 & 0 & 0 \\[4pt] -1 & 1 & 0 & r\tilde{\mu}^{-1} & 0 & 0 \\[4pt] 4\left(\tilde{\gamma}r^{-1} - \dfrac{g_0\rho_0}{\mu^*}\right) & \left(2r^{-1}\tilde{\gamma} - \dfrac{\rho_0 g_0}{\mu^*}\right)\nabla_1^2 & -4\tilde{\mu}\tilde{\beta}^{-1} & -\nabla_1^2 & 0 & 0 \\[4pt] \dfrac{\rho_0 g_0}{\mu^*} - 2r^{-1}\tilde{\gamma} & -r^{-1}[(\tilde{\gamma}+\tilde{\mu})\nabla_1^2 + 2\tilde{\mu}] & -\tilde{\lambda}\tilde{\beta}^{-1} & -3 & \rho_0 & 0 \\[4pt] -\dfrac{4\pi G\rho_0 r}{\mu^*} & 0 & 0 & 0 & 0 & r \\[4pt] 0 & \dfrac{-4\pi G\rho_0 \nabla_1^2}{\mu^*} & 0 & 0 & 0 & -r^{-1}\nabla_1^2 - 2 \end{bmatrix} \begin{bmatrix} \mu^*U \\ \mu^*V \\ P \\ Q \\ \phi_1 \\ g_1 \end{bmatrix}$$

$$\tilde{\gamma} = \frac{\tilde{\mu}(3\tilde{\lambda} + 2\tilde{\mu})}{\tilde{\lambda} + 2\tilde{\mu}} \qquad \tilde{\mu} \equiv \frac{\mu}{\mu^*}$$

$$\tilde{\beta} = \tilde{\lambda} + 2\tilde{\mu} \qquad \tilde{\lambda} \equiv \frac{\lambda}{\mu^*}$$

$$\text{(III-61RK)}$$

$$\mathbf{u}_0 = \begin{bmatrix} \mu^* U_E \\ \mu^* V_E \\ P \\ Q \\ \phi_1 \\ g_1 \end{bmatrix} = \begin{bmatrix} \mu^* \\ 0 \\ \rho_c g_c \\ 0 \\ 0 \\ 4\pi G \rho_c \end{bmatrix} A + \begin{bmatrix} 0 \\ \mu^* \\ 0 \\ 0 \\ 0 \\ 0 \end{bmatrix} B + \begin{bmatrix} 0 \\ 0 \\ \rho_c \\ 0 \\ 1 \\ \dfrac{n}{a} \end{bmatrix} E$$

The starting vectors are also discussed in Appendix IV.

The boundary conditions at the surface ($r = b$) are:

$$[\tau \cdot \hat{\mathbf{r}}]_-^+ = 0 \qquad\qquad\qquad\qquad\qquad\qquad \text{(II-32)}$$

$$g_1 = \frac{-(n + 1)}{b} \phi_1 - 4\pi G \sigma. \qquad\qquad\qquad \text{(II-39)}$$

If we apply a simple normal stress:

$$P = -g_0 \sigma$$
$$Q = 0$$

and let $\sigma \equiv$ redistributed load surface density $= \dfrac{1}{g_0}$ so that $P = -1$ dyne, the boundary conditions are

$$P = -1$$
$$Q = 0$$
$$g_1 = \frac{-(n + 1)}{b} \phi_1 - \frac{4\pi G}{g_0}.$$

These three equations are sufficient to determine A, B, and E.

Once self-gravitation is taken into account, two alternatives in loading the free surface present themselves: We could load the earth by some hypothetical pure stress (i.e., rocket engines). Alternatively we could load the surface by *redistributing* mass. The mass in this redistributed load would attract the earth the same time as the earth attracts the redistributed load. This attraction of the load for the earth turns out to be extremely important.

Figure III-6 shows the importance of these various terms for the case of a self-gravitating compressible elastic sphere with the homogeneous

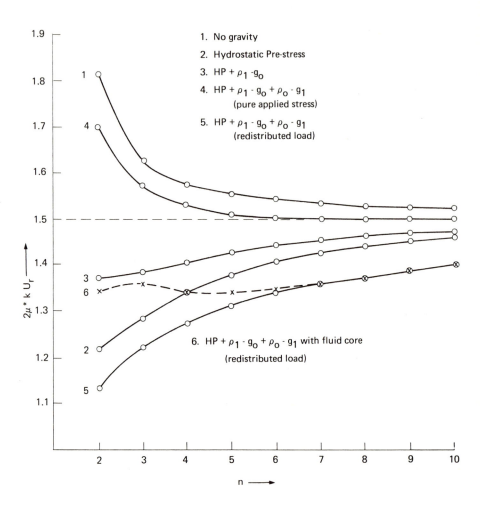

Figure III-6. The normalized surface displacement of a solid elastic sphere with various gravitational phenomena taken into account is shown as a function of order number. The first case (1) assumes $g_0 = g_1 = 0$ and was computed from equation (III-57RK). The second case (2) takes into account *H*ydrostatic *P*re-stress and was computed using (III-59RK). The third case (3) takes into account hydrostatic pre-stress and the $\rho_1 - g_0$ gravitational interaction (III-60RK). The fourth case (4) takes into account HP, $\rho_1 - g_0$, and $\rho_0 - g_1$, assumes a pure applied stress (III-61RK). Case (5) accounts for HP, $\rho_1 - g_0$, and $\rho_0 - g_1$ interactions for a redistributed load. Curve (6) is the same as (5) except the earth is assumed to have an inviscid fluid core with density $\rho_c = 10.52$ gm/cc. For this figure starting vectors (App. IV-24a) were used.

$\tilde{r}$	0.01	0.50	0.50	1.00
$\tilde{\mu}$	1.00	1.00	1.00	1.00
$\tilde{\lambda}$	1.00	1.00	1.00	1.00
ρ_0	12.50	10.52	7.52	3.76
g_0	0.00	1000.00	1000.00	1000.00

$r^* = 6371 \text{ km}$ $\qquad r = \tilde{r}r^*$

$\mu^* = 10^{12}$ $\qquad \mu = \tilde{\mu}\mu^*$

$\qquad\qquad\qquad \lambda = \tilde{\lambda}\mu^*$

Table III-3 Elastic property, density, and gravity profiles for the approximate model earth used for several heuristic calculations. The reason for the peculiar choice of density profile is that we assume in the propagation that $\partial_r g_0 = -4\pi G\rho_0 - 2r^{-1}g_0$, i.e., that $\nabla \cdot g_0 = 4\pi G\rho_0$. The relation between ρ_0 and g_0 shown satisfies this condition exactly in the mantle, and approximately in the core.

earth parameters given in Table III-3. As can be seen, hydrostatic pre-stress tends to diminish deformation. The increased density due to the deformation and the increased gravitational field produced by that density, g_1, almost restore the deformation to its initial non-pre-stress level. Gravitational attraction of the load for the earth changes the deformation substantially (the difference between curve 4 and curve 5).

As is to be expected as n becomes large, the deformation approaches that expected of a non-gravitating, homogeneous, elastic half-space.

1) Brief Physical Interpretation of Gravitational Effects. Following Backus, we have defined the gravitational potential ϕ so that it has the meaning usual to any other potential energy. This is contrary to the usual convention in gravity. $\phi \equiv$ work done by some outside agent *against gravity* in bringing a particle of unit mass from ∞ to r. $\mathbf{g} =$ gravitational attraction per unit mass $= \mathbf{F}_g/m$

$$\mathbf{g} = \frac{-GM_E}{r^2}\hat{r}$$

$$\phi = \int_{\infty}^{r} -\mathbf{g}\, dr = \frac{-GM_E}{r}.$$

Negative potential energy thus corresponds to an attractive force. One must do positive work to separate two gravitating particles. From the above equation it follows that

$$\mathbf{g} = -\partial_r\phi\hat{r} \underset{\substack{\text{more} \\ \text{generally}}}{=} -\nabla\phi_1.$$

From considering the divergence of g from a point mass, it follows

$$\nabla^2\phi = 4\pi G\rho; \quad \nabla^2\phi_0 = 4\pi G\rho_0; \quad \nabla^2\phi_1 = 4\pi G\rho_1.$$

It follows from the above equation that ϕ_1 is diminished near positive mass fluctuations ($\rho_1 > 0$) and augmented near negative mass fluctuations. We may now make the following comments as regards ϕ_1:

ϕ_1 DUE TO SURFACE LOADING:

Redistributed Load

Pure Stress

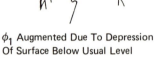

ϕ_1 Diminished Due To Positive Mass In Redistributed Load (Dominates Over Depression Of Surface Due To Load).

ϕ_1 Augmented Due To Depression Of Surface Below Usual Level (Negative Mass).

ϕ_1 DUE TO DEFORMATION OF CORE-MANTLE BOUNDARY:

ϕ_1 Augmented Due To Increase Of Distance To Denser Material.

Of course, if $\rho_c = \rho_m$ deformation of the core-mantle boundary does not perturb the potential.

For a redistributed surface load ϕ_1 is thus diminished at the surface and augmented at the core-mantle boundary. The gradient of ϕ_1, $\partial_r\phi_1$, is reduced, and therefore g is reduced.

The effects of gravitation at both the surface and the core are to diminish deformation. Direct computation confirms this reasoning as can be seen from the surface displacements for *Non-Gravitating* and *Self-Gravitating* boundary conditions at the surface and core-mantle interfaces.[11]

[11]Starting vectors (App. IV-24a) are used.   Values are not to be literally but only relatively interpreted. See Appendix IV.

SURFACE	CORE	$2\mu^{*}kU_{r}$
NG	NG	-2.00
NG	SG	-1.95
SG	NG	-1.37
SG	SG	-1.34

The other gravitational interactions are easy to understand. Application of surface pressure increases the density under the load, increasing $(\hat{\mathbf{g}}_1)$ and hence increasing the deformation through the $\rho_0 - g_1$ interaction. Similarly, the increase in density ρ_1 will increase the deformation through its interaction with g_0.

The decrease in deformation due to the introduction of the hydrostatic pre-stress term results from the resistance lines of equal hydrostatic pressure offer to being pushed together. (See Section III.A.1.a.)

2. Viscous Deformation of a Sphere

The equations for viscous flow (from Section II.A) are:

$$\nabla \cdot \tau - \rho_0 \nabla \phi_1 - \rho_1 \nabla \phi_0 = 0, \qquad (\text{II-23})$$

where

$$\rho_1 = -\nabla \cdot (\rho_0 \mathbf{u}) \equiv -(U\partial_r \rho_0 + \rho_0 \nabla \cdot \mathbf{u}), \qquad (\text{II-21})$$
$$\rho_1 = \text{incompressible case} = -U\partial_r \rho_0.$$

The simplest case is one which assumes gravity is constant: i.e., $\mathbf{g}_0 = \text{constant} = -\nabla \phi_0 = -g_0 \hat{\mathbf{r}}$, the body is not self-gravitating and $\rho_0 \nabla \phi_1 = 0$

$$\nabla \cdot \tau + g_0 U \partial_r \rho_0 \hat{\mathbf{r}} = 0 \qquad (\text{III-62})$$

The constitutive relations are the same as the *incompressible* elastic constitutive relations except strain is replaced by rate of strain. We can obtain the incompressible non-gravitating elastic Runge-Kutta system by letting $\lambda \rightarrow \infty$ in (III-57RK). The result is

$$r\partial_r \begin{bmatrix} \mu^{*}U \\ \mu^{*}V \\ P \\ Q \end{bmatrix} = \begin{bmatrix} -2 & -\nabla_1^2 & 0 & 0 \\ -1 & 1 & 0 & r\tilde{\mu}^{-1} \\ 12r^{-1}\tilde{\mu} & 6r^{-1}\tilde{\mu}\nabla_1^2 & 0 & -\nabla_1^2 \\ -6r^{-1}\tilde{\mu} & -2r^{-1}\tilde{\mu}(2\nabla_1^2 + 1) & -1 & -3 \end{bmatrix} \begin{bmatrix} \mu^{*}U \\ \mu^{*}V \\ P \\ Q \end{bmatrix}$$

The viscous Runge-Kutta equation for a non-self-gravitating fluid sphere with $\mathbf{g}_0 = -g_0\hat{\mathbf{r}} = $ constant is thus[12]:

$$
r\partial_r
\begin{bmatrix}
\eta^* V_U \\
\eta^* V_V \\
P \\
Q
\end{bmatrix}
=
\begin{bmatrix}
-2 & -\nabla_1^2 & 0 & 0 \\
-1 & 1 & 0 & r\tilde{\eta}^{-1} \\
12r^{-1}\tilde{\eta} & 6r^{-1}\tilde{\eta}\nabla_1^2 & 0 & -\nabla_1^2 \\
6r^{-1}\tilde{\eta} & -2r^{-1}\tilde{\eta}(2\nabla_1^2 + 1) & -1 & -3
\end{bmatrix}
\begin{bmatrix}
\eta^* V_U \\
\eta^* V_V \\
P \\
Q
\end{bmatrix}
+
\begin{bmatrix}
0 \\
0 \\
-g_0 U_V r\partial_r\rho_0 \\
0
\end{bmatrix}
$$

$$\text{(III-62RK)}$$

where $\eta^* = $ viscosity at some particular location
$\qquad \tilde{\eta} = \eta/\eta^*$

$$U_V = \int_0^t V_U \, dt \qquad \text{viscous radial displacement}$$

$\qquad \partial_r\rho_0 = $ non-adiabatic density gradient (which is the same as the actual observed density gradient in the incompressible case).

If $\partial_r\rho_0 = 0$ the viscous case is identical to the incompressible elastic case already depicted in Figure III-5 of the previous section with $\mathbf{u}$ replaced by $\mathbf{v}$. For a given initial load isostatic equilibrium will be approached exponentially just as in the flat-space case.

$$2\eta^* k v_r = -\rho_0 g_0 h f(k) \qquad v_r = \frac{\partial h}{\partial t}$$

$$\text{(III-63)}$$

$$h = h_0 e^{-\rho g t f(k)/2\eta k} \qquad k \equiv \frac{n + \frac{1}{2}}{b}$$

$f(k)$ goes from 1.31 at $n = 2$ to ~ 1 at $n = 7$ (Figure III-5). It is simply a factor expressing the "augmentation" of rate of decay at low order numbers due to spherical geometry. Augmentation is in quotes because whether it is an absolute increase or decrease depends on the definition of k.

---

[12]The placement of the buoyant term in the Runge-Kutta system falls out of Backus's (1967) technique naturally.

The exponential decay of the homogeneous viscous sphere may be conveniently simulated with a computer. The initial normal stress $(\tau_{rr})_{t=0}$ is given. We compute the instantaneous viscous velocity at $t = 0$. The initial load can then be relaxed

$$(\tau_{rr})_{t=\Delta t_1} = (\tau_{rr})_{t=0} + \rho_0 g_0 v_r \Delta t_1,$$

and an instantaneous velocity at $t = \Delta t_1$ computed. This process can

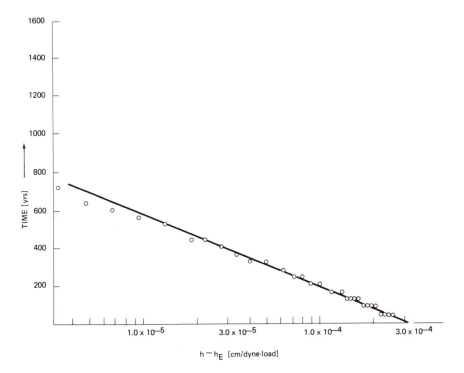

Figure III-7. Analytic isostatic adjustment of a homogeneous viscous sphere subject to a one dyne P_2 surface load (III-63) compared to decay calculated by successive relaxation of boundary conditions (o's) as described in the text.

$\eta = 3 \times 10^{21}$ poise $g_r = 1000$
$\rho_s = 3.313$ $h_E = 3.018 \times 10^{-4}$ cm

$r_s = 6371$ km
$n = 2$ (order number).

Stepping is due to quantization by printer plotter. A one dyne P_2 harmonic load is assumed applied at $t = 0$.

be repeated as often as necessary to obtain a decay history $h(0 + \Delta t_1 + \cdots) = (v_r)_{t=0}\Delta t_1 + (v_r)_{t=\Delta t_1}\Delta t_2 + \cdots$. This method is convenient as it may be used in cases where the adjustment of the surface is no longer exponential.

Figure III-7 shows the exponential decay of a homogeneous viscous sphere for a conveniently dense set of time intervals, and compares it to the theoretical decay (III-63).

Figure III-8 shows a case where the decay is not exponential. The "earth" is here assumed to be homogeneous but to have an inviscid fluid core of contrasting density. Buoyant forces arise at the core-

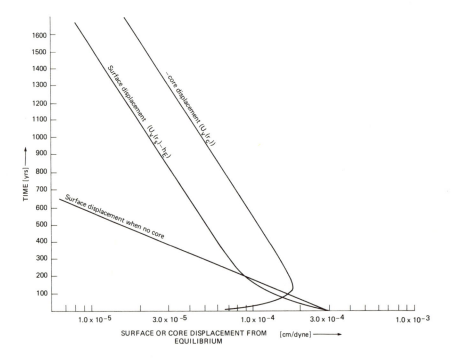

Figure III-8. The response of a homogeneous viscous sphere with an inviscid core to surface loading (at $t = 0$) with a P_2 harmonic. Surface displacement from ultimate equilibrium $h_E - U_v(r_s)$ and negative core displacement $-U_v(r_c)$ is shown. The surface displacement from equilibrium for the case of a homogeneous viscous sphere with no inviscid core is also shown for comparison. Boundary conditions were, at any instant:

$$P_s = -1 - \rho_s g_0 U_v(r_s) \qquad P_c = (\rho_c - \rho_m) g_0 U_v(r_c)$$
$$Q_s = 0 \qquad\qquad\qquad Q_c = 0$$

Other parameter values were taken: $g_r = \text{const} = 1000 \text{ cm/sec}^2$, $r_s = 6371$ km, $\rho_s = 3.313 \text{ gm/cc}$, $\rho_c - \rho_m = 4.4 \text{ gm/cc}$, $r_c = .5495\, r_s$, $\eta = 3 \times 10^{21}$ poise.

mantle boundary as well as at the surface, and these forces are augmented or relaxed according to $v_r(r_c, t)$ just as at the surface. The
changing boundary conditions drastically alter the character of the decay. Initial adjustment of the surface is more rapid than the coreless
case. Both the surface *and* the core-mantle boundary are depressed
at about the same rate. After a short time, however, buoyant forces
build up at the density discontinuity of the core-mantle boundary to a
magnitude large enough to first stop and then reverse further depression. The rising core-mantle boundary then slows the adjustment of the
surface, almost exactly as restricting flow to the channel between the
core-mantle boundary and the surface would do. Since the core-mantle
boundary is rising, the effect on the decay of the surface is slightly
more severe than restricting flow to a channel.

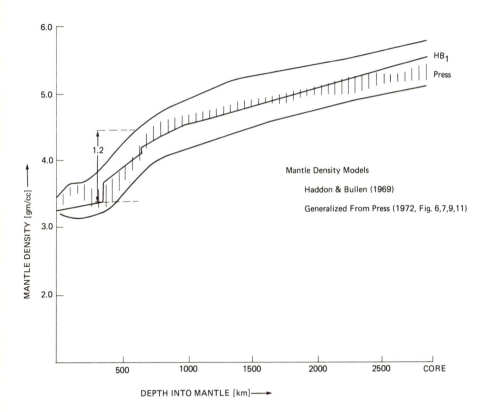

Figure III-9. Mantle density profiles from Haddon and Bullen (1969) and Press (1972,
Figures 6, 7, 9, 11) show non-adiabatic density gradients may be as high as 1.2 gm/cc
in the upper mantle between 400 and 700 km depth.

$\partial_r\rho_0$ may be unequal to zero. This is likely in the upper mantle between 350 and 700 km depth where several phase changes take place. A reasonable estimate of how large the non-adiabatic density gradient in this area might be may be inferred from Figure III-9. The total non-adiabatic increase in density could be as large as 1.2 gm/cc. The effect of such non-adiabatic density gradient will be similar to the effect of the density discontinuity at the core. Buoyant forces will build up as the density gradient is depressed, and flow will be encouraged in a layer above 350 km and discouraged in the lower mantle below 700 km.

Figure III-10 shows the effects of various density gradients between 350 and 700 km. The number beside each curve gives the total non-adiabatic increase in density. The non-adiabatic density gradient is assumed constant. As can be seen, the mantle gradients act to further

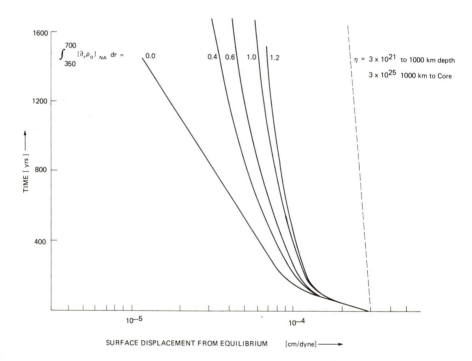

Figure III-10. Effect of non-adiabatic density gradients in the upper mantle on the fluid response to surface loading. Non-adiabatic gradients have integrated values between 350 and 700 km depth into the mantle as shown in the figure. The gradients are assumed constant in this interval. A P_2 load of 1 dyne amplitude is assumed. The response of a layered viscous mantle is shown for comparison.

break the decay (after it is about $\frac{1}{3}$ complete). In the case of

$$\int_{350}^{700} \nabla \rho \, dz = 1.2$$

the rate of decay of the surface P_2 harmonic after the first 300 years is very similar to the decay of a 1000 km thick layer of 3×10^{21} poise overlying a much more viscous substratum of 3×10^{25} poise, which continues to the core. The rate of adjustment continues to decrease till at the final stages the surface decays at a rate appropriate for flow limited to a 350 km thick channel.

This situation should cause no surprise. If non-adiabatic density gradients are really present, they may be initially deformed, but ultimately equilibrium must look:

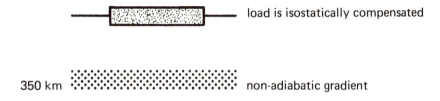

load is isostatically compensated

350 km — non-adiabatic gradient

The phase boundary may deform, enabling much more rapid initial response. At ultimate equilibrium, however, it is undeformed. Exactly as much fluid transport must occur in the layer above 350 km as if that layer had never deformed (assuming the phase boundary does not migrate at all). Thus, the time necessary for *complete* isostatic adjustment is similar to that required for a 350 km thick viscous channel, although the aspect of the decay is entirely different.

Our general method for calculating the effects of such non-adiabatic density gradients is subject to a simple verification. A very steep density gradient very near the core-mantle boundary should behave precisely as a density jump of the same magnitude at the core-mantle boundary. Figure III-11 shows this is indeed the case.

Finally, Figure III-12 shows the effect of low viscosity zones of various viscosities and thicknesses. Such low viscosity zones have little effect on the decay of low order harmonics, a result previously obtained by Anderson and O'Connell (1967).

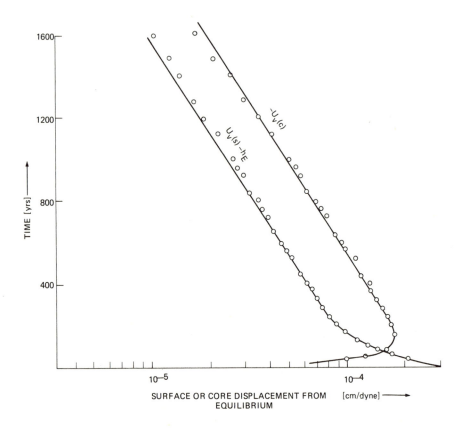

Figure III-11. Verification that a steep non-adiabatic density gradient near the core-mantle boundary behaves the same as a density jump at the core mantle boundary. o's are for the steep density gradient:

$$\int_{2800}^{2860} \nabla \rho_{NA}\, dr = 4.4,\ \rho_c - \rho_m = 0.0.$$

the solid lines are the case previously computed in Figure III-8; $\nabla \rho_{NA} = 0.0$ everywhere, $\rho_c - \rho_m = 4.4$. A one dyne P_2 load is assumed applied at $t = 0$.

3. Deformation of a Self-Gravitating Viscoelastic Sphere

a. The Viscoelastic Equations

The viscoelastic response may be adequately approximated as the sum of the elastic and viscous responses at any given stage of isostatic adjustment. This was demonstrated in Section III.A.2.g.2. In the self-gravitating case, we need only make sure the effects of elastic hydro-

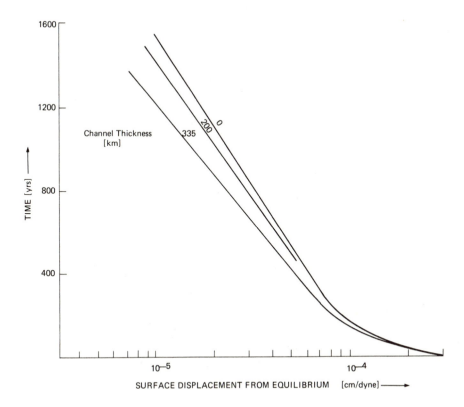

Figure III-12. Uppermost mantle low viscosity channels do not affect the adjustment of low order harmonic loads substantially. The channel thickness is as shown. Channels have a viscosity 3×10^{19} poise. Other parameters are as in Figure III-8. A P_2 harmonic load is assumed applied at $t = 0$.

static prestress, elastic density and gravity perturbations (body force changes), and buoyant forces arising from the viscous deformation are properly accounted for in both the viscous and the elastic systems. This is accomplished by adding a constant vector to the Runge-Kutta system in a fashion similar to what was done in equation III-62 RK.

This is to say we approximate the viscoelastic equation of motion:

$$(\nabla \cdot \tau)_{\text{viscoelastic}} - \nabla(\rho_0 g_0 U_E) - \rho_0 \nabla \phi_1 - \rho_1 g_0 \hat{\mathbf{r}} = 0,$$

where

$$\phi_1 = (\phi_1)_{\text{elastic}} + (\phi_1)_{\text{viscous}}$$

$$\rho_1 = (\rho_1)_{\text{elastic}} + (\rho_1)_{\text{viscous}} = U_v(\partial_r \rho_0)_{NA} - \rho_0 \nabla \cdot \mathbf{u}_E - U_E \partial_r \rho_0$$

$$\hat{g}_1 = (\hat{g}_1)_{\text{elastic}} + (\hat{g}_1)_{\text{viscous}}$$

by two coupled equations of motion:

$$(\nabla \cdot \tau)_{\text{elastic}} - \nabla(\rho_0 g_0 U_E) - \rho_0 \nabla \phi_1 - \rho_1 g_0 \hat{\mathbf{r}} = 0, \qquad \text{(III-64a)}$$

$$(\nabla \cdot \tau)_{\text{viscous}} - \nabla(\rho_0 g_0 U_E) - \rho_0 \nabla \phi_1 - \rho_1 g_0 \hat{\mathbf{r}} = 0. \qquad \text{(III-64b)}$$

These equations are put into Runge-Kutta form

$$r \partial_r \mathbf{u} = \mathbf{A}(u)\mathbf{u} + \mathbf{b},$$

in equation III-64RK (p. 93).

Since $\mathbf{A}_{\text{elastic}}$ becomes identical in form to $\mathbf{A}_{\text{viscous}}$ as $\lambda \rightarrow \infty$, a good check that $\mathbf{b}_{\text{viscous}}$ contains all the terms it should is that it contain just those terms the viscous system does not have in common with the elastic system when $\lambda = \infty$.

We discussed the P component of $\mathbf{b}_{\text{elastic}}$ previously. It is just the buoyant force arising from viscous deformation of non-adiabatic density gradients and of course is present in both the viscous and elastic propagators. The viscous deformation of non-adiabatic density gradients will also cause gravitational perturbations which must be input in $\mathbf{b}_{\text{elastic}}$. The form for these terms will be discussed in a moment.

If we assume the core to be an inviscid fluid, the boundary conditions at the core-mantle boundary follow from (II-31), (II-34), (II-41), and (II-42):

$$\hat{g}_1 \big|_{r=a} = 4\pi G(\rho_c - \rho_m)(U_V + U_E) + \frac{n}{a}\phi_1$$

$$Q\big|_{r=a} = 0 \qquad\qquad\qquad\qquad \text{(III-65)}$$

$$P\big|_{r=a} = \rho_c g_0 U_E + (\rho_c - \rho_m)g_0 U_V + \rho_c \phi_1$$

Thus the starting vectors in the elastic system are given in equation III-66 (p. 94).

The starting vectors for the viscous system are given in equation III-67 (p. 94).

$(\phi_1)_V$ and $(\hat{g}_1)_V$ are determined by $U_V(r)$. The potential due to the deformation of the core-mantle boundary is:

$$(\phi_1)_V \big|_{r=a} = -4\pi a G(\rho_c - \rho_m) U_V \left(\frac{1}{2n+1}\right). \qquad \text{(III-68)}$$

The Viscoelastic Propagator System III-64RK

$$
\tilde{r}\partial\tilde{r}
\begin{bmatrix}
\mu^* U_E / r^* \\[2pt]
\mu^* V_E / r^* \\[2pt]
P \\[2pt]
Q \\[2pt]
\phi_1 \\[2pt]
r^*\hat{g}_1
\end{bmatrix}
=
$$

$$
\begin{bmatrix}
-2\tilde{\lambda}\tilde{\beta}^{-1} & -\tilde{\lambda}\tilde{\beta}^{-1}\nabla_1^2 & \tilde{\beta}^{-1}\tilde{r} & 0 & 0 & 0 \\[6pt]
-1 & 1 & 0 & \tilde{r}\tilde{\mu}^{-1} & 0 & 0 \\[6pt]
4\left(\tilde{r}^{-1}\tilde{\gamma} - \dfrac{\rho_0 g_0 r^*}{\mu^*}\right) + \dfrac{4\pi G\rho_0^2 r^*}{\mu^*} & \left(2\tilde{\gamma}\tilde{r}^{-1} - \dfrac{\rho_0 g_0 r^*}{\mu^*}\right)\nabla_1^2 & -4\tilde{\mu}\tilde{\beta}^{-1} & -\nabla_1^2 & 0 & +\tilde{r}\rho_0 \\[10pt]
-2\tilde{r}^{-1}\tilde{\gamma} + \dfrac{\rho_0 g_0 r^*}{\mu^*} & -\tilde{r}^{-1}[(\tilde{\gamma}+\tilde{\mu})\nabla_1^2 + 2\tilde{\mu}] & -\tilde{\lambda}\tilde{\beta}^{-1} & -3 & +\rho_0 & \tilde{r} \\[8pt]
0 & 0 & 0 & 0 & 0 & \tilde{r} \\[8pt]
\dfrac{-4\pi G r^{*2}}{\mu^*}[\tilde{r}\partial_{\tilde{r}}\rho_0] + 4\tilde{\mu}\tilde{\beta}^{-1}\rho_0] & \dfrac{-8\pi G\rho_0 r^{*2}}{\mu^*}\tilde{\mu}\tilde{\beta}^{-1}\nabla_1^2 & \dfrac{-4\pi G\rho_0 r\tilde{\beta}^{-1}r^*}{\mu^*} & 0 & -\tilde{r}^{-1}\nabla_1^2 & -2
\end{bmatrix}
\begin{bmatrix}
\mu^* U_E / r^* \\[2pt]
\mu^* V_E / r^* \\[2pt]
P \\[2pt]
Q \\[2pt]
\phi_1 \\[2pt]
r^*\hat{g}_1
\end{bmatrix}
$$

$$
+
\begin{bmatrix}
0 \\[4pt]
0 \\[4pt]
-g_0\tilde{r}[\partial_{\tilde{r}}\rho_0]_{NA}\,U_V \\[4pt]
0 \\[4pt]
\tilde{r}(\phi_1(r))_V \\[4pt]
-4\pi G\tilde{r}r^*[\partial_{\tilde{r}}\rho_0]_{NA}\,U_V + n(\phi_1(r))_V
\end{bmatrix}
$$

$$
\text{note } \tilde{\lambda}\tilde{\beta}^{-1} - 1 = -2\tilde{\mu}\tilde{\beta}
$$

$$
\tilde{r}\partial\tilde{r}
\begin{bmatrix}
\eta^* V_U' / r^* \\[2pt]
\eta^* V_V / r^* \\[2pt]
P \\[2pt]
Q
\end{bmatrix}
=
\begin{bmatrix}
-2 & -\nabla_1^2 & 0 & 0 \\[4pt]
-1 & 1 & \tilde{r}\tilde{\eta}^{-1} & 0 \\[4pt]
12\tilde{r}^{-1}\tilde{\eta} & 6\tilde{r}^{-1}\tilde{\eta}\nabla_1^2 & 0 & -\nabla_1^2 \\[4pt]
-6\tilde{r}^{-1}\tilde{\eta} & -2\tilde{r}^{-1}\tilde{\eta}(2\nabla_1^2 + 1) & -1 & -3
\end{bmatrix}
\begin{bmatrix}
\eta^* V_U / r^* \\[2pt]
\eta^* V_V / r^* \\[2pt]
P \\[2pt]
Q
\end{bmatrix}
$$

$$
+
\begin{bmatrix}
0 \\[4pt]
0 \\[4pt]
r\rho_0\hat{g}_1 - g_0 U_V\tilde{r}[\partial_{\tilde{r}}\rho_0]_{NA} - 4\rho_0 g_0 U_E \\[4pt]
-\rho_0 g_0\nabla_1^2 V_E + 4\pi G\rho_0^2 U_E r \\[4pt]
\rho_0\phi_1 + \rho_0 g_0 U_E
\end{bmatrix}
$$

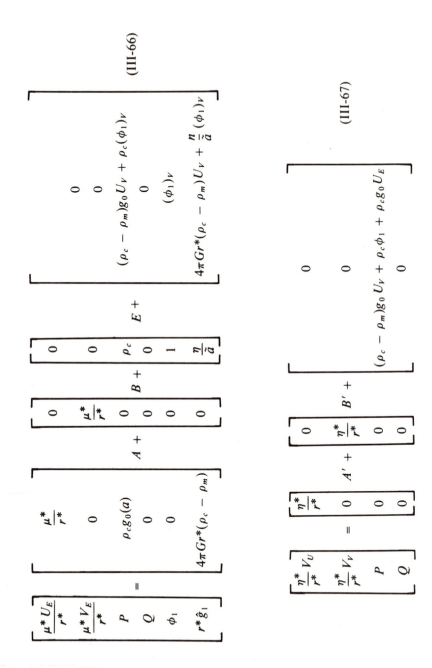

$$(\text{III-66})$$

$$(\text{III-67})$$

This is just the potential of a harmonic surface density $\sigma_n = (\rho_c -$ $\rho_m)U_V$. Upward, away from $r = a$, the potential decays as $\left(\dfrac{a}{r}\right)^{n+1}$, i.e., similar to (II-36), so:

$$(\hat{g}_1)_V \big|_{r=a} = \left.\frac{+\partial\phi_1}{\partial r}\right|_{r=a} = +\left(\frac{n+1}{2n+1}\right)4\pi G(\rho_c - \rho_m)U_V. \qquad \text{(III-69)}$$

The same result would have been obtained if we had substituted (III-68) into the expression for $(\hat{g}_1)_V$ given in (III-65). It is important to realize that $(\hat{g}_1)_V$ communicates to the propagator that $(\phi_1)_V$ is due entirely to a surface mass distribution at $r = a^-$. With this information, the propagator can correctly interpolate what $(\phi_1)_V$ will be at radii larger than $r = a$.

In the viscous system, (III-67), $\phi_1(a)$ is known from the elastic solution.

Now return to the problem of the gravitational terms in $(b)_{\text{elastic}}$. We want to introduce the effects of the viscous distortion of the gravitational potential ϕ_1 and the gravitational field strength $\hat{g}_1$ into the elastic propagating system. Consider just the depth r_α. The potential due to the deformation of a lamina of thickness $\Delta\tilde{r}$ about $\tilde{r}_\alpha$ may be written:

$$\Delta\phi_1(\tilde{r}_\alpha) = 4\pi G r_\alpha [\partial_{\tilde{r}}\rho_0]_{NA} \Delta\tilde{r} U_V(\tilde{r}_\alpha)\left(\frac{1}{2n+1}\right),$$

$$\Delta\hat{g}_1(\tilde{r}_\alpha) = \frac{-(n+1)}{2n+1} 4\pi G[\partial_{\tilde{r}}\rho_0]_{NA} \Delta\tilde{r} U_V(\tilde{r}_\alpha).$$

Once again we have selected $(\hat{g}_1)_V = \partial_r\phi_1$, such that ϕ_1 decays away from its source $(r > r_\alpha)$. The vector

$$\tilde{r}\partial_{\tilde{r}}\begin{pmatrix}\phi_1 \\ r^*\hat{g}_1\end{pmatrix}$$

may now be approximated by dividing $\Delta\phi_1(\tilde{r}_\alpha)$ and $\Delta g_1(\tilde{r}_\alpha)$ by $\Delta\tilde{r}$ (to obtain an approximation to the derivative) and multiplying by $\tilde{r}$. $\Delta\hat{g}_1$ must also be multipled by r^*. If this is done, (III-70) results, where the P term has been added for completeness.

$$(\mathbf{b}(r))_{\text{Elastic}} = \begin{bmatrix} 0 \\[6pt] 0 \\[6pt] -g_0[\partial_{\tilde{r}}\rho_0]_{NA}\,U_V\tilde{r} \\[6pt] 0 \\[6pt] 4\pi G r\tilde{r}[\partial_{\tilde{r}}\rho_0]_{NA}\,U_V(r)\left(\dfrac{1}{2n+1}\right) \\[6pt] \dfrac{-(n+1)}{(2n+1)}\,4\pi G r^*\tilde{r}[\partial_{\tilde{r}}\rho_0]_{NA}\,U_V \end{bmatrix} \qquad \text{(III-70)}$$

From (III-68) we see (III-70) is entirely parallel to (III-66). Effectively $(\mathbf{b})_{\text{elastic}}$ may be thought of as a series of new boundary conditions, each of which must be propagated, in turn, to the surface $\tilde{r} = 1$. Notice, though, that only viscously deformed *non-adiabatic* density gradients act as sources for perturbation of the gravitational field. After propagation to the surface, the surface boundary conditions determine A, B, E in the elastic system and A', B' in the viscous system.[13]

$$\hat{g}_1\big|_{r=b} = \frac{-(n+1)}{b}\phi_1 - 4\pi G[\rho_0(U_V + U_E) + \sigma]$$

$$P\big|_{r=b} = -g_0\sigma - \rho_0 g_0 U_V \qquad \text{(III-71)}$$

$$Q\big|_{r=b} = 0$$

We have shown how the elastic and viscous parts of (III-64RK) are solved. The procedure by which the viscoelastic solution is obtained from the solution of each of these parts is as follows:

At $t = 0$ no viscous flow has taken place and $U_V(r) = 0$. The deformation is thus simply that of a self-gravitating elastic sphere. Given the boundary conditions at the surface, we can solve as we have previously, taking care to store (at a suitable number of radii) the quantities we need to communicate to the viscous system: $\hat{g}_1(r)$, $U_E(r)$, $V_E(r)$, $\phi_1(r)$. Given these quantities, we use the viscous system to compute the value of $V_U(r)$, $V_V(r)$ at $t = 0$. We then pick a Δt_1, and compute a new value of $U_V(r) = V_U(r)\Delta t_1$. This is communicated to the elastic system which computes new elastic displacements for the "relaxed"

[13]Calculation of ϕ_1 due to deformation of a non-adiabatic density gradient between 350 and 650 km using the inhomogeneous propagator (III-64RK) agrees well with approximate calculations done by hand. Details of the method of calculation are given in Cathles (1971, Appendix VIII).

boundary conditions. The elastic results are communicated to the viscous system which uses the relaxed boundary conditions, and $U_V(r, \Delta t_1)$ to compute new viscous velocities, and hence a new $U_V(r, \Delta t_2) = V_U(r, 0)\Delta t_1 + V_U(r, \Delta t_1)\Delta t_2$. This process may be repeated so as to examine as lengthy a decay history as desired.

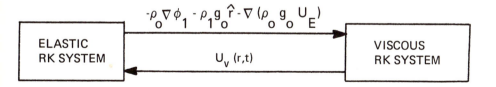

1) The Theorem of Correspondence as an Alternative Approach. We may use the correspondence theorem, derived in Section II.B.2, to obtain a solution to the viscoelastic problem directly from the elastic equation II-12. The formalism of the Theorem of Correspondence leads to all the results we have obtained by more physical reasoning. The non-adiabatic density gradient, for example, is deduced by the formalism from the elastic parameters and the density profile $\rho_0(r)$. Instead of stepping forward in time, a boundary condition problem is simply solved for a number of different S values. The inverse Laplace transform may easily be taken. Such an approach would not neglect the solid-elastic → fluid-elastic transition as we have done; but, as we have shown, this effect is small, and limited to the early stages of load removal. The rate of viscous adjustment will automatically slow as flow takes place and the boundary conditions change. With the exception of the solid-elastic → fluid-elastic transition, our method and the method of the Theorem of Correspondence should be entirely equivalent. Parsons (1972) has used this approach.

b. Specific Cases.

Figure III-13 shows the effect of self-gravitation for a homogeneous, uniform earth with and without fluid core. We see that self-gravitation in both cases slows the adjustment. The rate of adjustment appears to have been diminished more than might be suggested by the elastic example (Figure III-6). If we look at the stress configuration, the explanation is clear. Gravitation has altered the mode of flow, diminishing it in the deep interior. As in the pure viscous case, decay is slowed still further by introducing a non-adiabatic density gradient between 350 and 650 km.

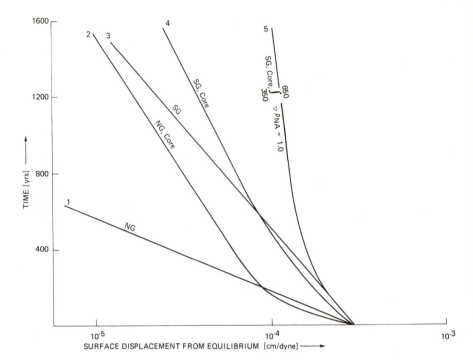

Figure III-13. Comparison of the adjustment of a self-gravitating viscoelastic sphere with and without core to the adjustment of a simple viscous sphere. P_2 harmonic load applied at $t = 0$.

(1), (2) Earth model as in Figure III-8.

$$g_0 = 1000 \qquad \qquad \rho_m = 3.313$$
$$\rho_c - \rho_m = 4.4 \qquad \qquad \eta = 3 \times 10^{21} \, \text{p}$$

(3) Pseudo-homogeneous E model designed so gravity approximately O.K.

$r = .01$	.549	1.0	$\times 6.356 \times 10^9$ cm
$\rho_0 = 15$	7.52	3.313	
$\lambda = 1.$	1.	1.	$\times 10^{12}$ dyne-cm^{-2}
$\mu = 1.$	1.	1.	$\times 10^{12}$ cyne-cm^{-2}
$\eta = 1.$	1.	1.	$\times 3 \times 10^{21}$ poise
$g_0 = 0$	1000	1000	

(4) Pseudo "E" model

$r = .549$	1.	$\mu = 3.0$	.7
$\rho_0 = 5.527$	3.313	$\eta = 1.$	1.
$\lambda = 4.4$	1.0	$g_0 = 1000$	1000

(5) Haddon Bullen HB_1 model with $[\nabla \rho]_{NA}$ as indicated. (Haddon and Bullen, 1969).

z	2863	2185	985	635	635	335	335	0.0	km
$\mu \times 10^{-12}$	2.954	2.555	1.837	1.462	1.396	.749	.697	.709	dyne-cm^{-2}
$K \times 10^{-12}$	6.349	5.407	3.490	2.665	2.698	1.863	1.706	1.017	dyne-cm^{-2}
ρ_0	5.527	5.188	4.539	4.200	4.150	3.700	3.441	3.313	gm/cc
g_0	1073.0	1010.2	995.8	998.7	998.7	995.5	995.5	983.2	cm-sec^{-2}

$$\rho_{core} = 9.927 \text{ gm/cc}$$
$$\eta = 3 \times 10^{21} \text{ poise at all radii}$$

As remarked in the non-self-gravitating section, decay with a non-adiabatic density gradient in the upper mantle persists for times of the order of that necessary for a 350 km viscous channel to adjust to the load applied. Figure III-14 shows the adjustment history of the $n = 2$ mode of a Haddon and Bullen (1969) earth model with a non-adiabatic density gradient between 350 and 650 km and a fluid core of density 9.927 ($\rho_c - \rho_m = 4.4$). As can be seen, even after 10,000 years the surface displacement is still only 75% of what it would be at complete equilibrium.

Again, this may be thought of as the result of three decay times: First, that of a hollow earth; second, that of flow restricted to the mantle; and last, that of flow restricted to a channel above about 350–400 km. Roughly speaking, vertical displacement for $n = 2$ is constant from the surface to 400 km depth. Flow is restricted to the last layer when the load at the surface is balanced by the buoyant forces in the deformed density gradient at 350 km $\left(\int_{350}^{650} \nabla \rho_{NA} = 1.0 \right)$. Suppose we place a harmonic load of density 3.3 and unit amplitude at the free surface. When $3.3(1 - x) = 1.0x$ flow is restricted to the upper 350 km layer. $x = .77$. Hence, the 75% decay after 10,000 years.

The upper channel will introduce a certain hysteresis into loading and unloading phenomena. Since the wavelengths of interest to us are much greater than the thickness of the channel, we may use flat earth theory and (III-25) to estimate this effect.

$$\tau = \frac{3\eta}{\rho g k^2 D^3}.$$

If we use

$$k = \frac{n + \frac{1}{2}}{r_E}$$

$$\eta = 3 \times 10^{21} \text{ poise}$$
$$D = 350 \times 10^5 \text{ cm}$$
$$g = 10^3$$
$$\rho = 3.0 \text{ gm/cc}$$

$$\tau_n = \frac{897 \times 10^3 \text{ yrs}}{(n + \frac{1}{2})^2}$$

$$\tau_2 = 143 \times 10^3 \text{ yrs}$$

We see the decay is strongly dependent on n and D. $D = 350$ km is

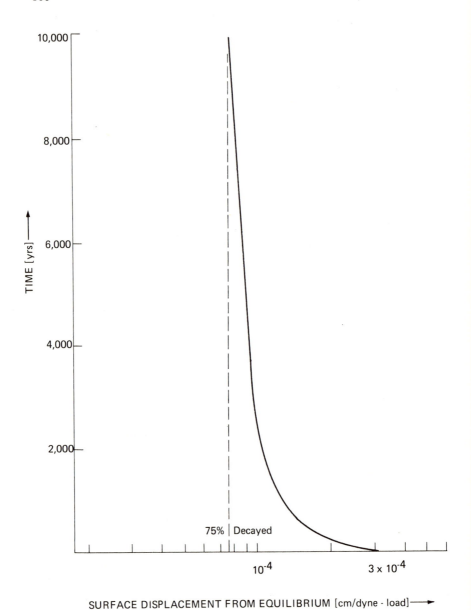

Figure III-14. Adjustment of an HB$_1$ model earth (see Figure III-13) with uniform 3×10^{21} poise viscosity throughout the mantle and a constant non-adiabatic density gradient in the upper mantle between 350 and 650 km depth with integrated value of 1.0 gm/cc^3. Note adjustment to a P_2 harmonic load is about 75% complete in 10,000 years.

really too small. An average D of 450 km would not be an unreasonable choice.

$$(\tau_2)_{D\,=\,450\,km} = 67 \times 10^3 \text{ yrs.}$$
$$(\tau_{10})_{D\,=\,450\,km} = 3.8 \times 10^3 \text{ yrs.}$$

Flow above the non-adiabatic density gradients in the upper mantle will thus not contribute significantly to the decay of low order load harmonics. Higher order harmonics may be affected, however. Some residual effects may be observable. Gravity anomalies might show some correlation to the continental margins, for example, if there were substantial non-adiabatic density gradients in the upper mantle (or a substantial low viscosity channel). Non-adiabatic gradients would produce a dipolar potential source: Positive at the surface (excess load), but negative at 350–550 km depth where lighter material is depressed to support the surface load. At long wavelengths (low n) this dipolar character will substantially mask any observation of a deformed geoid, even though the sphere might be far from ultimate isostatic equilibrium.

If the viscosity above 450 km is 3×10^{22} poise, the significance of the upper channel is greatly reduced.

Phase boundary migration will tend to reduce the effect.

4. Harmonic Loads of Order Number One

One other consequence of self-gravitation should be discussed: Self-gravitation permits $n = 1$ deformation of the earth by redistributed surface loads. It is clear that we may transport material from one side of the earth to the other and that this redistribution will deform the earth to some extent. Until recently (Cathles, 1971; Farrell, 1972) no one calculated load deformation coefficients of order one. This was perhaps due to an association of load deformation coefficients and Love numbers. A Love number of order one would correspond to a displacement of the whole earth in its celestial orbit, and thus have no direct geophysical interest.

To find the $n = 1$ deformation coefficient, one may proceed from either of two viewpoints: One viewpoint focuses on the necessity of the center of mass of an object not acted upon by unbalanced outside forces to remain fixed in space. The other focuses on making sure all forces are in fact balanced. We shall briefly present both viewpoints and then concentrate on the second, which seems to the writer more direct, less complicated, and more immediately physically interpretable.

All harmonic surface loads except $n = 1$ have a center of mass coincident with the center of mass of the undisturbed earth. The center of mass of a P_1^0 load, however, is located a distance $\frac{2}{3}b$ from the center of mass of the earth (b = radius of the earth). Of course, the center of mass of the earth-load system will remain fixed in space (provided no unbalanced outside forces act in redistributing the surfacial load). The center of mass of the *system* will thus shift relative to the core-mantle boundary. The sense of this shift is shown in Fig. III-15b. It is easy to show that *for a perfectly rigid earth* the center of mass of the system will shift about 40 m relative to the undeformed core-mantle boundary if 200 m of water (amplitude) is shifted from one hemisphere to the other. Since buoyant forces are calculated from the center of mass of the system, it is clear the shift in the center of mass relative to the density discontinuity at the core-mantle boundary (and relative to the density stratification in the mantle) will have a substantial effect on any elastic or viscous deformation of the earth. In particular, there will be a substantial buoyant pressure pushing the core-mantle boundary up under the load, *even if* the mantle has not deformed either elastically or viscously.

Figure III-15a presents vector diagrams of the body forces acting at the center of the earth. As can be seen, the net body force at the center of the core is zero for all harmonic load redistributions except $n = 1$. Even for $n = 1$, the body force vectors occur only in balanced pairs. As depicted by the second set of vectors in Figure III-15a, the P_1^0 surface load attracts the earth with *exactly* the same intensity as the earth attracts the load. The load is plastered to the surface of the earth by gravitational forces just as a small piece of lint is plastered to a larger one by electrostatic forces.

Such a viewpoint lends itself to easy incorporation into our theoretical structure. For $n = 2, 3, \ldots$ our propagator technique calculates the body force attraction of the load for the mantle as a matter of course. We have seen this is of considerable significance (Figure III-6). We have commonly taken into account the attraction of the surface load and deformed mantle for the core by adding a term P (core) = $\rho_c \phi_1$ (See II-42).

In the case of $n = 1$, the load, in a sense, is "supported" entirely by body forces, some deformation of the earth occurring in the process. Some of the balancing body forces derive from the mantle, the rest derive from the core. If we use the usual Runge-Kutta technique to calculate the full effect of the mantle, whatever stress remains at the core-mantle boundary must be balanced by the body force attraction of the core by the deformed mantle and the surface load. This result is

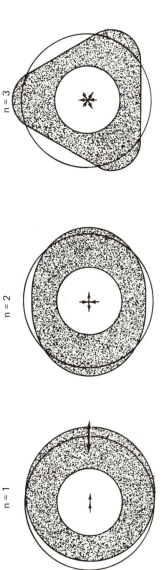

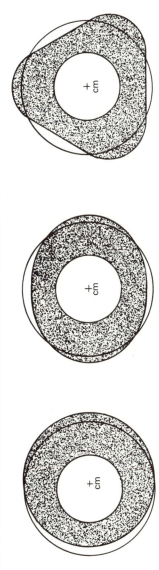

n = 1   n = 2   n = 3

A. GRAVITATIONAL FORCE VECTORS ARISING FROM REDISTRIBUTION OF SURFICIAL MATERIAL

B. SHIFT OF CENTER OF MASS OF THE EARTH-LOAD SYSTEM RELATIVE TO CORE-MANTLE BOUNDARY

Figure III-15. Illustration of two viewpoints on $n = 1$ static load deformation. The first viewpoint concentrates on the balance of body forces. The surface load redistribution is seen to attract the earth with exactly the same force as the earth attracts the load redistribution. It can also be seen for all higher order load redistributions ($n > 1$), no unbalanced body forces act on the core. The second viewpoint concentrates on the location of the center of mass of the earth relative to, for example, the core-mantle boundary. A P_1 load redistribution shifts the center of mass relative to this reference surface and calls buoyant forces into play. Higher order load redistributions do not.

given to us automatically if we require the core-mantle boundary to be fixed in space. This new boundary condition is easy to introduce.

It should be pointed out the gravitational viewpoint may lead to different results than the center-of-mass viewpoint depending on how the center-of-mass calculations are implemented. Assume, for example, a perfectly rigid earth. Referring to Figure III-15b, we see that a P_1^0 redistribution of mass will cause a shift in the center of mass relative to the core-mantle boundary *and* relative to the undeformed surface boundary as well. Thus our point of observation, $r = b$, will shift as the load is redistributed. Even with an infinitely rigid earth, the surface will appear depressed in the strict Eulerian description.

The strict Eulerian description is not the most meaningful system, however. An observer situated on the earth could keep track of deformations of the earth with a suitable array of strain meters. Such strain would be meaningful to him in deducing the physical properties of the earth. To keep track of the movements of the center of mass of the earth (or the surface) through space, however, the observer would need to make careful astronomical measurements. Even with these measurements, he would discover only that which he already knew; namely, that the earth had shifted slightly to compensate for the shift in surface load and to keep the center of mass of the system fixed in space. Also, since the magnitude of the shift of the center of mass would be proportional to the magnitude of the redistributed load, any Love numbers defined on the basis of this viewpoint would be dependent on the magnitude of the load and this is not a satisfactory situation.

The second approach gives deformation and gravity perturbations relative to the position of the surface of the earth after load redistribution but before any deformation has taken place. We take cognizance of the shift of the center of mass only insofar as it affects boundary conditions at the core-mantle boundary. We ignore the movement of the earth through space. The load deformation coefficients are independent of load magnitude.

In this way we find load deformation coefficients for $n = 1$ and for an HB_1 earth to be[14]

$$n = 1 \qquad h' = -.246 \qquad k' = 0.0$$
$$n = 2 \qquad h' = -.980 \qquad k' = .312$$

The sign convention is such that the above coefficients may be compared to those listed in Appendix IV and with those commonly cited in the literature. The difference between an HB_1 earth and a Gutenberg

---

[14]In a recent calculation using a slightly different approach Farrell (1972) obtained $h' = -.290$ and $k' = 0.0$ for $n = 1$.

model earth may be observed by comparing the load deformation co-efficients for $n = 2$ listed above with those given in Appendix IV.

h' as given above corresponds to an elastic response of $-.4535 \times 10^{-4}$ cm to a 1 dyne surface load. The elastic response to a 1 dyne P_2 load is $-.1083 \times 10^{-3}$ cm.

5. Viscoelastic Decay Spectra

Figures III-16–20 give the viscous decay curves at a selection of order numbers for an HB_1 elastic earth with various viscosity and non-adiabatic density structures.

As can be seen in Figure III-16, for a uniform viscous earth the adjustment to a P_3^0 load is more rapid that P_1^0 or P_2^0. This is because

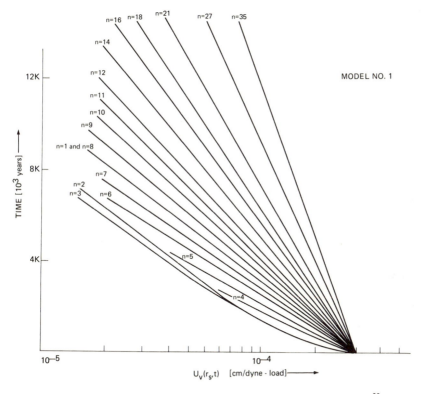

Figure III-16. Viscous Decay Spectrum ($U_V(r_s, t)$ v.s.t) for a uniform 10^{22} poise earth with $\nabla \rho_{NA}$ everywhere zero. The surface load is assumed redistributed at $t = 0$. Elastic and density profiles were taken to correspond to the HB_1 model (see Figure III-13). This model is designated Model #1 in Table IV-1 and is also described there. Notice that the decay of the $n = 4$ and 5 harmonics are faster than the lower order harmonics after about 4000 years. Total displacement $U(r_s)$ is the sum of this displacement $U_V(r_s, t)$ and the elastic displacement, $U_E(r_s, t)$, which is not shown.

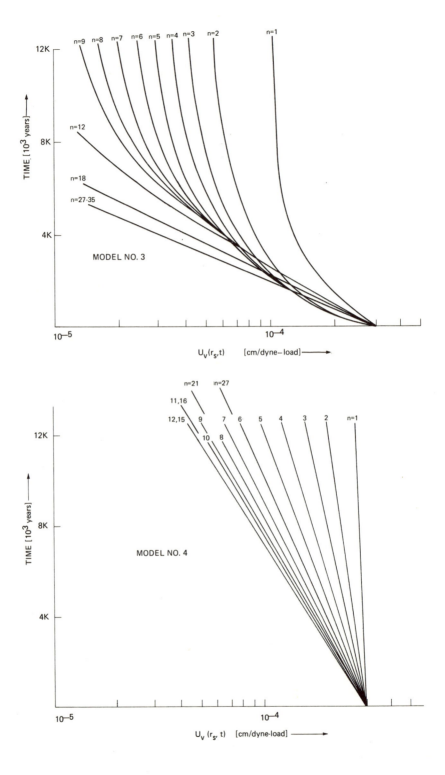

Figure III-17. Viscous decay spectrum for Model #3 (see Table IV-1). $\eta = 10^{21}$ poise between 0 and 335 km depth into the mantle and 10^{22} poise from 335 km to the core-mantle boundary. There is a uniform non-adiabatic density gradient between 335 and 635 km depth with integrated magnitude of .5 gm/cc. The effects of the non-adiabatic density gradient can be clearly seen in the slowing of decay after a certain point for all the lower order harmonics (fishtailing).

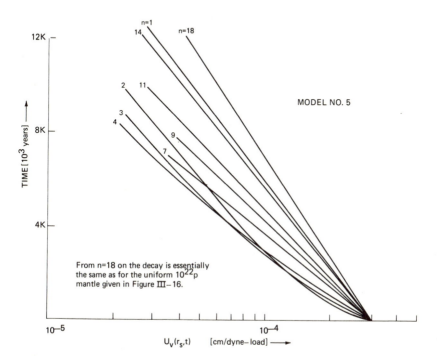

MODEL NO. 5

From n=18 on the decay is essentially the same as for the uniform 10^{22}p mantle given in Figure III—16.

$U_v(r_s,t)$ [cm/dyne—load] ⟶

TIME [10^3 years] ⟶

Figure III-19. Viscous decay spectrum for earth Model #5 (Table IV-1). The viscosity is 10^{22} poise to 985 km depth, then increases linearly to 2×10^{22} p at 2185 km depth and 3×10^{22} p at 2863 km depth (core-mantle boundary). $\nabla \rho_{NA} = 0$ everywhere. The lower order harmonics see the increase in viscosity distinctly (compare to Figure III-16) but past $n = 18$ the decay spectrum is the same as in a uniform 10^{22} poise mantle (Figure III-16).

Figure III-18. Viscous decay spectrum for earth Model #4 (Table IV-1). The viscosity of the mantle below 985 km depth is taken as 10^{24} poise, the viscosity above as 10^{22} poise. $\nabla \rho_{NA} = 0$ everywhere. The high viscosity lower mantle drastically slows adjustment of the lower order harmonics but does not affect harmonics with $n > 27$. The decay of these harmonics is the same as in Figure III-16.

of the interference of the core. By $n = 8$, the decay is again as slow as for $n = 1$. The rest of the decay spectrum is regular.

Addition of a low viscosity channel and a non-adiabatic density gradient (Figure III-17) slows decay of the low order numbers, after a certain point. At the same time, decay of the higher order numbers is much more rapid due to the 335 km thick low viscosity channel.

Decay of the lower order numbers is *drastically* slowed if a high viscosity is given to the mantle beneath 1000 km depth (1000 km from the surface of the crust, 985 km from the surface of the mantle). Only at $n = 27$ does the decay become the same as with a low viscosity lower mantle. This is shown in Figure III-18.

Figures III-19 and 20 give the decay spectra for earth models we shall consider later.

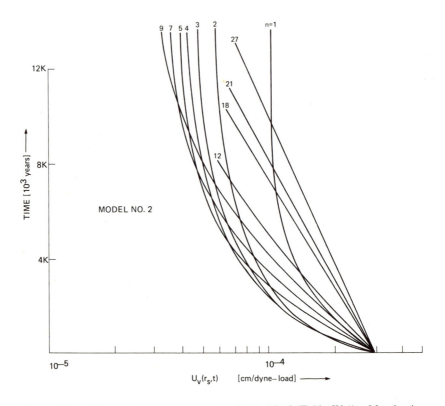

Figure III-20. Viscous decay spectrum for earth Model #2 (Table IV-1). Mantle viscosity is everywhere 10^{22} poise, but there is a uniform non-adiabatic density gradient between 335 and 635 km depth. Comparison with Figure III-17 highlights the influence of an upper mantle non-adiabatic density gradient on the decay of low order harmonics. Past $n = 27$ decay is similar to Figure III-16.

FOUR

Application of Theory: The Earth's Viscosity Structure

WE HAVE CALCULATED decay spectra that describe how various earth models would adjust to the sudden application of one dyne harmonic load of different order number. For convenience these earth models are summarized in Table IV-1. If these models are combined with a model of the melting of the late Wisconsin ice and the attendant filling of the ocean basins, i.e., a model of the post-Wisconsin load redistribution, isostatic adjustment for the various earth viscosity models can be predicted. Comparison of these predictions to the adjustment observed geologically can indicate which mantle viscosity model is most appropriate.

The general approach can be described as follows: Suppose that at some point τ_i years in the past, a distributed load increment $\Delta L(\theta, \phi, \tau_i)$ dynes was suddenly applied to the surface of the earth. The manner of adjustment of the earth to that load increment can be computed from the harmonic constituents of that increment.

$$\Delta L(\theta,\phi,\tau_i) = \sum_{m=1}^{r} \sum_{n=m}^{r} (\Delta^c L_n^m(\tau_i) \cos m\phi + \Delta^s L_n^m(\tau_i) \sin m\phi) P_n^m(\theta),$$

$$(\text{IV-1})$$

then

$$U^{\#}(\theta,\phi,t) = \sum_{m=1}^{r} \sum_{n=m}^{r} (\Delta^c L_n^m(\tau_i) \cos m\phi$$

$$+ \Delta^s L_n^m(\tau_i) \sin m\phi) P_n^m(\theta) U^{\#}(n, t - \tau_i) F_n.$$

$U^{\#}(0, \phi, t - \tau_i)$ describes the worldwide uplift that follows the load redistribution $\Delta L(\theta, \phi, \tau_i)$. It is the total uplift response consisting of the sum of the elastic, $U_E^{\#}$, and fluid, $U_V^{\#}$, uplift at any time, $t - \tau_i$. $U_V^{\#}$ are shown in Figures III-16–20. F_n is a set of filter weights we shall discuss in a moment.

If the load is removed in several incremental steps (say N of them), then the adjustment of the earth's surface to the combination of these

Model #1
   HB_1 Elastic Earth (see Figure III-13)
   $\nabla \rho_{NA} = 0$ everywhere
         $\eta = 1.0 \times 10^{22}$ poise
      $\rho_{core} = 9.927$ (for all models)
   $\rho_0(r)$ as in HB_1 model (for all models)

Model #1'
   Same as 1 except glaciers removed so that ice levels at $t = 6.5$ KBP
   were same as those at 7 KBP before, and ice was totally removed at
   4.5 KBP rather than 5 KBP.

Model #2
   HB_1 Elastic Earth
   $\nabla \rho_{NA} = 0$ everywhere except between 335 and 635 km into the
         mantle, where it assumes a constant value whose integrated
         magnitude is:

$$\int_{335}^{635} \nabla \rho_{NA} dr = .5 \text{ gm/cc}$$

   $\eta = 1.0 \times 10^{22}$ poise

Model #3
   HB_1 Elastic Earth
$$\int_{335}^{635} \nabla \rho_{NA} dr = .5 \text{ gm/cc, elsewhere zero}$$
   $\eta = 1.0 \times 10^{21}$ poise 0–335 km
   $\eta = 1.0 \times 10^{22}$ poise 335 km-core

Model #4
   HB_1 Elastic Earth
         $\eta = 1.0 \times 10^{22}$ poise 0–985 km into the mantle
         $\eta = 1.0 \times 10^{24}$ poise 985 km-core
   $\nabla \rho_{NA} = 0$

Model #5
   HB_1 Elastic Earth
   $\nabla \rho_{NA} = 0$
         $\eta = 1.0 \times 10^{22}$ poise 0–985 Km
         $\eta = 2.0 \times 10^{22}$ poise at 2185 km
         $\eta = 3.0 \times 10^{22}$ poise at 2863 Km (core)
   η is linearly interpolated between these values

Table IV-1   Earth models for which worldwide isostatic adjustment was computed. For
all earth models except #1' the load removal described in Section IV.A.4 was assumed.

incremental unloadings can be described by the numerical convolution integral:

$$U^{\#}(\theta,\phi,t) = \sum_{i=1}^{N} \sum_{m=1}^{r} \sum_{n=m}^{r} (\Delta^c L_n^m(\tau_i) \cos m\phi + \Delta^s L_n^m(\tau_i) \sin m\phi)$$

$$\cdot P_n^m(\theta) U^{\#}(n, t - \tau_i) F_n. \quad \text{(IV-2)}$$

It is assumed that $U^{\#}(n, t - \tau_i < 0) = 0$.

The rate of uplift was computed as in (IV-2) except that $V^{\#}$ instead of $U^{\#}$ was used. In this case, correcting the rate of uplift for the effects of elastic response was achieved:

$$V^{\#} = V_V^{\#}\left(1 - \frac{U_E^{\#}(t - \tau_i = 0)}{U_V^{\#}(t - \tau_i = \infty)}\right).$$

It should be noticed that m runs from 1 to r in (IV-2). This automatically assures us that we deal only with redistributed loads. A load with $m = 0$ would be a uniform DC load that could only be produced by a net influx (or efflux) of material from outside the earth (another planet, meteorites, etc.). It would not, of course, be geologically reasonable to presume that any significant such transport of material has occurred in the last ten or twenty thousand years.

The transformed load removal data were filtered by multiplying the transform spectrum by a suitable set of filterweights, F_n. Filtering is necessary to eliminate the Gibbs phenomenon. If one tries to represent a function with sharp edges by a Legender (or Fourier) series that is truncated at order r, that representation will overshoot the function near the sharp edges and produce troughs just outside the edges. These troughs and ridges have no physical meaning but are artifacts of an imperfect mathematical representation. Clearly they could be easily confused with physical phenomena that have significance to the viscosity of the mantle.

The Gibbs phenomena's troughs and ridges may be eliminated if the frequencies higher than r are removed from the load description by filtering. For the loads presented in this chapter, a .5% Gaussian filter was used. The procedure used is described in detail in Appendix V. Use of this filter assures us that an ice load 3000 m thick will have mathematical side lobes less than 15 m amplitude. Equivalently, if the ultimate viscous uplift in an unloaded region will be 700 m, any change in elevation even in the immediately surrounding regions of greater than about 3.5 m is physically meaningful. Actually the load representation appears to be somewhat better than this.

In all cases presented, load redistributions were described at 72
equally spaced points on each of 72 meridian circles symmetrically distributed about the equator. Such a grid permits Legender coefficients
to be determined out to order number 36. Thus r in equation IV-2 is
taken equal to 36. As discussed in Appendix V, Ellsaesser's weights are
used to improve the accuracy of the Legender transformation. Those
coefficients are only rigorously valid out to $n = r = 18$, but the error
involved in the extension of their use to $r = 36$ should be negligible for
our purposes.[1]

An elastic model for the earth as given by Hadden and Bullen is
assumed (see Fig. III-13).

The coefficients of the incremental load removal, $\Delta^c L_n^m(\tau_i)$ and
$\Delta^s L_n^m(\tau_i)$, were obtained by Legender transforming the load distributions defined in the next section. For the computation of model uplift in equation IV-2, five equal removal steps were assumed between
each of six Legender transformed, filtered, load stages at $T = 17, 12,$
10, 8, 7, 5 thousand years before present (see Table IV-2 and discussion
following).

The landward edges of both the Canadian and Fennoscandian
glaciers were tapered, and water was permitted to load both Hudson's
Bay and the Gulf of Bothnia as meltwater was added to the oceans after
these areas were uncovered.

Figure IV-1 illustrates the results of computation of the isostatic
adjustment of one of the earth models (uniform 10^{22} poise mantle,
Model #1) to the load redistribution chosen to represent the melting of
the Wisconsin glaciers. The sinking of the oceans is evident from the
increasing "relief" of the continents. Uplift in Canada can be seen to
follow glacial retreat and to be quite cycle-shaped in the early stages of
uplift. Other features of interest can be noted, including the early flow
of the suboceanic mantle material under continents such as Australia
for temporary storage. We shall discuss these matters in more detail
later.

For small-scale load redistributions, higher resolution with no decrease in validity can be obtained using the simple half-space non-self-
gravitating theory developed in the first part of Chapter III. In the next
section, after assessing the duration of the last glacial load and the
manner of its removal and checking the load magnitude by computing
eustatic sea level, we use this simple theory to assess in order: (1) the

---

[1]All Legendre transforms were made using Legendre polynomials generated by a
program made available by the Geophysical Fluid Dynamics Laboratory of the
Princeton Forrestal Campus and written originally by Dr. F. Baer.

influence of the lithosphere on isostatic adjustment, (2) the viscosity of the uppermost mantle, and (3) the viscosity of the mantle above 1000 km. The self-gravitating model with spherical geometry is then used to assess the viscosity of the lower mantle.

Necessary geological evidence is collected in each section. Whenever possible, physical interpretations are given to supplement the computational results. Critical data bearing on glacial uplift, such as gravity anomalies, are reviewed. At the end of the chapter, initial assumptions (such as the assumption of a layered Newtonian viscosity) are reassessed. Suggestions for further work are made.

The sections we have chosen are not all mutually exclusive. For example, models of eustatic sea level have implications not only for the volume and mode of ice removal but also for the viscosity of the lower mantle. The Fennoscandian uplift probes primarily the upper mantle (<1000 km depth), but it also is affected by the viscosity of the lower mantle and places some restraints on its viscosity. We handle these inconsistencies by making cross-references between sections.

A. The Duration and Magnitude of the Ice Load: Meltwater Curves and Eustatic Sea Level

Before the response of the earth to Pleistocene load redistribution can be calculated, we must know the nature of the load redistributions. How long was the last ice load applied to Canada and Fennoscandia? How large was it? How, in detail, did the ice retreat? At least tentative answers to these questions are provided by geological investigation of past climates, radioactive isotope studies, ice mechanics, and studies of glacial end moraines. Such studies indicate that the maximum Wisconsin ice load was probably applied about as long as it has now been removed, with substantial ice loads applied about twice as long. Ice mechanics and end morain data permit the magnitude of the ice load to be assessed. A meltwater curve, showing the increase in depth of the world's oceans that resulted from the melting of Wisconsin glaciers, can be computed in the process. When this meltwater curve is combined with the computed increase in depth of the oceans resulting from isostatic adjustment to the meltwater load, a eustatic curve results. Comparison of the eustatic sea level curves computed for the various earth models (Table IV-1) to eustatic sea level curves geologically observed provides a check on the size of the estimated ice load and its estimated mode of removal as well as indicating the relative appropriateness of the various earth models.

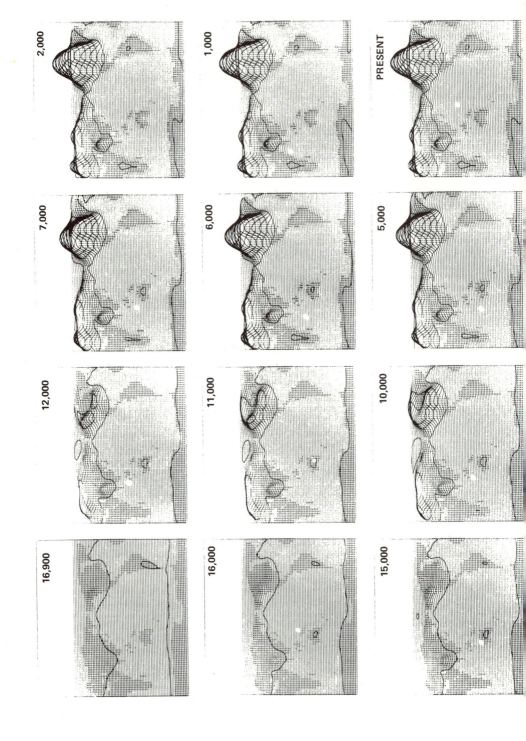

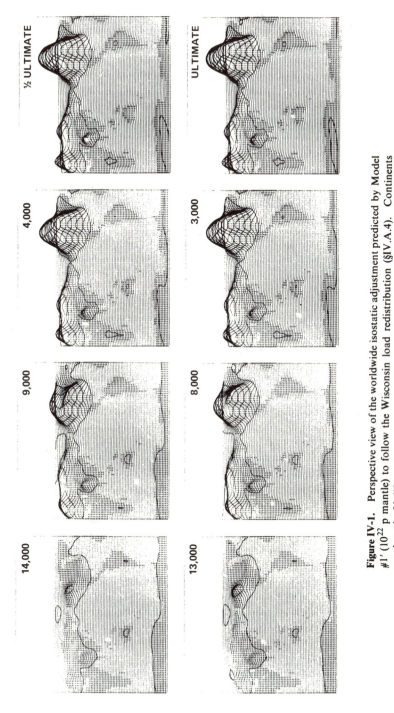

Figure IV-1. Perspective view of the worldwide isostatic adjustment predicted by Model #1' (10^{22} p mantle) to follow the **Wisconsin** load redistribution (§IV.A.4). Continents are dotted. Uplift contours are every 100 m starting at zero. Times for each figure are in thousands of years **BP**. This figure shows the uplift in formerly glaciated areas, the sinking of the loaded ocean basins and an initial uplift of Australia followed by sinking, indicating that the continents can act as temporary depositories for mantle material squeezed out from under the loaded oceans.

1. Pleistocene Ice Fluctuations in General

Beginning in the Eocene, average yearly temperatures in both Europe and North America began a gradual but not strictly uniform decline. By the end of the Pliocene, average temperatures in western Europe had fallen about 12–15°C, and those in the United States a similar amount (Dorf, 1960, 1970; Woldstedt, 1958, Vol. I, p. 8). London, England, had seen a transformation from a climate, flora, and fauna similar to today's tropical rain forests of Indonesia (10°N Latitude) to a climate, flora and fauna similar to its current one (50°N Latitude). The alligators and crocodiles that once abounded were preserved along with the remnants of the tropical rain forests in the Eocene London clay (Dorf, 1960). From fossil evidence, Dorf (1970) draws the climate curves of Figure IV-2.

Paleomagnetism as well as rough calculations of expected continental drift indicate a change in latitude of less than 10° for Europe and America since the Eocene (Holmes, 1965, p. 1211). Consequently, a major change in worldwide climate is implied.

As the climate cooled, glaciers developed in Alaska, the Sierra Nevada, Iceland, Antarctica, and presumably in Greenland. Denton and Armstrong (1969b) have demonstrated, by dating lava flows inter-

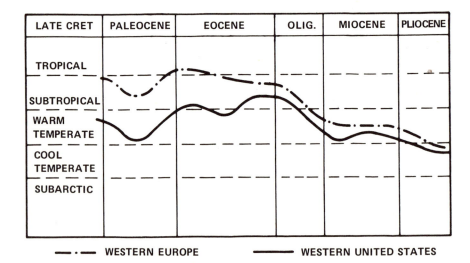

Figure IV-2. Climatic regimes inferred from fossil evidence for western Europe and western United States. The figure, reproduced with the author's permission, is from E. Dorf, Symposium North American Paleontological Convention, 1969, © the Allen Press, Inc.

bedded with till, that extensive glaciation of the Wrangell Mountains of southern Alaska occurred 9–10 million years before the present and intermittently at least 16 times since then. They find evidence of a pronounced climatic deterioration in Alaska of about 7°C in the Miocene. This evidence is corroborated by planktonic fauna in the marine tillites of the Yakataga Formation studied by Bandy et al. (1969) which indicate an influx of cold water foraminifera 13 million years before present. These foraminifera suggest a decrease in surface water temperature of 10–15°C. Herman (1970) found evidence of ice-rafted material in paleomagnetically dated marine cores from the Arctic Ocean indicating continental glaciation commenced prior to 6 million years before present. Interestingly, she suggests that the Arctic remained free of permanent pack ice till about 0.7 million years before present. The Sierra Nevada Mountains were glaciated 3 million years before present. Iceland was glaciated about the same time (Curry, 1966).

Denton and Armstrong (1968, 1969a) were able to show by potassium-argon dating lava flows overlying glacial till in the Taylor Valley, where the ice spills through the Transantarctic Mountains, that the east Antarctic ice sheet had *fully developed* at least by 4 million years ago. Since the morphology of the region had been only slightly modified since that time, there is an indication that this date is to be regarded as a minimum. The ice in the Taylor Valley has undergone at least five fluctuations since 4 million BP but has never disappeared. The fluctuations do not correlate in any obvious way with the Pleistocene glaciations of the Northern Hemisphere. There is direct evidence of glaciation in the Jones Mountains ~7m years BP (Denton et al., 1971).

It appears then that, as the climate cooled, ice developed in the most suitable locations first, gradually increasing in volume in an oscillating fashion until, at the beginning of the Pleistocene (2 million years before present), it had attained approximately the volume it has today. This trend is reflected in Tertiary sea levels as shown in the graph of Figure IV-3, which makes no attempt to portray small fluctuations. The fall in sea level is about what would be expected for construction of the present Antarctic (and Greenland) ice cap.

During the Pleistocene itself, whose boundary with the Pliocene generally is taken as 2 ± 0.2 million years BP (Fairbridge, 1968, *Encyclopedia of Geomorphology,* p. 918), there were at least five major glaciations of the Northern Hemisphere. The interglacial periods were slightly warmer than our present climate (Toronto was 2–3° warmer than it is today, Dorf, 1960), but not warm enough to melt the Green-

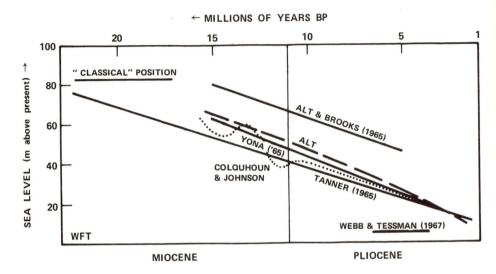

Figure IV-3. Tertiary sea levels (meters above present) as a function of time (10^6 years BP). The "classical" pre-glacial position is ~ 80 m. Other curves were compiled from the literature by Tanner. The figure, reproduced with the author's permission, is from W. F. Tanner, *Palaeogeography, Palaeoclimatology, Palaeoecology, 5* (Special Issue), p. 12, 1968, © the Elsevier Publishing Co.

land glacier, although probably enough to cause it to recede. Glacial advances in Europe, Asia, and North America appear in phase. The last ice age is called Wisconsin (North America), Wurm (Alps), Weichsel (Northern Europe), or Varsovian (Poland, U.S.S.R.), depending upon geographic location.

Recent work (see especially Broecker and Van Donk, 1970) is providing increasing support for the theory that the various ice ages are correlated with perturbation in the orbital parameters that govern the amount of solar energy received by a given hemisphere in a year (the Milankovitch Theory). It might be reasoned that, as the temperature of the Northern Hemisphere dropped, and as glaciations of various extent began, any minor fluctuations in the amount of solar radiation reaching the Northern Hemisphere, particularly in the summer, might be reflected in glacial fluctuations. Such minor fluctuations in solar radiation occur due to changes in the eccentricity of the earth's orbit (period 90,000 years), changes in the tilt of the earth's mean axis of rotation relative to the ecliptic (period 41,000 years), and precession of the earth's axis of rotation (period 21,000 years). These fluctuations are governed by orbital mechanics and thus can be calculated. The 90,000 year cycle of orbital eccentricity seems particularly important

(Broecker, 1966). It can be seen from Figure IV-4 that the glacial stages follow a rough 90,000 year cycle.

Fluctuation of ice volume is reflected in sea levels. The species composition and total number of foraminifera living in surface waters is temperature dependent. The temperature of the surface water environment, the continental ice volume, or some combination of the two is reflected in the O^{18}/O^{16} ratio of foraminifera skeletons (the more ice or the colder it is the larger the δO^{18} values).[2] Figure IV-4 illustrates the agreement between these various types of evidence.

The form of the δO^{18} curves in Figure IV-4 suggests that a gradual build-up of ice or decrease in temperature is commonly followed by a very sudden melting of the ice or rise in temperature. If the δO^{18} curves reflected ice volume, such a form would be in accord with ice mechanical calculations. Weertman (1963) has calculated that ice sheets should take 15 to 30 thousand years to grow, but could dissipate in 2 to 4 thousand years. Broecker et al. (1960) summarize evidence for an abrupt change in climate about 11,000 years before present. They point out that changes in the rate of marine sedimentation, in the clay content of the Mississippi River, the drying of lakes in the American Great Basin, and sharp changes in the type of pollen in European varves all indicate an abrupt change in climate. End morain data indicate the Wisconsin ice dissipated extremely rapidly, especially from Fennoscandia, after about 10,000 BP.

From this discussion we may conclude:

(1) Antarctica was continously and fully glaciated for at least the last 4 million years and probably at least the last 7 million years. (See also Denton et al., 1971.) It is virtually certain that Greenland was continuously glaciated for the last 2 million years, although direct proof is lacking.

(2) The Northern Hemisphere was glaciated a substantial fraction of the last 600,000 years, but was not continuously glaciated.

(3) Deglaciation in general and in particular after the last glacial

---

[2]Dansgaard and Tauber (1969) have recently suggested that Emiliani's marine $\dfrac{O^{18}}{O^{16}}$ curves may be interpreted to reflect the worldwide volume of continental ice. Broecker and Van Donk (1970) do not take a stand on this issue in their review article but point out that whether the curves are interpreted as temperature or as ice volume fluctuations really makes little difference as changes in temperature may be taken to imply ice fluctuations. Unfortunately, the relation between temperature and ice volume is undoubtedly not a linear one. Whatever the resolution of this question, agreement between species' proportion, total number of foraminifera, and δO^{18} in foraminifera skeletons appears excellent. (Broecker and Van Donk, 1970; West, 1968, Fig. 10.2, 10.9).

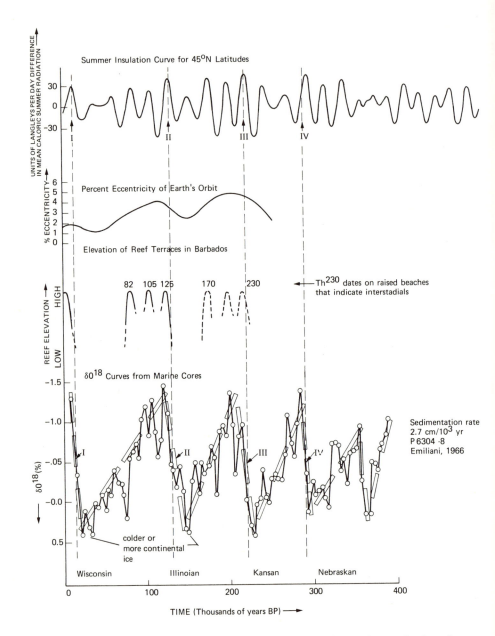

Figure IV-4. The correlation between peaks in summer insulation, the termination of glacial stages, and high sea level appears quite good. High orbital eccentricity appears quite important. Islands, like Barbados, that are tectonically rising provide high resolution records of sea level fluctuations. Curves are from Broecker and Van Donk (1970, Figures 2a, 10) and Mesolella et al. (1969, Figure 9).

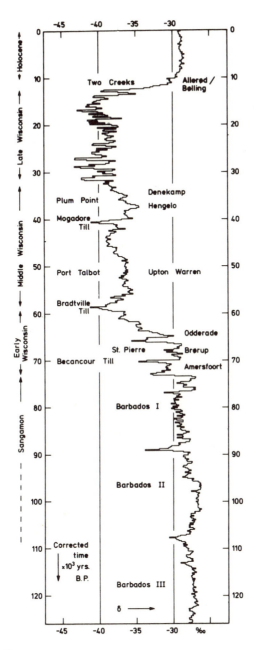

Figure IV-5. δO^{18} profile of the Greenland Camp Century ice core. Measurements made on core in 200 year intervals on a corrected time scale deduced from ice flow model. Tentative interpretation of European and American glacial events is given by Dansgaard et al. and the division of the Wisconsin was proposed by them (far left). The figure, reproduced with the first author's permission, is from W. Dansgaard, H. B. Johnsen, and C. C. Langway, in *The Late Cenozoic Glacial Ages*, K. K. Turekian, ed., p. 52, © Yale University Press.

stage proceeded very rapidly, an order of magnitude more rapidly than the time necessary to build up the continental ice sheets.

2. The Last Stage of the Last (Wisconsin) Glaciation

The last glacial stage (the Wisconsin) is of greatest importance to our study of recent isostatic adjustment phenomena. High resolution δO^{18} data are provided by studies of ice cores through the Greenland ice cap. These cores do not suffer the ambiguity of δO^{18} measurements in foraminifera skeletons. There seems to be general agreement that δO^{18} measurements in ice cores reflect primarily the temperature at which precipitation occurred. Dansgaard et al. (1969a, 1971) have interpreted a Greenland core assuming a constant precipitation rate and making simple assumptions regarding horizontal ice flow. The resulting "climate" curves (an example of which is shown in Figure IV-5) are remarkably detailed, reflecting even very brief warm periods such as the Alleröd and the Bölling. The agreement between the Greenland core and the δO^{18} analyses of marine cores summarized by Broecker and Van Donk is good. The Port Talbot and Plumb Point interstadials (see Figure IV-5) can be recognized: (1) geologically in the field (formation of soils, moraines, etc.), (2) in δO^{18} studies, and (3) in sea level curves. Our problem is to determine how great a removal of the Wisconsin glacial load occurred during these interstadials. The best tool at our disposal for answering this question is a study of sea level prior to 19,000 years BP.

Figure IV-6 shows that available data from continental shelves (which were largely water free during the Wisconsin Maximum) suggest that sea level was higher 25 to 30 thousand years BP than it was 15 thousand years BP. Sea level was minimum 15 to 19 thousand years BP. These are the conclusions of Emery et al. (1971).

It is not clear how close sea level came to its present level 35 thousand years ago. Shepard (1963, p. 5ff) briefly summarizes some C^{14} dated material from Mexico and Texas indicating a high stand of sea level of about -50 ft at 25,000 years BP. This would correlate with low stands of lakes in the Great Basin of the United States and pollen indicating an interstadial in the Lake Erie region (Farmingdale or Plumb Point). Hoyt et al. (1965) have suggested, on the basis of radiocarbon dates from Georgia, that sea levels during the Port Talbot and Plumb Point-Farmdalian interstadials were as high or higher than present and that North America and Europe were therefore ice-free at these times. They caution that their radiocarbon dates may be subject to question, however. Milliman and Emery (1968) present a sea level curve for

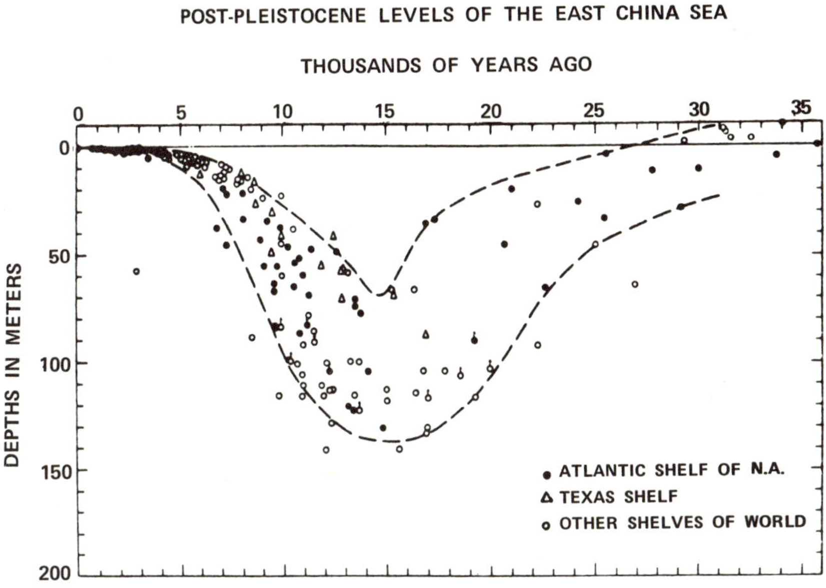

POST-PLEISTOCENE LEVELS OF THE EAST CHINA SEA

THOUSANDS OF YEARS AGO

Figure IV-6.   Plot of published sea level data versus radiocarbon age.   Dates from shallow-water shells, oolites, salt-marsh peat, wood, coralline algae, and coral.   Note that near-present sea levels are suggested for times ~30–35 thousand years before present.   The figure, reproduced with the first author's permission, is from K. O. Emery, H. Niino, and B. Sullivan, in *The Late Cenozoic Glacial Ages,* K. K. Turekian, ed., p. 383, 1971, © Yale University Press.

the past 35,000 years.   Their curve shows sea level approached its present level about 32,000 years BP.

Broecker and Van Donk (1970, p. 184), in reviewing sea level data, have questioned the validity of radiocarbon dates older than 35,000 years. They feel it is "quite likely" sea level reached within 18 m of its present level. Guilcher (1969) reviews the problem of possible high sea level stands during the Wisconsin and concludes that it is necessary to remain cautious because some evidence supports and some contradicts a Wisconsin interstadial.   For contradicting evidence, he points to the fact no great increase in temperature is indicated by most δO^{18} curves. This objection was recognized by Shepard (1963, p. 7).   Guilcher also cites evidence that the Adriatic Sea was 100 m below its present level 20,000 BP.   Numerous data points on Milliman's and Emery's curve support this.

Of course, it could be argued that glaciers grew very rapidly between

25 and 20 thousand years BP. Indeed many sea level curves are drawn
with very sharp drops in this time interval (Hoyt et al., 1965, p. 391;
Shepard, 1963, p. 6; Curray, 1961, p. 1708). As we have seen, however, Weertman (1963) has calculated that ice sheets should take 15 to
30 thousand years to grow, but can dissipate in 2 to 4 thousand years.
In this light, such rapid and large glacial growths as 100 m sea level in
5,000 years become suspect.

Evidence of a marine transgression 50 to 35 thousand years BP is
provided by New Guinea (Veeh and Chappell, 1970). This area gives
high resolution data because of its rapid tectonic uplift (3 mm/yr). The
tectonic uplift may be subtracted out if sea level is assumed to be
known from other areas at particular times. The authors assume that
sea level was between -13 and -16 m 80,000 years BP and between
-5 and -12 m 6300 BP. On this basis they infer the sea level curve
shown in Figure IV-7. As can be seen in Figure IV-7, New Guinea indicates sea levels reached -25 m about 30,000 years BP.

We conclude there is substantial evidence of a significant rise in sea
level (to between -25 and -15 m) during a substantial fraction of the
60,000 year span of the Wisconsin glaciation. Also, it appears that the
last glacial load was applied to the continents for about as long a time
as it has now been removed. Thus, if the glaciated areas are not near
equilibrium at present, we must account for the brief application of the
glacial load in our modeling.

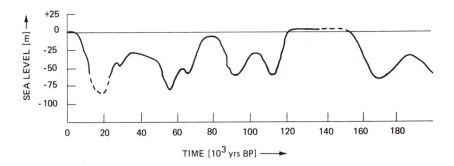

Figure IV-7.   Sea level curve indicated by New Guinea. The 20 to 40 thousand-year BP
record is particularly distinct in this area, providing perhaps the best indication to date
of Wisconsin interstadial sea levels. The figure, reproduced with the first author's permission, is from H. H. Veeh and J. Chappell, *Science, 167,* p. 864, © 1970 by the
American Association for the Advancement of Science.

3. Wisconsin Ice Thickness

Direct geological evidence of glacial thickness is sparse, especially for North America. It can be inferred from erratics of the Bothnian lowlands carried over the Scandinavian mountains that the ice sheets in Fennoscandia may have had thicknesses in excess of 2500 m (Flint, 1957, p. 368). General conclusions may be drawn from ice mechanics, however. Ice flows as if it were almost a perfectly plastic fluid (Weertman, 1963; Nye, 1959). Almost all the shear takes place very near the bottom of the glacier. The most important variable in determining the thickness of a glacier is thus its areal extent (and the yield stress, τ_0, of the ice at the base of the glacier). The thickness of a glacier only depends on the accumulation rate to the one-sixth power (Nye, 1963). The maximum thickness H and half width L of a glacier scale:

(IV-3)

$$H = \left(\frac{2L\tau_0}{\rho g}\right)^{1/2} \qquad H_{AVE} = \frac{2}{3}H$$

where ρ = density of ice, g = gravitational field strength (Weertman, 1963, p. 149).

τ_0 is the biggest uncertainty in using equation IV-3. It is generally thought to have a range between 0.5 and 1.0 bar (Moran and Bryson, 1969, p. 98) depending on the temperature at the base. Continental glaciers in warm latitudes might behave differently from glaciers in exceedingly cold regions such as Antarctica. Variation in τ_0, as a result of frictional heating is presumably the cause of glacial surges (Wilson, 1964). Topography is also a source of uncertainty. Greenland is not the unbiased example it might be, as the glacier there is substantially dammed in by peripheral mountain belts. The current Antarctic glacier, which is roughly the same areal extent as the Wisconsin glaciation in North America, has a general maximum thickness of 3000 m with small domes up to 3500 m thick (Bently et al., 1964, Plate 3; Denton et al., 1971). The thickness of ice in central Greenland today is 3100 m (Moran and Bryson, 1969, p. 98).

The sophistication of equation IV-3 can be increased by considering various modifications. The thickness is augmented if isostatic adjustment to the ice load is taken into account (Weertman, 1963). Flow laws other than perfect plasticity and other boundary conditions, such as calving, active growth, or stagnation, can be considered (Paterson, 1972). We assume an initial maximum thickness for the Canadian glacial system of 3750 m. Such a thickness could be computed from (IV-3) with a τ_0 of .33 bars. An ice thickness of 3750 m for the most recent Canadian glacier is in good accord with Paterson's (1972) estimate of 3.6 km. Equation IV-3 is used to scale the glacial thickness as the ice-covered area diminishes (or to obtain the thickness of glaciers smaller than the Canadian one) and to compute ice volumes. This approach avoids some of the difficulties inherent in the uncertainties of τ_0, and is shown in the following section to introduce negligible errors relative to the more sophisticated approaches of Paterson (1972).

4. Computation of a Meltwater Curve

The retreat of the Wisconsin ice has been studied extensively. It is well understood in Europe. There remain greater uncertainties in North America. No attempt will be made here to review the types of information available or the techniques used. The end result of such studies is an isochron map showing the retreat (or fluctuation) of ice margins with time.

DeGeer's (1954, Fig. 1, p. 304) isochron map is given in Figure IV-8. (This map is also reproduced and discussed in Woldstedt (1958, Vol. II, p. 150)). Recent investigation of a 6500-year-old pumice horizon in the Arctic shows uplift increases from West Spitsbergen to East Spitsbergen. This may suggest that there was thick ice in at least the North Barents Sea and that the Fennoscandia ice sheet was much larger than suggested by DeGeer (Schytt et al., 1968). Arguing against a large ice volume in the Barents Sea is evidence that land uplift increases from northern Norway south (Sauramo, 1939). We assume that there was not much ice in the Barents Sea. However, the filtering we are obliged to do smears out the ice load, so in fact our model assumes ice in the seas neighboring Fennoscandia.

Similar isochron maps for North America by Bryson, Wendland, Ives and Andrews (1969) and Prest (1969) are shown in Figures IV-9, 10. The differences between these two maps probably reflect the uncertainty in field knowledge of the ice retreat in North America. The most important differences between the two are: (1) the rate at which the Arctic archipelago was deglaciated and, by inference, the thickness of the ice load there, (2) the time of final deglaciation (7000 or 5000 years BP), and (3) Prest's map covers the deglaciation of the coast ranges as well as the central and eastern portion of the continent.

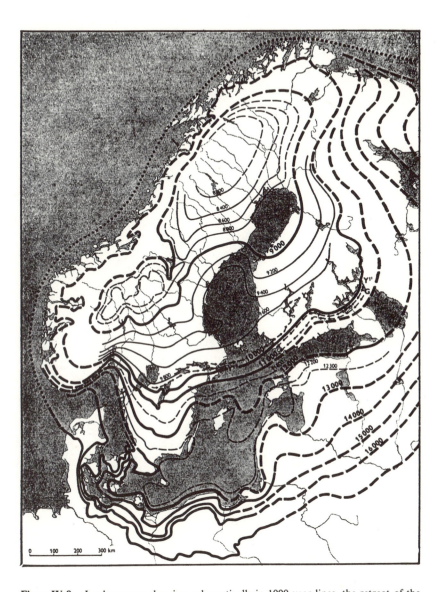

Figure IV-8. Isochron map showing, schematically in 1000 year lines, the retreat of the last ice in Fennoscandia. The figure, reproduced with the publisher's permission, is from E. H. De Geer, *Geologiska Föreningens I. Stockholm Förhandlingar, 76*, p. 304, 1954, © Geologiska Föreningen. *Note:* figure also appears in Woldstedt (1958).

Figure IV-9. Isochron map showing the retreat of the last ice in North America. Contours are in thousands of radiocarbon years BP. The figure, reproduced with the first author's permission, is from R. A. Bryson, W. M. Wendland, J. D. Ives, and J. T. Andrews, *Arctic and Alpine Research, 1,* p. 2, 1969, © The Institute of Arctic and Alpine Research.

Unless the Fennoscandian ice sheet was, in fact, much larger than is commonly thought, the major contributor to sea level rise was North American ice. Paterson (1972) estimates that it accounted for at least $\frac{2}{3}$ of the total fall in sea level. It is generally assumed that Greenland's contribution was negligible (Donn et al., 1962; Bryson et al., 1969). Antarctica may have contributed 6 m of eustatic sea level rise, mainly

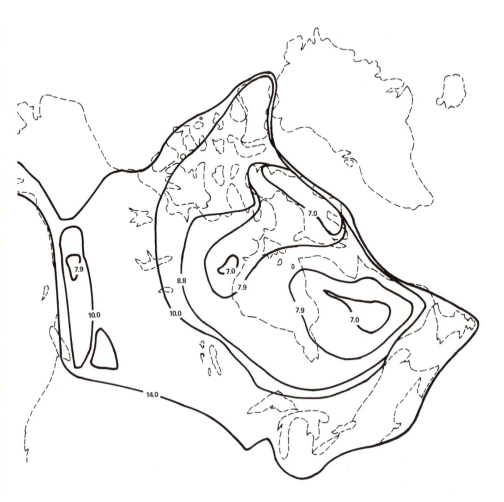

Figure IV-10. Isochron map showing the retreat of the last ice in North America sketched from a more detailed map by Prest (1969). Comparison to Figure IV-9 may suggest uncertainties in isochron determination. Isochrons are in thousands of radio-carbon years BP.

due to calving at its perimeters due to the sea level rise dictated by the Northern Hemisphere (Hollin, 1962).

Table IV-2 gives the values picked for the characteristic half-lengths of the glaciers in Canada and Fennoscandia and the corresponding average glacial heights calculated from equation IV-3. The average height assuming a low value for τ_0 (.33 bars) is summarized for the various glaciers of the world as a function of time. This, as we have previously commented, is the same as assuming maximum thickness of

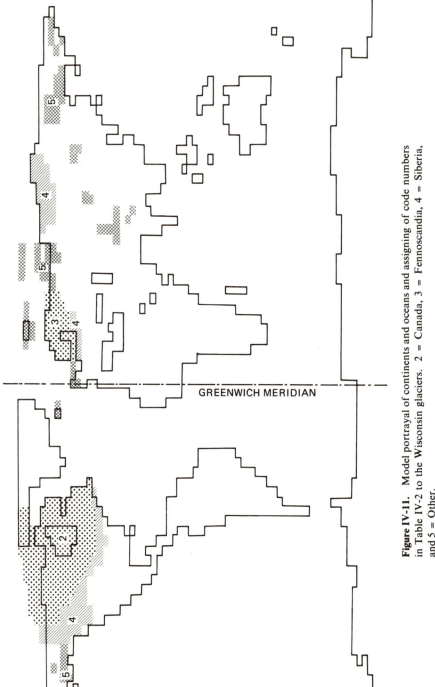

GREENWICH MERIDIAN

Figure IV-11. Model portrayal of continents and oceans and assigning of code numbers in Table IV-2 to the Wisconsin glaciers. 2 = Canada, 3 = Fennoscandia, 4 = Siberia, and 5 = Other.

Figure IV-12. Digital representation of the coastline of North America and of the fashion in which the late Wisconsin ice melted. Numbers represent thousands of radiocarbon years before present that the ice stand indicated existed. This figure is abstracted from Figure IV-9 and so is based on the isochron map of Bryson, Wendland, Ives, and Andrews.

3750 m for the Canadian glacier and scaling thicknesses for the other glaciers and the Canadian glacier at later times according to (IV-3).

Figure IV-11 shows the initial heights (see Table IV-2) assigned to each glacial system. (This figure also gives a good indication of the resolution of modeling.) The code number assigned is kept the same. Figures IV-12 and 13, direct abstractions of Figures IV-8 and 9, show how the area covered by ice was assumed to shrink with time. The ice thickness (from Table IV-2) times this area then gives the ice volume, which, when multiplied by .9174 gives the equivalent volume of water. Dividing by the area of the world's oceans then gives the meltwater curve listed in Table IV-3. The computed meltwater increment of 108.9 m is in good agreement with Donn et al.'s (1962) estimate of 105 to 123 m. It is also in good agreement with Paterson's (1972) maximum estimate of 114 m.[3] The approach we have used is discussed by Bloom (1971), although he does not compute actual ice volumes. The form of our meltwater curve (Table IV-3) is closely similar to that of Bloom's meltwater curves.

We might point out specifically: (1) The 14,000 BP ice stand in Fennoscandia was taken to have occurred at 12,000 years BP. (2) As the ice melted, the Gulf of Bothnia and Hudson's Bay were opened to the influx of sea water. From this point account was taken of the changes in load on these areas that resulted from further sea level increases. (3) A uniform retreat from 10,000 years BP to no ice at 8,000 BP was assumed in Fennoscandia. (4) The ice was assumed to be totally gone in Canada 5,000 BP.

5. Computation of a Eustatic Sea Level Curve

Eustatic sea level may be conveniently defined as the distance from the earth's centre to the average sea level at any given time. The meltwater increment calculated in the last section does not translate directly into a change in eustatic sea level because the ocean basins may isostatically adjust to the meltwater load.

---

[3]Paterson's maximum estimate is 114 m. Our meltwater increment is larger than that considered probable by Paterson. This is primarily because we use the isochron map of Bryson et al. (1969, Figure 7) rather than that of Prest (1969). The method used introduces no significant errors itself. Had we used the same glacial areas as Paterson, our volume estimates would have been within a few percent of his until 11.8 thousand years BP, when he changed from a steady state glacial model to a mean between a steady state and a stagnant glacial model. Thereafter our estimates of ice volumes would be 13–18% larger than Paterson's, which is within Paterson's estimates of accuracy for ice volume of 20% (not counting uncertainties in ice margin position).

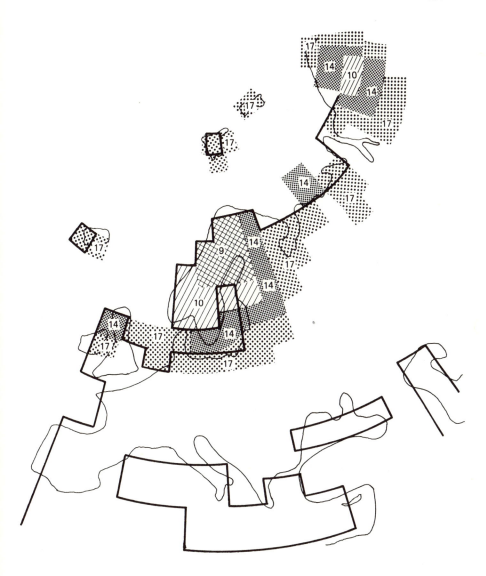

Figure IV-13. Digital representation of coastline of Fennoscandia and of the fashion in which Fennoscandia ice melted. Numbers represent thousands of radiocarbon years before present that the ice stand indicated existed.

Figure IV-14 shows the depression of the ocean basins that is predicted by the earth models of Table IV-1. (The method of calculation is summarized in equations IV-1, 2, and is discussed in that section.) Figure IV-14 shows substantial difference between the various earth models. A model mantle of reasonably uniform 10^{22} poise viscosity

Canadian Glacial System

Time	Estimated L	H_{avg}
		$\tau_0 = .33$ bars
14,000	2060 km	2.62
12,000	1760	2.42
10,000	1375	2.14
9,000	1100	1.91
8,000	1000	1.82
7,000	410	1.16
5,000	Gone	

Fennoscandia Glacial System

17,000	960	1.79
14,000	825	1.65
10,000	550	1.37
9,000	275	.95
8,000	Gone	

Average Height of Glaciers Used in Computations, Based on Above Calculations

		H_{avg}	H_{max}
$T = 17,000$ BP	(2) Canada	2.5 km	3.75
	(3) Fennoscandia	1.76	2.63
	(4) Siberia	1.52	2.28
	(5) Other	1.0	1.50
$T = 12,000$ BP	(2) Canada	2.42 km	
	(3) Fennoscandia	1.55	
	(4) Siberia	1.25	
	(5) Other	.6	
$T = 10,000$ BP	(2) Canada	2.15 km	
	(3) Fennoscandia	1.37	
	(4) Siberia	0.8	
	(5) Other	0.0	
$T = 8,000$ BP	(2) Canada	1.85 km	
	(3) Fennoscandia	0.0	
	(4) Siberia	0.0	
	(5) Other	0.0	
$T = 7,000$ BP	(2) Canada	1.2 km	
$T = 5,000$ BP	All glaciers melted		

Table IV-2 Calculation of average glacial thicknesses based on Figures IV-8 and 9. The maximum thickness of the Canadian glacial system was assumed to be 3750 m. Equation IV-3 was then used to determine the heights of other glaciers and of the Canadian glacier at subsequent times. This is equivalent to assuming $\tau_0 = .33$ bars. The numbers in parentheses next to the glacial identification refer to Figure IV-11. Bryson, Wendland, Ives and Andrews isochron map is used (Figure IV-9).

Time (Thousands of Years)	Average Sea Depth (m below Present)
BP	
17	108.9
12	81.6
10	52.5
8	20.1
7	5.0
5	0.0

Table IV-3 Wisconsin and post-Wisconsin meltwater curve. (Average depth of sea in meters less than present depth at various times in the past.) Calculated from Table IV-2 and Figures IV-11–13 by summing area elements 5° in latitude and 2.5° in longitude, multiplying ice thickness, coverting to water volume, and dividing by the area of the world's oceans. Meltwater curve is based on Bryson, Wendland, Ives, and Andrews Isochron map.

(Model #1) predicts a flat ocean depression that is nearly complete at present. The depression is the magnitude that would be anticipated:

$$33m = \frac{108.9}{(3.313 = \text{density of the upper mantle})}$$

For a high viscosity lower mantle (Model #4), however, most of the increased capacity of the oceans comes from deep troughs peripheral to regions of former ice load.

Table IV-4 shows average increase in ocean depth as a function of time predicted by the models in Table IV-1 for the load redistribution of Table IV-2. Some reduction in the expected increase in ocean capacity results because filtering has smeared the edges of the ocean load and the resulting depression. Thus only 26.2 m of increased capacity is ultimately predicted for any model, instead of ~33 m.

Combining the isostatic adjustment curves of Table IV-4 with the meltwater curve of Table IV-3 results directly in predicted eustatic sea level curves.

Geologically, a eustatic sea level curve may be inferred by the history of dated shorelines at a tectonically stable locality or by averaging dated sea levels at a large number of localities so as to average out tectonic or isostatic movements. Isostatic adjustment to the ocean load fluctuations suggests that stable platforms may be rare (see Figure IV-14, also Mörner, 1969, Bloom, 1967). Averaging a large number of localities reduces resolution. Mörner (1969) has suggested that resolu-

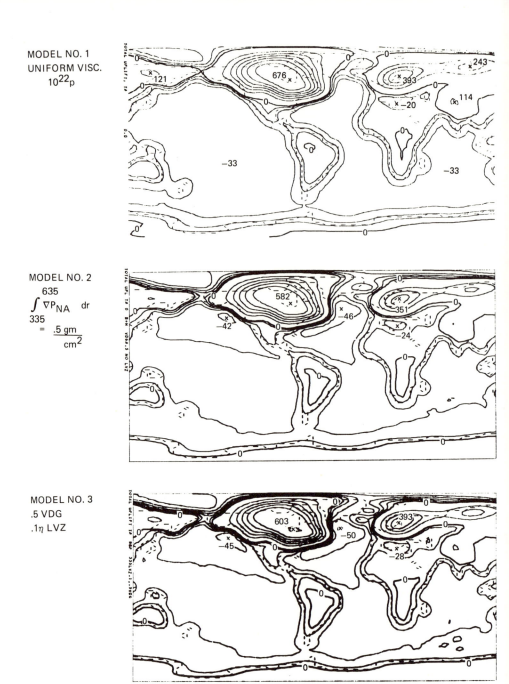

Figure IV-14. Model uplift at $T_{\text{Model}} = 0\text{BP}$ for the earth models of Table IV-1 displayed on a "mercator" projection with undistorted latitude scale. Thick contour = zero change in elevation. Positive uplift contours every 100 m. Contours of depression every 10 m. Peripheral "channel" troughs are clearly evident.

MODEL NO. 5
$\eta = 1,2,3$
$\times 10^{22}$p

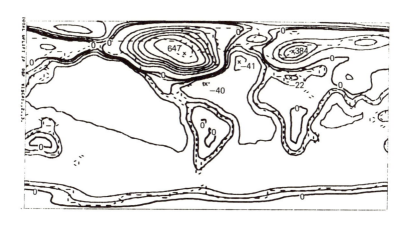

MODEL NO. 4
10^{24}p

LOWER MANTLE

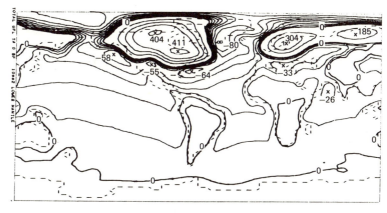

Model Time	Uniform Viscosity		$\nabla \rho_{NA} \neq 0$		High Viscosity Lower Mantle	
				LVZ		
	Model #1	Model #1	Model #2	Model #3	Model #5	Model #4
15		−0.8				
11		−5.4				
10	−7.8		−7.5	−8.1	−7.0	−4.1
9		−10.2				
7		−15.9	−15.0	−16.0	−14.7	−8.1
6	−19.0	−18.6	−17.1	−18.2	−17.1	−9.2
5		−20.8				
4		−22.6				
3	−23.9		−20.2	−21.3	−22.1	−12.0
2		−24.4				
0	−25.4	−25.3	−21.1	−22.0	−24.2	−13.8
−2		−25.7	−21.3	−22.1	−24.9	−14.8
−∞	−26.2	−26.2	−26.2	−26.2	−26.2	−26.2

Table IV-4   The isostatic adjustment of the ocean basins for the earth models listed in Table IV-1.

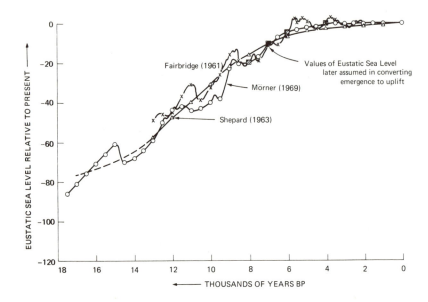

Figure IV-15. A comparison of observed eustatic sea level curves of various workers. It can be seen that the curves vary in detail but not in general form. Boxes indicate positions of eustatic sea level used for calculating isostatic uplift from sea level curves. It is assumed eustatic sea level was 20 m below present 8000 BP, −10 m at 7000 BP, −5 m at 6000 BP, and at present elevation in the last 4000 years. Eustatic sea level is assumed to have changed smoothly between these points. The dashed extension to Shepard's curve has been added for easy comparison to calculated curves.

tion may be regained by refining an average eustatic curve by subtracting it from the sea level curve of an isostatically uplifting locality, such as the Swedish coast, and requiring the resulting isostatic adjustment curve to be smooth and sensibly related to load redistributions. Eustatic curves determined in various ways by various workers do not differ dramatically. Figure IV-15 shows the eustatic curves of Fairbridge, Shepard, and Mörner. Also shown are the particular values of eustatic sea level we shall later use to obtain isostatic curves from sea level curves in Canada.

Figure IV-16 compares the eustatic curve computed for Model #1 to the eustatic curves of Figure IV-14. Curve A at the top is the meltwater curve taken directly from Table IV-3. The lowest curve in Figure IV-16 shows the increased capacity of the ocean basins that resulted from their isostatic adjustment under the meltwater load (from Table IV-4). The calculated eustatic curve is the difference between these two curves. All three eustatic curves are required to coincide at present time.

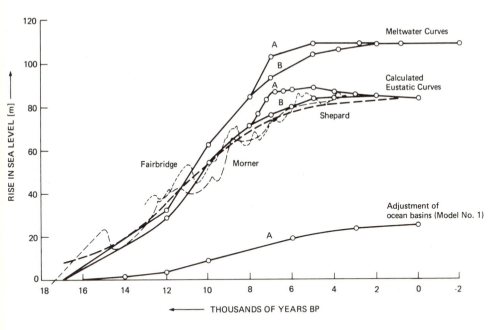

Figure IV-16. Comparison of the eustatic sea level curve computed on the basis of Model #1 (uniform 10^{22} poise mantle) to eustatic curves determined geologically.

The agreement between the calculated and observed curves is remarkably good both in form and in the magnitude of eustatic sea level change indicated (~ 80 m). The calculated curves could be shifted slightly toward the present (500 and perhaps 1000 years), but the quality of the agreement would be diminished if they were shifted any toward the past. The glaciers might have melted slightly less rapidly than we have modeled, but probably did not melt more rapidly. This indicates Prest's (1969) isochron map would not produce good agreement with observed eustatic sea level curves. Prest's map indicates dissipation of nearly all the ice by 7000 BP (see Figure IV-10). A meltwater curve going to zero near 7000 BP could not produce a eustatic curve in good agreement with observed eustatic curves. This conclusion was also reached by Moran and Bryson (1970).

Figure IV-17 shows the eustatic sea level curves that would be expected if the ocean basins adjusted instantaneously to the meltwater load. The agreement with the eustatic curves of Figure IV-14 is slightly better than that of Figure IV-16. Part of the improvement is due to the fact that isostatic adjustment of the ocean basins is larger since they were computed directly by taking 1/3.313 of meltwater Curves A

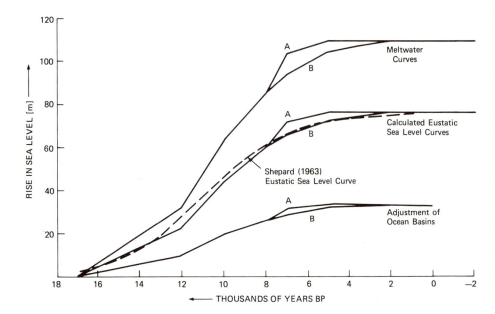

Figure IV-17. Eustatic sea level curve computed assuming oceans isostatically adjust immediately to any meltwater increment compared to Shepard's geologically determined eustatic sea level curve.

and B. Thus no smearing of the load due to filtering was involved. Part of the improvement, however, is due to a slight alteration in shape of the calculated eustatic curves. This suggests the agreement between calculated and observed eustatic curves might be slightly improved if the lower mantle were less viscous than the upper mantle.

The general quality of the match in both Figures IV-16 and IV-17 can be improved (and the Holocene high sea level indicated eliminated) if the last part of the meltwater curve is altered slightly (from curve A to curve B). This small, late alteration in the meltwater curve will not affect the adjustment of the ocean basins noticeably. Their adjustment is dominated by the earlier large influxes of meltwater. Thus, the eustatic curve corresponding to meltwater curve B can be calculated with negligible error from the ocean basin adjustment curve appropriate for A. The result (eustatic curve B) is in better agreement with the geological eustatic curves of Figure IV-16. It is clear that the existence of a Holocene high sea level is substantially dependent on the details of the melting of the last portions of the glacial ice unless the lower mantle is very fluid.

Figure IV-18 (Model #4) shows that a thousand-kilometer-thick

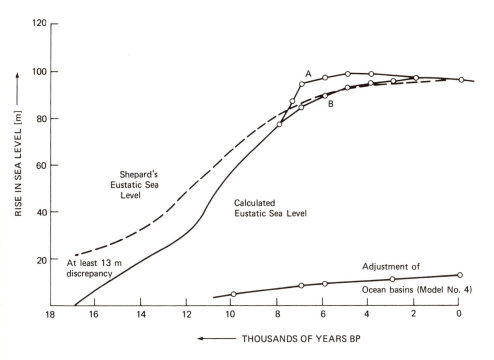

Figure IV-18. Comparison of eustatic sea level curve computed on the basis of Model #4 (10^{24} poise lower mantle) to Shepard's geologically determined eustatic sea level curve. There is at least a 13 m discrepancy.

channel of 10^{22} poise viscosity underlain by a high viscosity lower mantle (10^{24} poise) would not adjust rapidly enough to generate a eustatic curve in good agreement with that geologically observed. Here no appeal can be made to the smearing of the ocean's load. Most of the increased capacity of the ocean results from the generation of the channel depression surrounding formerly glaciated areas (see Figure IV-14). The eustatic curves could be brought into accord if the ice volume were reduced about 13%.

From Table IV-4 it can be seen the eustatic curves of all the earth models with near 10^{22} p mantle viscosities would be in acceptable agreement with observed eustatic curves.

Figures IV-16 and IV-17 make a case for a lower mantle of low ($\sim 10^{22}$ p) viscosity in the sense that they reconcile morain data (Figures IV-8, 9, 11), ice mechanics, and eustatic sea level observations in a straightforward, uncomplicated fashion. We are not forced to a double domed ice model to reduce the meltwater volume as were

Moran and Bryson (1969), who took no account of the isostatic adjustment of the ocean basins. However, mantles of higher viscosity could accommodate the eustatic data if the meltwater volume were reduced. Reduction in the meltwater volume might seem difficult in the light of possible additional ice in the Barents Sea and in light of the fact that we have already chosen (effectively) a rather low value for the yield stress of ice ($\tau_0 = .33$ bars). However, others (Paterson, 1972), have calculated smaller meltwater volumes by assuming smaller glaciated areas. Sea level variations recorded by islands compared to those recorded by continents and the suggested lack of a broad (long duration) Holocene high sea level provide additional evidence suggesting the viscosity of the lower mantle may be $\sim 10^{22}$ p and thus, indirectly, that our calculated ice volume is appropriate.

6. Bloom's Test of Isostasy

Bloom (1967) has pointed out that, if the oceans adjust rapidly to the imposition of the post-Pleistocene meltwater load, but the coasts remain fixed (being midway between the loaded oceans and the continents, which will rise initially to conserve mantle mass), then islands and continental shelf areas will act as "dip sticks" and record the full (meltwater) increase in ocean depth, whereas the continents will tend to record only the eustatic change in sea level. Since the density of the upper mantle is about 3.3 gm/cm^3, the change in eustatic sea level will be about two-thirds the meltwater increase. Given a layer of meltwater of some thickness added to the oceans, we may construct Table IV-5.

Donn et al. (1962) draw attention to accumulating evidence for a -109 m sea level off the New Jersey coast. This has been recognized

	Observed Coastal Shelf or Island Sea Level	Expected Continental Sea Level. $\frac{2}{3}$ of Island Sea Level	Morner's Sea Level
Late Wisconsin	109 m	77 m	70–90 m
Early Wisconsin	130 m	91 m	85–90 m
Illinoian	145 m	101.5 m	

Table IV-5 Comparison of sea levels on continents and islands. Figures show that, if the ocean basins are assumed to adjust rapidly isostatically, near-shore sea levels and sea levels from islands and continental margins are in good agreement. The additional entries in this table are adapted from Donn *et al.'s* (1962) estimates of 114–133 m meltwater in early Wisconsin maximum and 137–159 m meltwater for the Illinoian maximum.

previously by Flint (1957, pp. 261–271). Emery and Garrison (1967) present evidence for late Wisconsin shorelines currently at −130 m and deeper off the east coast of the United States. Donn et al. (1962) cite evidence supporting a −130 m sea level on the California shelf, from islands off California and from terrain in the Pacific and Far East, and to accumulating evidence of a −145 m shoreline. Garrison and McMaster (1966) have geomorphically identified wave cut terrain at −145 m and −118 m on the continental shelf off New England. In all cases, the lowest sea levels appear to be farthest from the present coast. Much of these data are recent and were obtained from oceanographic vessels.

Mörner (1969, p. 438) draws attention to the fact that his isostatic curves for the Swedish west coast place eustatic sea level at about −70 m 13,000 B.C. (see also Figure IV-15). He indicates the Early Wisconsin minimum sea level "must have been −85 to −90 m." A eustatic sea level of this magnitude fits better with the general European view of −80 to −100 m. Mörner cites Daly, Woldsted, Flint, and others, as supporting sea levels in this range. It is probably fair to assume that most of these early studies were based on accessible near-shore measurements.

In general, then, there is a suggestion that the older, near-shore sea levels are accommodated within the 80–100 m eustatic range. Shorelines at lower present elevations are found offshore. These fall generally within the "dip-stick" range of 110–145 m, although there is some indication the "dip stick" may overestimate the amount of melt-water added in some cases (see discussion in Section IV.C.3.d). The agreement between the ranges of sea level observed on continents and shelves and the range which might be predicted assuming rapid isostatic adjustment of the ocean basins suggests that the ocean basins may adjust relatively rapidly with respect to the rate at which they were filled following post-Pleistocene glaciation.

7. The Question of a Holocene High Sea Level: Wellman's Test of Isostasy

In discussing Figures IV-16 and 17, we commented that the existence of a Holocene high sea level is sensitively dependent on the details of the melting of the last glacial ice and on the viscosity of the lower mantle. It is clearly important to our investigation of mantle viscosity. Unfortunately there is some disagreement regarding the existence of Holocene high sea levels. As can be seen from Figure IV-15, Fairbridge believes there is evidence for such highs, whereas Mörner and

Shepard believe not. Recent evidence suggests there was *not* a Holocene high meltwater level. No evidence of Holocene high sea levels was
found during an expedition to certain ("dip stick") islands in the South
Pacific (Curray et al., 1970; Newell and Bloom, 1970; Bloom, 1970a, c;
Shepard, 1970; Scholl et al., 1969).

It is possible to have a eustatic sea level high in the absence of a
meltwater maximum if the ocean basins adjust slowly enough. Wellman (1964) has shown that unless the adjustment of the ocean basins is
more rapid than a decay time of 2000 years would indicate, a higher
than present sea level should be generally observed on continental
coasts. This effectively means P_2 and P_3 relaxation times must be of
the order of 2000 years or lower if a Holocene high sea level is to be
avoided. From Figures III-16–20 it can be seen this in turn required
a lower mantle viscosity near 10^{22} poise.

It can be seen from Figure IV-18 that the adjustment of the ocean
basins with a high viscosity lower mantle is smooth. Thus any Holocene high sea level that occurs in the absence of a meltwater high, and
because of delayed isostatic adjustment of the ocean basins will be
broad and smooth. Fairbridge's Holocene high is too sudden and
jagged to be considered the sort expected. For this reason, and also
in general, present evidence seems to weigh against the existence of a
Holocene eustatic sea level high.

The discussion to this point might be summarized: A comparison of
calculated and observed eustatic sea level curves, sea level variations
on islands as compared to continents, and the probable lack of a
Holocene sea level high (of the right form), all tentatively, but non-
conclusively, suggests a relatively fluid ($\sim 10^{22}$ poise) lower mantle.

B. The Lithosphere

1. Review of the Past Literature

The effects of an elastic crust or lithosphere in altering the form of
isostatic adjustment has been the subject of considerable discussion.
Daly (1934) appealed to the crust in his "down punching" hypothesis
both to avoid formation of a bulge of squeezed out mantle material
peripheral to the glacial ice load (for which he could find no geological
evidence) and to explain "hinge" lines (areas where old shorelines or
"strand lines" are bent up sharply as if hinged) "observed" in North
America and Scandinavia. Niskannen (1949) concluded at the end of a
series of analyses that crustal effects dominate the recent Fennoscan-

dian uplift and make determination of the upper mantle viscosity difficult if not impossible. Heiskanen and Vening Meinesz (1958, pp. 357ff.)
concluded the opposite: that the effects of the elastic crust on the
Fennoscandian uplift are quite negligible, and a moment 1/3600 that
of the ice load would be sufficient to neutralize the resistance due to
elastic downbending of the crust and cause upwarping. They suggested
that the hinge lines noticed by Sauramo and discussed by Daly developed over preexisting fault planes. McConnell (1968), in an analysis
similar to the one we shall present, showed that the effects of the crust
were only important for short wavelength harmonic load components;
the uplift of central Fennoscandia was not affected by even a substantial crust (250×10^{23} N-m flexural rigidity).

Seismologists have made estimates of lithosphere (crust plus rigid
layers of the uppermost mantle) thickness from seismic velocity variations. Shear waves show an abrupt drop in velocity at about 70 km
depth in oceanic areas (Kanamori and Press, 1970) and at about twice
that depth under continental land masses (Press, 1972). This shear
wave decrease has been interpreted to indicate the transition from rigid
material to comparatively "fluid" asthenosphere beneath. Press (1972)
associates the base of the lithosphere with the solidus of a phase transition. High heat flow areas such as the Basin and Range seem to have
anomously thin lithosphere thicknesses (see White, 1971; Archambeau,
Flinn and Lambert, 1969, Helmberger and Wiggins, 1971).

Seismic waves used in most analyses have periods ranging from a few
seconds (short period) to a minute or so (long period). Glacial uplift
phenomena are measured over tens to thousands of years. Thus there is
a frequency difference of at least eight orders of magnitude between the
two phenomena. Seismic velocity variations may not be directly related
to changes in long term (10–100 years) rigidity. For this reason we shall
assess the strength of the lithosphere by non-seismic means. The reader
may be interested to compare the determination summarized here to
the seismic results of others. A rough correlation is suggested. The
possible effects of phase transition at the base of the lithosphere are
assessed at the end of this section.

2. Effects from Elastic Strength of the Lithosphere

a. Background Theory

The foundations for determining the elastic effects of the lithosphere
were laid in Section III.A.2.e and are summarized in Figure III-3 and
the parameter α. It was shown that an elastic crust (lithosphere) will
reduce the total ultimate isostatic adjustment to a harmonic load by

α^{-1} and augment the rate of adjustment toward isostatic equilibrium by α. It was further shown (footnote to Equation III-33) that:

$$\alpha = 1 + \frac{k^4 D}{\rho g} \qquad\qquad \text{(IV-4)}$$

where D is the flexural rigidity of the lithosphere, ρ is the density of the mantle under the lithosphere, and g is the gravitational constant. k equals $\frac{2\pi}{\lambda}$ where λ is the wavelength of the harmonic load. α^{-1} was plotted in Figure III-3 as a function of λ. Because of the dependence on k^4 (or λ^{-4}) there is quite a sharp cutoff between loads of small wavelengths that are largely supported by the elastic crust and loads of slightly longer wavelength that are scarcely affected.

In Appendix VI it is shown that a load of circular cross section can be represented, so far as the central elastic or fluid response is con-

Rigidity, Thickness, Viscosity of Lithosphere

Data	Region	Apparent Flexural Rigidity, N-m	Characteristic Time, Years
Lake Bonneville	Basin and Range province	5×10^{22}	10^4
Caribou Mountains	Stable continental platform	3×10^{23}	5×10^6
Interior Plains	Stable continental platform	4×10^{23}	5×10^6
Bothnia uplift	Stable continental platform	7×10^{22}	5×10^8
Lake Algonquin	Stable continental platform	6×10^{24}	10^3
Lake Agassiz	Stable continental platform	9×10^{24}	10^3
Hawaiian archipelago	Oceanic lithosphere	2×10^{23}	10^7
Island arcs	Oceanic lithosphere	2×10^{23}	10^7

Table IV-6 The flexural rigidity of the lithosphere. The table, reproduced with the author's permission, is from R. I. Walcott, *Journal of Geophysical Research, 75,* p. 3947, © American Geophysical Union.

Flexural Rigidity of the Crust	Corresponding Crustal Thickness, H, assuming $\lambda = \mu = 3.34 \times 10^{11}$ dyne cm^{-2}	$\bar{\lambda}_{50\%}$	$R_{50\%}$	Radius of Relative Stiffness $\ell = \dfrac{D}{\rho g}^{1/4}$	Comments (see also Table IV-6)
$.038 \times 10^{23}$ N-m	8 km	115 km	22 km	18 km	Oceanic crust (Brotchie and Silvester, 1969)
$.5 \times 10^{23}$	19	220	42	35	Basin and Range (Walcott, 1970)
1×10^{23}	23	262	50	42	
5×10^{23}	41	392	73	62	Long-term crustal rigidity, about value taken as typical of continental crust by Brotchie and Silvester (1969)
10×10^{23}	52	466	90	74	
50×10^{23}	88	697	134	111	
100×10^{23}	110	828	159	132	Short term crustal rigidity suggested by lakes Agassiz and Algonquin, Walcott (1970).
500×10^{23}	188	1238	238	197	

Table IV-7 The radius of a circular square-edged load whose central portion would undergo only 50% of the expected isostatic adjustment due to the existence of a lithosphere with flexural rigidity indicated. $\lambda_{50\%}$ is the equivalent harmonic wavelength. $\lambda_{50\%}$ and $R_{50\%}$ were calculated using equation (IV-4) and (IV-5) and $\rho = 3.3$ gm/cm^3 and requiring $\alpha = 2$. Due to the dependence of α on k^4 the central regions of a load with radius 50% bigger than $R_{50\%}$ will undergo 94% the expected isostatic adjustment whereas a load 50% smaller in radius than $R_{50\%}$ will undergo only 6%. ℓ is the radius of relative stiffness. Large ice loads will induce crustal depression a distance of about 2ℓ from the ice edge.

cerned, by a harmonic load with wavelength $\bar{\lambda}$, where

$$\bar{\lambda} = 5.2\, R_0 \tag{IV-5}$$

R_0 is the radius of the circular (cylindrical square-edged) load.

The major uncertainty in assessing the influence of the lithosphere is the value of its flexural rigidity. Walcott (1970) has made a number of field measurements of flexural rigidity. Minimum values of the flexural rigidity of the lithosphere may be inferred from the long term support of mesas (Caribou Mountains) and horsts (Boothia uplift), and observed flexure of the earth's crust either under localized loads (Hawaiian Islands) or bending into a trench (positive seaward crustal flexure). Walcott's findings are summarized in Table IV-6. The flexural rigidity inferred from short term loads (10^3–10^4 years) is considerably higher than the flexural rigidity determined from longer term loads. Walcott interpreted this to indicate the lithosphere is viscoelastic (a Kelvin-Voight solid). He found a short term flexural rigidity of near 10^{25} N-m.

Table IV-7 has been constructed to show the size loads whose central regions would be about 50% supported by lithospheres with flexural rigidities covering the range indicated by Walcott. Due to the inverse fourth power dependence of α on λ, loads just 50% bigger in radius than $R_{50\%}$ will attain 94% of the expected isostatic equilibrium in their central regions while loads 50% smaller in radius will be only 6% supported by buoyant forces beneath the lithosphere. (For loads 25% bigger and smaller, these figures are 76% and 29%.) It can therefore be seen from Table IV-6 that the central areas under loads the size of Fennoscandia ($R_0 = 550$ km) will not be affected by the elastic strength of the lithosphere even if the lithosphere under Fennoscandia had a flexural rigidity several times larger than Walcott's largest estimate.

Brotchie and Silvester (1969) have investigated the influence a large diameter load has on neighboring regions. The load, even at complete isostatic equilibrium tends to cause depression past its edge because of the rigidity of the lithosphere. It can be seen from Figure IV-19 that the depression resulting from application of a square edged load extends a distance of between 2ℓ and 3ℓ km from the edge of the load. ℓ is the Radius of Relative Stiffness and is listed for different flexural rigidities in Table IV-7. Outside of the region of depression is an area of uplift. This is caused by the crustal flexure and can amount to about 3% of the maximum (isostatic) depression. More distributed loads smear the depression less ($\sim 2\ell$ km away). The magnitude of flexure is diminished to less than 1% of the maximum (isostatic) central depression. The maximum flexural uplift occurs about 3ℓ km from the

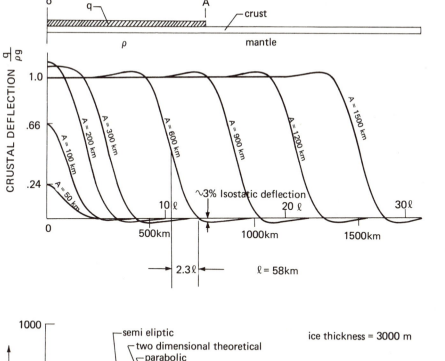

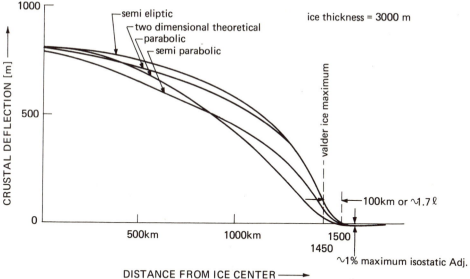

Figure IV-19. Deflections caused by variously shaped loads on a lithosphere of flexural rigidity $D = 3.76 \times 10^{23}$ N-m overlying in inviscid fluid substratum with density 3.73 gm/cm^3. For a square-edged load deflection is induced $2l$ to $3l$ km beyond the edge of the load. Tapered loads extend less than $2l$ km from the load edge. The notation of % deflection and extent of induced deflection has been added to the figures. The figures, reproduced with the first author's permission, are from J. F. Brotchie and P. Silvester, *Journal of Geophysical Research, 74*, pp. 5243, 5247, © 1969 by the American Geophysical Union.

ice edge and might thus be about 8 m in magnitude for a 3 km thick glacier. For $\ell = 58$ the flexural bulge would lie about 170 km from the ice edge, depression would be produced 116 km from the ice edge (see Figure IV-19). If the crust of North America had a short term flexural rigidity 50–100 × 10^{23} N-m, as suggested by Walcott, the depression due to the ice load would extend 200 to 250 km from the ice edge.

This last estimate of the extent of downwarping due to the ice load might be considered a maximum estimate. Brotchie and Silvester (1969, p. 523) point out that the stresses near the edge of a large glacial load are sufficient to cause surface cracking. These stresses attain a value of about 103 bars tension at the edge of a 3000 m thick parabolic shaped ice cap. The tensile strength of granite is about 40 bars (Jaeger, 1964, p. 75). Surface cracking will reduce the flexural rigidity of the crust and may be related to the hinge lines observed in Canada and Fennoscandia.[4]

b. The Case of Lake Bonneville

The framework developed above can be used to determine the flexural rigidity of the lithosphere. For example: Lake Bonneville, the ancestor of the present Great Salt Lake, had a radius of about 95 km and a depth of about 305 m (Crittenden, 1963). From Section III.A.2.f, equation III-46, the expected elastic response to such a load can be computed:

$$\bar{u}_z = \frac{-\sigma^* \rho g h_0}{(\lambda^* + \mu^*) 2\mu^* \bar{k}} = 2.6 \text{ m}$$

ρ = density of water
h_0 = depth of water load = 305 m
$\sigma^* = \lambda^* + 2\mu^*$
$\lambda^* = \mu^* = 7 \times 10^{11}$ dyne – cm^2 (Lamé parameters)
$\bar{k} = 2\pi/\bar{\lambda}$; $\bar{\lambda} = (5.2)(95 \times 10^5 \text{ cm})$

The once level shorelines of Lake Bonneville are today unbowed about 65 m at the center of the lake. This is considerably more than

---

[4]Some caution is required in hinge line interpretation. Flint (1971, p. 354) states: "In some areas strandlines are so numerous and so closely spaced that piecing together a single one from the discontinuous remnants is very difficult. In others some strand lines are so faint that they can lead to considerable error in determining the positions of former water planes." Thus the abrupt "hinge lines" are not necessarily indicated by the geological data, although areas of marked curvature in upwarped shorelines may be.

2.6 m, so some isostatic adjustment must have taken place. If isostatic equilibrium were attained under the water load and now fully recovered, the central upbowing should be $\frac{305}{3.3}$ = 92.5 m. This is more than the 65 m uplift observed. If the discrepancy is attributed entirely to the elastic crust

$$\alpha = \frac{92.5}{65} = 1.42.$$

Thus from (IV-4):

$$\frac{\bar{k}^4 D}{\rho g} = .42,$$

and $D = 5.2 \times 10^{23}$ N-m.

This is in agreement with the calculations of Brotchie and Silvester (1969), who found a model consisting of a 95 km radius of 310 m water load and an adjacent 65 km radius 130 m water load produced a maximum central depression of 61 meters with a crustal flexural rigidity of 3.76×10^{23} N-m. The error in our approximate method is about 20%.

This analysis of Lake Bonneville provides only an estimate of crustal flexural rigidity, since there is no assurance the water load was applied for long enough to attain isostatic equilibrium. From other data Walcott (1970) estimated a rigidity of $.5 \times 10^{23}$ N-m for the Lake Bonneville lithosphere.

c. Gravity Anomalies in Fennoscandia

Another relevant example is the free air gravity anomalies in Fennoscandia. Figure IV-20 shows they may be characterized by a harmonics with wavelength between 400 and 660 km. If we suppose these wavelengths are at least 50% supported by the lithosphere, Table IV-7, the flexural rigidity of the Fennoscandian shield is between 10 and 50 × 10^{23} N-m. In Section IV.E we shall find this estimate in good accord with Fennoscandian uplift data analyzed by McConnell. The estimates also agree favorably with Walcott's estimates of the flexural rigidity of the North American shield (see Table IV-6), Lake Algonquin and Agassiz entries).

Honkasalo's (1964) gravity map shows positive gravity anomalies up

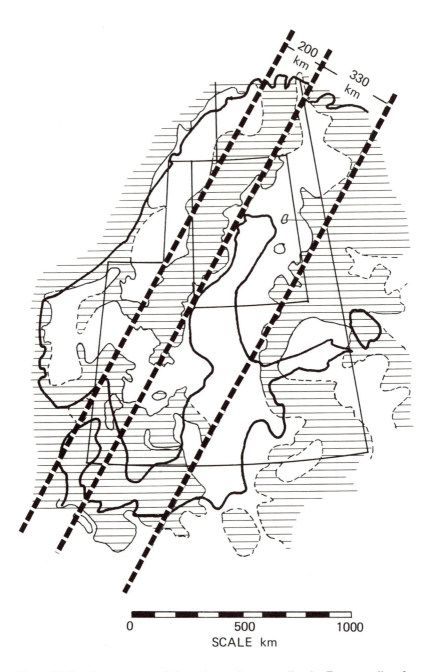

Figure IV-20. Zero contour of free air gravity anomalies in Fennoscandia after Honkasalo (1964). Positive anomalies are shaded. Note that anomalies can be described by a harmonic with wavelength between 400 and 660 km. From Table IV-7, this suggests a long term flexural rigidity between 10 and 50 × 10²³ N-m. Also shown are blocks over which gravity was averaged. The average anomaly in the larger block is −3.4 mgal; the average in the smaller block is −3.5 mgal. This might be taken to suggest about 25 m of isostatic uplift remain in Fennoscandia (7 m/mgal).

to +40 mgal and negative anomalies as large as −60 mgal. However, if the anomalies are averaged over an area large enough to remove those anomalies that could be supported by the elastic rigidity of the lithosphere, a negative anomaly of about 3.5 mgal remains (see Figure IV-20.). A one-milligal gravity anomaly could correspond to about 7 m of remaining isostatic uplift; so this suggests about 25 m of glacial uplift remain in Fennoscandia.

The reader should be cautioned that this is not the conventional interpretation of gravity in Fennoscandia. There does seem to be a trend in this direction, however. Niskannen (1939) estimated 200 m of remaining uplift from the −30 mgal anomaly indicated by his gravity map in the Bothnian Sea. Kääriäinen (1953, p. 72) in reviewing the subject of gravity anomalies was a bit more cautious. He pointed out that the negative anomalies appear to parallel the Finnish coast and that the boundary between the −40 mgal anomaly at the eastern end of the Gulf of Finland and the +28 mgal anomaly west of Helsinki was known to correspond to a geological contact (contrasting rock types). Kääriäinen nevertheless averaged things suitably and found about 200 m remaining uplift coupled with the appropriate caveat that correlation between gravity and uplift could not be proved or disproved. Honkasalo (1964) compiled a gravity map for all of Fennoscandia (see Figure IV-20), and he stated more categorically: "... we see that there is apparently no kind of correlation between the gravity anomalies and the present land rise as computed on the basis of geological, mareograph and precise leveling investigations. On the contrary, the correlation between the geologic map of rocks on the one hand and the gravity anomalies on the other is indisputable" (Honkasalo, 1964, p. 23). Honkasalo reiterated his conclusions more recently (1966).

Sea level data off the East Coast of the United States suggest 2ℓ may be closer to 100 km than 200 km (see next section). From Table IV-7 this indicates a low lithosphere flexural rigidity of around 1.0×10^{23} N-m for that coastal area. Detailed discussion of this case is deferred to the next section, since effects from an upper mantle low viscosity channel are also prominent in these data.

We conclude that, although the lithosphere's rigidity seems to vary from location to location in a not unreasonable fashion (rigidity of thermally hot areas, such as the Basin and Range, or oceanic crust being less than older continental shield areas), the magnitude of the lithosphere's flexural rigidity is not sufficient to affect the central uplift of the larger Pleistocene loads (Fennoscandia or larger). Because of the thin Basin and Range lithosphere, the uplift of Lake Bonneville is not dominated by the elastic lithosphere either. The lithosphere will produce edge effects. It will smear out the glacial load perhaps in some cases by as much as 250 km. Surface cracking may reduce the extent of smearing, but flexural uplift outside the load is of negligible magnitude. A formalism has been developed to take into account the effects of the crust. We shall use it when evaluating the uplift of Fennoscandia or smaller sized load redistribution.

3. Effects of a Possible Phase Change at the Base of the Crust

Migration of a phase boundary at the base of the crust might complicate isostatic adjustment and glacial uplift interpretation. Table III-2 indicates that in 30,000 years a phase boundary should migrate only 6.7% of the way to equilibrium if L = 50 cal/gm or 22% if L equals 15 cal/gm. For 10,000 years these figures are 3.9% and 12%. Thus for load cycles of 30,000 to 10,000 years' duration, phase boundary migration could cause surface uplifts with magnitudes between 2.2% and 11% of the isostatic adjustment.[5] These estimates are rough, but sufficiently accurate for us to conclude that it is unlikely crustal phase boundary migration will contribute significantly to surface deformation in the time scale appropriate for glacial uplift. We shall assume that this is in fact the case and ignore crustal phase boundary migration in our analysis. Results do not indicate this decision is inappropriate.

Analysis of isostatic depression in Greenland and Antarctica might help resolve the question of whether the base for the lithosphere represents a phase transition. In Section IV.A we concluded that the present ice load has been applied continuously to Antarctica for at least 4 million years and probably at least 7 million years, and to Greenland for at least 2 million years. From Section III.A.3.a.2, Table III-2, we see that this time duration is ample to allow a phase boundary at the crust-mantle boundary with L = 15 cal/gm to migrate at least 85% of the way to its ultimate equilibrium position. This is also sufficient time to expect complete isostatic equilibrium.[6] Full phase boundary migration can be expected to produce a depression 53% as large as the isostatic depression, so for a three kilometer thick ice load $(.85)(.53)(885) = 400$ m should be added to the 885 m of isostatic depression expected from a density 3.4 gm/cm^3 mantle under a 3300 m thick ice load. Thus, we arrive at the conclusion that, *if* the boundary between the mantle

----

[5]$2.2\% = (53\%)(3.9\%)$; $11\% = (53\%)(22\%)$. Complete phase boundary migration will cause a surface depression 53% as large on the isostatic depression (see Tables III-1 and 2).

[6]Woollard (1962) has verified that Antarctica is currently in isostatic equilibrium.

and the crust is a phase boundary with L as low as 15 cal/gm, the depression of surfaces of Greenland and Antarctica which were once at present sea level will now be 1285 m below present sea level. If, however, the crust-mantle boundary is not a phase boundary (or if $L \geq 50$ cal/gm, which would migrate only 28% to equilibrium in 1 myr), the expected depression of such surfaces is only 885 to 1000 m. Since sea level was lowered about 70 m by the build-up of Antarctic and Greenland ice, surfaces once at sea level should be 1215 or 800 m below present sea level, depending on whether the crust-mantle boundary is a phase boundary or not.

Geologically, one should be able to distinguish areas of Greenland or Antarctica which were once at or very near sea level (flora, fauna, salt marsh remnants, etc.). If such areas lie 1215 m below present sea level, the boundary between the mantle and the crust is likely a phase boundary. If they lie about 815 m below present sea level, it is not.

To my knowledge no one has made the sort of geological determinations suggested above. Unfortunately little can be concluded from examination of existing bedrock contour maps such as presented by Bently (1962) and Kupsch (1967).

C. The Viscosity of the Uppermost Mantle

1. Introduction and Review of the Literature

Seismologists have observed a subcrustal low velocity channel (Archambeau et al., 1969; Hales, Cleary et al.; 1968; Helmberger and Wiggins, 1971; Press, 1972) and find it to be quite variable both in thickness and magnitude from one locality to another. Variations in the velocity channel appear to correlate both with heat flow and with geological province (Archambeau et al., 1969). Hales, Cleary et al. (1968), for example, find a pronounced low velocity channel under the Basin and Range provinces of the western United States, which is also a high heat flow region, but find the channel pinches out under the Great Plains. The low velocity channel also appears to correlate with a channel of high electrical conductivity (Porath, 1971).

Generally the low velocity channel is taken to be about 100 km to 150 km thick. It is believed that it may be a zone of partial melting, although evidence of partial melting is not conclusive. Elsasser (1971b) has pointed out that it probably corresponds to the region of the mantle in which the temperature (with depth) profile approaches closest to the melting point curve (melting point as a function of temperature and pressure). Such a region of closest approach must exist since the temperature gradient is greater than the melting point gradient at the earth's surface.

Again, because of the eight orders of magnitude that separate the time scales of glacial uplift and seismic wave propagation, one cannot presume that the seismic low velocity channel will manifest itself as a low viscosity channel, although this may be suggested.

Several authors have proposed that flow in a low viscosity channel is responsible for nearly all observed isostatic adjustment. A channel flow model for glacial uplift was first quantitatively proposed by Van Bemmelen and Berlage (1935). It has been appealed to recently by Lliboutry (1971), Artyushkov (1971), and Walcott (1971) to explain all or most of the glacial uplift of Fennoscandia and Canada.

Uplift data from Utah, Greenland, the Arctic Archipelago, and sea level data off the east coast of the United States definitely indicate the existence of a low viscosity channel beneath the lithosphere. We show that it must be about 75 km thick and 4×10^{20} poise (or the equivalent). The next section will discuss whether this channel is underlain by a fluid ($\sim 10^{22}$ p) or rigid mantle.

A low viscosity channel beneath the lithosphere and overlying a fluid ($\sim 10^{22}$ p) mantle might have an appealing genetic explanation related to mantle convection.

2. Theoretical Description of Channel Flow

Calculation of the response of a layered viscosity structure is easily achieved using the propagator techniques discussed in Section III.A.2.a. A program can be written to compute the effects of a low viscosity channel and an elastic crust. Results of such computation are given in the Section IV.D. It is useful, however, to have a rule of thumb to indicate the extent to which a particular channel of particular thickness and relative viscosity will be significant.

In Section III.A.2.d we showed that, if H were the thickness of a fluid channel and kH were small compared to unity such that $(kH)^2$ was small, then

$$\bar{v}_z \big|_{z=0} = \frac{\rho g h k^2 H^3}{3\eta} \tag{IV-6}$$

In other words, each harmonic component of load, of initial amplitude h, decays exponentially:

$$h(t) = h_0 \exp\left(\frac{-\rho g H^3}{3\eta} k^2 t\right) \qquad\qquad (\text{IV-7})$$

k is the wave number of the harmonic load and equals $\dfrac{2\pi}{\lambda}$, where λ is the wavelength of the load harmonic. In Section III.A.3.a.2, equation III-56, we showed a harmonic temperature perturbation will decay:

$$T(t) = T_0 \exp\left(\frac{-K}{\rho c} k^2 t\right) \qquad\qquad (\text{IV-8})$$

Thus, if $kH \ll 1$, i.e., if the wavelength of the harmonic load is large compared to the thickness of the channel, the load will adjust just as a similar temperature anomaly would adjust, by conduction, in an infinite, thermally conducting medium. We may thus use our intuitive feeling about how heat conducts to infer how a large load over a small viscous channel will adjust.

For example, Figure IV-21, shows the response of a viscous channel to the sudden unloading of a square-edged load. By analogy to the case of suddenly pushing a cool rod in hot sand, it can be seen that the edges of the rod warm first (uplift) cooling the neighboring sand (causing a peripheral sinking) in the process. Only when the thermal gradient to the center, $r = 0$, becomes linear will the center of the depression begin to uplift at all. The center will then approach the equilibrium surface $z = 0$ exponentially with time. The peripheral areas are never elevated above their equilibrium positions.

Response to loading follows by analogy to a hot rod in cool sand. As the edges of the rod cool, the neighboring medium heats up (positive peripheral bulge). This continues until there is approximately a linear temperature gradient from the center of the rod outward. The center and the linear gradient then cool exponentially. Some depression occurs outside the load edge that would not occur in the heat conduction. This is because the channel appears infinitely deep to short load harmonics. For a half-space, adjustment is more rapid to long wavelengths than to shorter ones. Thus depression will be induced a distance from the load edge even in the channel flow case. (The lithosphere will also tend to cause such an effect.)

This phenomenon indicates that, if the load is not large with respect to the channel thickness or if we are interested in uplift near the edge of the load (where kH is not small), then a more complete description is needed. The quantitative behavior of a channel for all wavelength

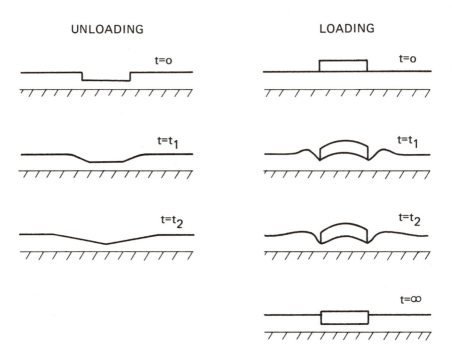

Figure IV-21.   Isostatic adjustment of a viscous channel in response to unloading and loading. Unloading case is inferred by direct analogy to heat conduction. The loading case is a bit more sophisticated near the edges and takes into account the fact that the channel will look infinitely deep to very short wavelength load harmonics. Thus some depression is induced beyond the load edge and inside the peripheral bulge.

loads (not just large ones), follows from the exact expression for flow restricted to a channel of thickness H derived in Section III.A.2.d (equation III-27).

$$h(t) = h_0 e^{-t/\tau} \qquad\qquad\qquad\qquad\qquad\text{(IV-9)}$$

$$\tau = \frac{2\eta k}{\rho g}\, \mathcal{L} \qquad\qquad\qquad\qquad\qquad\text{(IV-10)}$$

$$\mathcal{L} = \frac{C^2 + k^2 H^2}{CS - kH}, C = \cosh kH, S = \sinh kH \qquad\text{(IV-11)}$$

For kH large, $\mathcal{L} = 1$, the channel looks infinitely thick to the harmonic load, and the response is the same as that of a viscous half-space. At small kH, $\mathcal{L}$ becomes very large, indicating a load with a large harmonic wavelength will take much longer to adjust isostatically if flow is restricted to a channel than it would if flow were permitted in the whole half-space. Table IV-8 gives $\mathcal{L}$ for various λ and H, and may be used to evaluate the magnitude of channel phenomena where a low viscosity channel overlies a higher viscosity substrata. The use of Table IV-8 is illustrated below.

$\mathcal{L}$ is the number of times more slowly a given load harmonic will adjust toward isostatic equilibrium if it is applied to a viscous channel of depth H rather than to an infinite half-space of the same viscosity. Thus Table IV-8 shows that a harmonic load of wavelength $\bar{\lambda} = 185$ km will adjust 1.89 times more slowly if it is applied to a 50 km thick viscous channel than it would if it were applied to a viscous half-space of the same viscosity. Table IV-8 shows that the 185 km wavelength load would adjust as fast if applied to a 200 km thick channel as it would if it were applied to a half-space (the entry in Table IV-8 is

						Central Fennoscandia
	W =	50 km	100	200	500	775 (R_0 = 550)
	$\bar{\lambda}$ =	185 km	370	740	1850	2860
H	$\dfrac{2860}{\bar{\lambda}}$ =	15.5	7.73	3.86	1.54	1.00
50 km		1.89	5.67	30.4	313	1290
		(30)	(44)	(117)	(480)	
75 km		1.25	2.29	9.78	145	381
		(19)	(18)	(38)	(223)	
100 km		1.13	1.89	5.67	46.0	153
		(18)	(15)	(22)	(71)	
200 km		1.0003	1.13	1.89	8.71	24.3
		(1.6)	(9)	(7)	(13)	

Table IV-8 A table for the computation of asthenosphere effects. Values of $\mathcal{L}$, computed from equation IV-11, are given for various values of harmonic load wavelength $\bar{\lambda}$. The width of channel that corresponds to $\bar{\lambda}$ is approximately $W = \dfrac{\bar{\lambda}}{3.7}$ (see Appendix VI, equation 10). The channel thickness is H. $\mathcal{L}$ is the number of times slower a harmonic load of wavelength $\bar{\lambda}$ will approach equilibrium if it induces flow in a channel of thickness H rather than in a half-space of the same viscosity. The numbers in parentheses were obtained by multiplying $\mathcal{L}$ by $\dfrac{2860}{\bar{\lambda}}$. As explained in the text, these entries give the ratio of the rate of adjustment of a load harmonic of wavelength applied to a channel of depth H to the rate of adjustment of a load harmonic of wavelength 2860 km applied to a viscous half-space of the same viscosity (i.e., to the rate of adjustment of central Fennoscandia).

1.0003). Thus a 185 km wavelength harmonic load will not see the bottom of a 200 km channel. This agrees with our discussion in Chapter III. The top caption of Table IV-8 shows that a 185 km wavelength load is equivalent to a square-edged load strip of width, W, equal to 50 km. Table IV-8 shows explicitly the response of central Fennoscandia. As discussed in Appendix VI, the response of central Fennoscandia can be described by a harmonic wavelength 5.2 times the radius of the ice load of 550 km or 2860 km.

A question of great relevance can now be answered by using Table IV-8. Suppose we have a thin, viscous channel overlying a fluid half-space of some viscosity. Under what conditions will the channel begin to influence significantly the central uplift? How wide a strip on each side of the load edge will be influenced by channel flow?

If the channel has a viscosity 25 times less viscous than the underlying half-space, division of the entries of the last column of Table IV-8 by 25 gives the number of times slower central Fennoscandia uplifts by inducing flow in the channel than by inducing flow in the underlying half-space. It can be seen immediately that the uplift of central Fennoscandia will be significantly affected only by a 200 km thick channel of this sort.

If the channel is thinner than 200 km and we wish to determine the range of influence the channel has from the load edge, we must consider the relative rates of adjustment via channel and half-space flow for various harmonic wavelengths. This is done by using the figures in parentheses in Table IV-8. The numbers in parentheses compare the channel adjustment of a load harmonic of indicated wavelength, $\bar{\lambda}$, to the half-space adjustment of central Fennoscandia. Division of the entries in parentheses by 25 gives the number of times slower a harmonic wavelength $\bar{\lambda}$ will adjust by inducing channel flow than central Fennoscandia will adjust by inducing half-space flow. The width of influence of the channel is roughly that at which these two rates are equal. Thus Table IV-8 indicates that a channel 25 times less viscous than the underlying mantle in Fennoscandia will influence uplift in a band about 300 km from the ice edge if the channel is 100 km deep, a band about 150 km broad if the channel is 75 km deep, and less than 50 km broad if the channel is 50 km deep. In each case the criterion used is that the numbers in parentheses divided by the number of times more fluid the channel should be about equal to unity, that is that the channel trough develop about as rapidly as central Fennoscandia adjusts toward isostatic equilibrium.

Cross-checks on Table IV-8 are provided by the computation of the earth models of Table IV-1 and by the computation described in Section IV.D.

3. Determination of the Viscosity of the Uppermost Mantle from Uplift Data from Various Geographical Localities

a. Lake Bonneville

From the loading history of Lake Bonneville, Crittenden (1963) calculated that the viscous decay time of the center of the load must be less than 4000 years if the observed 65 m central uplift of shorelines is to be produced at present (see Figure IV-22). This implies a maximum viscosity of about 10^{21} poise, as shown below:

$$h(t) = h_0 e^{(-\rho g t)/2\eta \bar{k}}; \quad \tau = \frac{2\eta k}{\rho g} = 4000 \text{ years} \tag{IV-13}$$

$$\bar{k} = \frac{2\pi}{\bar{\lambda}}$$

$$\bar{\lambda} = 5.2 \, R_0 \, (\text{as shown in Appendix VI}) = 520 \text{ km} \tag{IV-14}$$

$$R_0 = 100 \text{ km}$$

$$\rho = 3.3 \text{ gm/cc}$$

$$g = 10^3 \text{ cm-sec}^{-2}$$

Thus $\eta = 1.7 \times 10^{21}$ poise $\tag{IV-15}$

The decay time could well be half the value quoted by Crittenden, so this viscosity is a maximum estimate. It is a slightly higher viscosity than quoted by Crittenden due to our use of relation (IV-14) rather than the Heiskanen-Vening Meinesz approximation used by Crittenden.

We still need to take into account the effect of the lithosphere in speeding the isostatic recovery.

If the flexural rigidity of the lithosphere is as high as 5.2×10^{23} N-m (the maximum it could be and still permit 65 m central displacement under the 305 m water load—see page 151:

$$\alpha = 1.42.$$

If the flexural rigidity is as low as suggested by Walcott ($D = .5 \times 10^{23}$) in the Bonneville area:

$$\alpha = 1.03.$$

Thus, augmenting $\eta_{\text{Bonneville}}$ by α, we see

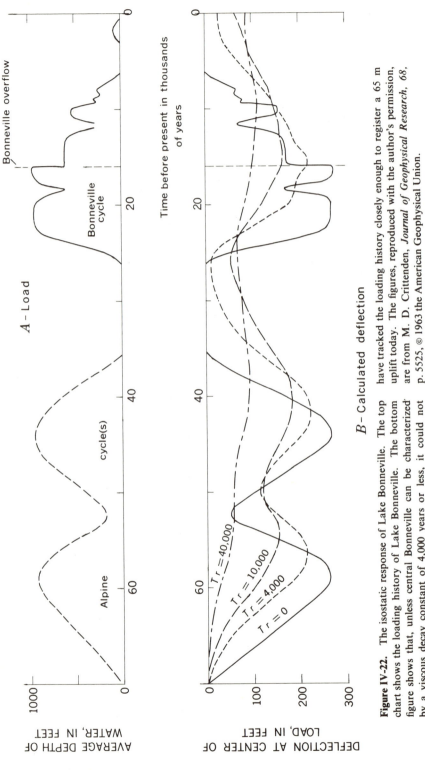

Figure IV-22. The isostatic response of Lake Bonneville. The top chart shows the loading history of Lake Bonneville. The bottom figure shows that, unless central Bonneville can be characterized by a viscous decay constant of 4,000 years or less, it could not have tracked the loading history closely enough to register a 65 m uplift today. The figures, reproduced with the author's permission, are from M. D. Crittenden, *Journal of Geophysical Research, 68,* p. 5525, © 1963 the American Geophysical Union.

$$1.75 \times 10^{21} < \eta_{\text{Maximum}} \atop \substack{\text{uppermost mantle} \\ \text{viscosity under Bonneville}} < 2.4 \times 10^{21} \text{ poise} \qquad \text{(IV-16)}$$

Appendix VI shows that half the horizontal transport of material occurs below a depth equal to 1.4 R_0 beneath the center of a circular load, and that 0.2 the horizontal transport occurs below a depth of 2.5 R_0. Consequently, if the viscosity of the mantle under Lake Bonneville is 10^{21} poise, it must have this viscosity to close to 250 km depth. This can also be seen from Table IV-8. It cannot have a viscosity greater than 10^{21} poise, but the viscosity could be less. If it is less, the thickness of the low viscosity channel immediately under the crust may be correspondingly thinner.

b. Greenland

The Meisters Vig area of northeast Greenland has been studied by both Washburn and Stuiver (1962) and Laska (1966) with excellent agreement between the two studies. Washburn and Stuiver show (and it can be inferred from Figure IV-23) that the uplift at Meisters Vig can be described accurately by an exponential decay time of 1000 years. Deglaciation occurred 9 to 8.5 thousand years ago, and there has been between 60 and 120 m of uplift in the area. The initial uplift was very rapid (9 m/100 years). The remarkable thing is that at present there are sizable glaciers only 8 km away!

The uplift in Greenland must thus be due to a small retreat of the glacier there. The glaciers have been stable at their present position for the last 8000 years. It is probably a bit generous to assume that the uplift followed the unloading of a 60 km wide strip.

We may estimate the maximum viscosity and thickness of a channel which will produce a 1000 year decay time for a 60 km wide unloaded strip in the same fashion that we estimated the viscosity under Lake Bonneville on page 163. From Appendix VI:

$$\bar{k} = 1.7/60 \times 10^5 \text{ cm}; \quad \bar{\lambda} = 320 \text{ km}.$$

$$\tau = (1000 \text{ yrs}) (3.15 \times 10^7 \text{ sec/yr}) = \frac{2\eta\bar{k}}{\rho g},$$

provided

$\rho = 3.313 \text{ gm/cc}$,

$g = 10^3$,

$\eta_{LVZ} = 2.0 \times 10^{20} \text{ poise}$.

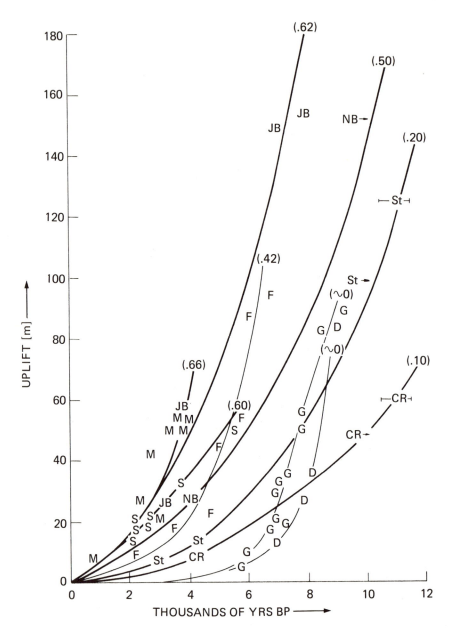

Figure IV-23. Isostatic uplift curves of Arctic Canada. The emergence curves compiled by various workers have been converted to uplift curves using Morner's eustatic sea level curve. Locations are identified in Figure App. VII-6. M = E. Melville Penin. (Farrand, 1962); JB = James Bay (Farrand, 1962); S = Southampton (Farrand, 1962); NB = North Bay (Farrand, 1962); F = N. Fox Basin (West Baffin Island); (Andrews, 1966; Ives, 1964); St = Sault (Farrand, 1962); CR = Cape Rich (Farrand, 1962); G = Greenland (Laska, 1966); D = Devon Island (Müller and Barr, 1966). The present rate of uplift estimated from the initial part of each curve is given in m/100 yrs in parentheses at the top of each curve.

From Table IV-8 it can be seen this viscosity must extend to between 75 and 100 km or the "channel" viscosity must be less.

The lithosphere will have a substantial, but not completely dominant effect on the uplift in Greenland. Assuming a thin lithosphere with flexural rigidity of .5 × 10²³ N-m (such as Walcott suggested was appropriate for the Lake Bonneville area),

$$\alpha = 1 + \frac{\bar{k}^4 D}{\rho g} = 2.$$

Thus the viscosity derived above should be increased by a factor of two. Half the load is supported by the lithosphere so only about half the isostatic depression one would calculate from the ice load thickness should have been recovered.

$$(\eta_{LVZ})_{\text{Greenland}} = 4 \times 10^{20} \text{ poise;} \qquad\qquad \text{(IV-17)}$$
$$H_{LVZ} \geq 75 \text{ km.}$$

c. The Arctic Archipelago

Figure IV-23 gives a compilation of uplift curves converted from emergence[7] curves from the literature using Mörner's eustatic sea level curve (any of the other eustatic curves would have done as well). It shows the uplift that has occurred at various sites (see Figure App. VII-6) over the last few thousand years. The curves are divided into two groups, light-lined and heavy-lined. The light curves are for sites deglaciated particularly late or at the edge of the main ice load. The solid curves are for localities inside the main ice load.

The uplifts in Melville Island and northeast Spitzbergen are exceedingly similar to the uplift curves of Greenland and Devon Island (see Müller and Barr, 1966, Figure 2). The isostatic adjustment for the light-lined curves is extremely rapid. With the exception of Fox Basin, all the curves are located at the edge of the main glaciation. Fox Basin looks, at least in part, like a channel response to late deglacia-

---

[7]In Canada uplift data come primarily from the radiocarbon dating of shells, driftwood, and bone which are thought for various reasons to lie at (or some known distance below) what was the mean sea level at the time of deposition. An emergence curve is obtained when, for a particular location, the elevations of a number of such fossil shorelines relative to present sea level are plotted with respect to their radiocarbon age. An uplift curve may be obtained from the emergence curve by subtracting the (eustatic) elevation of sea level at the time of formation of each shoreline.

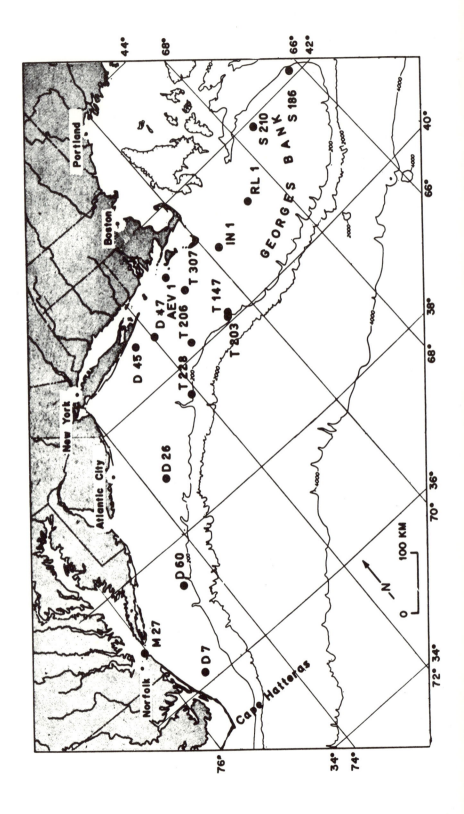

Table 1. Radiocarbon dates for indicators of ancient sea levels of the continental shelf off northeastern United States.

Symbol	Ship and station	Material	Location	Depth (m)	Source	Age	Lab. No.
IN 1	Invader 1	C. virginica*	40°59'N,69°44'W	45	(26)	7,310 ± 300	W-1981
D 7	Delaware 7-1	C. virginica*	36°09'N,75°20'W	33	(7)	8,130 ± 400	W-1402
M 27	Bridge boring	C. virginica*	36°59'N,76°06'W	21	(5)	8,135 ± 160	ML-196
T 307	Trident 307	E. directus†	40°50'N,70°52'W	55	(27)	9,150 ± 220	I-2475
AEV 1	A. E. Verrill 1	C. virginica*	41°18'N,71°00'W	34	(28)	9,300 ± 250	W-2013
D 60	Delaware 60-7	C. virginica*	37°24'N,74°39'W	64	(29)	9,600 ± 600	L-948
D 26	Delaware 26	C. virginica*	38°49'N,73°39'W	55	(7)	9,780 ± 400	W-1403
D 45	Delaware 45	C. virginica*	40°43'N,72°25'W	37	(7)	9,920 ± 400	W-1400
S 210	?	C. virginica*	41°55'N,67°35'W	46	(30)	10,300 ± 150	S-210
S 186	?	C. virginica*	42°05'N,67°15'W	53	(6)	10,600 ± 130	S-186
D 47	Delaware 47	C. virginica*	40°40'N,71°59'W	51	(7)	10,850 ± 500	W-1401
T 206	Trident 206	Astarte sp. M. arcticum‡	40°10'N,71°26'W	86	(27)	10,850 ± 150	I-2474
RL 1	Ruth Lea 1	Peat, partly salt-marsh	41°09'N,68°43'W	59	(8)	11,000 ± 350	W-1491
T 228	Trident 228	P. magellanicus§	39°37'N,72°07'W	147	(27)	13,200 ± 210	I-2545
T 147	Trident 147	M. arcticum‡	40°09'N,70°29'W	122	(27)	13,420 ± 210	I-2473
T 203	Trident 203	M. arcticum‡	40°06'N,70°32'W	130	(27)	14,850 ± 250	I-2544

* *Crassostrea virginica* (Gmelin). † *Ensis directus* Conrad. ‡ *Mesodesma arctatum* (Conard). § *Placopecten magellanicus* Gmelin.

Figure IV-24. Figure showing locations of samples indicative of past sea levels and table showing the depths and ages of these samples. Recent detailed charts (Emery and Uchupi, 1972, pp. 31–33) permit submerged terraces, which were formed by Pleistocene sea level still-stands, to be traced for hundreds of kilometers. The present depths of the deepest Wisconsin terrace (the Franklin Terrace) are similar to the deepest samples show above (i.e. 140 m). In addition, as the terrace is followed from the northeast to the southwest, the depth increases and then decreases, suggesting a "channel flow" loss of material to the glacially rebounding area to the north.

The figure and table, reproduced with the first author's permission, are from K. O. Emery and L. E. Garrison, *Science, 157,* p. 685, © 1967 by the American Association for the Advancement of Science.

tion. Notice that the tail of that curve occurs at comparatively recent times and (Figure III-9) that the deglaciation was recent in that area. Thus, all the light curves in Figure IV-23 are in some sense near the edge of the (most influential) ice load. Their uplift will be most heavily influenced by flow in the uppermost mantle since their Fourier load description consists dominantly of high frequencies. Their rapid adjustment requires a fluid uppermost mantle and suggests a thin low viscosity channel. The viscosity they suggest is similar to that suggested by Greenland: 4×10^{20} poise.

d. Shorelines off the United States East Coast

If one examines a particular coastline in detail, some complications become evident. Consider for example the data of Emery and Garrison (1967) given in Figure IV-24. Emery and Garrison's data are plotted against Mörner's eustatic sea level curve in Figure IV-25. The tails of the arrows in Figure IV-25 indicate where the data would fall if present elevations were taken literally and no correction made for hydro-isostatic adjustment. The arrows indicate the hydro-isostatic adjustment correction that one would make if one assumed full isostatic adjustment had occurred under the water load now present over each location.[8] For the older, lower shorelines this is not a strictly reasonable correction since, on the basis of 110 m of meltwater added, one could expect at most 33 m of ultimate isostatic adjustment.

Three observations can be made from Figure IV-25:

(1) The agreement between Mörner's eustatic curve and sea level data from the shelf off the East Coast is better if the sea level data are corrected for hydro-isostatic adjustment than if they are not.

(2) The agreement would be improved further if less or no hydro-isostatic adjustment were applied to D47, S186, S210, D45, AEV-1, and M27. From Figure IV-25 it can be seen that these are exactly those locations which lie closest to the present coastline.

(3) The hydro-isostatically corrected elevations of locations closest to the edge of the continental shelf (T203, T147, T228, T206, and D60) all fall below the eustatic curve. They must have sunk an additional amount since their formation.

---

[8]For example, T203 is moved up 39.4 m from its present elevation of -130 m, since 130 m of sea water load might cause $\dfrac{130}{3.313} = 39.4$ m of isostatic adjustment.

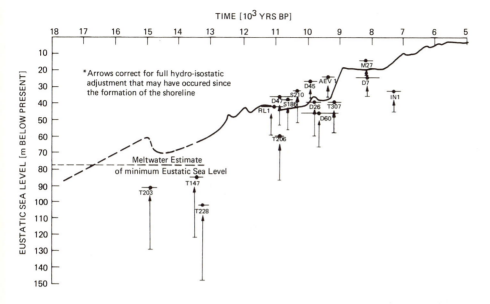

Figure IV-25. Emery and Garrison's sea level data from Figure IV-24 corrected for hydro-isostatic adjustment and compared to Morner's eustatic sea level curve. Arrows correct for the full hydro-isostatic adjustment that could have occurred since the formation of the shoreline.

The first point, particularly if the next two points can be accounted for, may be taken as evidence that hydro-isostatic adjustment of the ocean basins has occurred.

The second point, that locations nearest the present coast need less hydro-isostatic correction, is probably a consequence of the rigidity of the elastic crust. Full hydro-isostatic adjustment appears to be approached beyond about 60 km from the present coast. From our discussion in Section B of this chapter we see this would imply a rather low radius of relative stiffness for the crust off the east coast of about 25 km. (This would correspond to flexural rigidity more appropriate for oceanic lithosphere than continental or shelf lithosphere.) Thus, the behavior of near-shore localities can be accounted for, provided the lithosphere near the east coast and under the shelf is relatively weak and incompetent. We referred to this result in Section IV.B.

The fact that localities at the edge of the shelf are at present lower than they should be once isostatic equilibrium is attained indicates either:

(1) these localities are at present out of isostatic equilibrium, having sunk since their formation past their equilibrium point; or

(2) the shorelines observed now were formed when the localities were elevated. They are now near equilibrium, but because of their elevated condition at the time of formation appear to have sunk further than they should.

Both these conditions would occur if there is an uppermost mantle low viscosity channel. The near-shore, earliest formed localities would be depressed upon loading of the oceans if there is a low viscosity channel. Material near the coasts would be squeezed out of such a channel by the ocean load and pushed under the continents. Similarly, material from such a low viscosity channel would be sucked out by unloaded regions in their quest for mantle material with which to achieve isostatic equilibrium.

By the same token, during the Pleistocene, channel material would be squeezed out from under the ice load, and regions near the coast and especially near the ice-loaded areas would have been above isostatic equilibrium. Shore lines would be carved at anomalously low stratographic levels in these areas. Thus both explanations for the observed shoreline data outlined above are probable if a thin low viscosity channel exists. A 75 km thick 4×10^{20} poise (25 times less than the underlying mantle) channel would (see Table IV-8) have a maximum about 200 km from the edge of the former glacial load or between 100 and 150 km from edge of the effective glacial load (taking into account smearing of load by a crust of low-flexural rigidity (10^{23} N-m) (see Section IV.B).

e. Conclusions from Geological Evidence

Greenland suggests a low viscosity channel of 4×10^{20} poise and at least 75 km thick. The Bonneville uplift would require such a channel to be about 75 km thick. This is shown in Table IV-9. If we assume the viscosity of the upper mantle is about 10^{22} poise (we shall find this is appropriate), the low viscosity channel is about 25 times less viscous than the mantle that affects the Fennoscandian uplift. From Table IV-8 we see that channel troughs a bit less than 200 km wide should develop in a time similar to that characterizing the Fennoscandian uplift (divide numbers in parentheses in the 75 km thick channel row by 25 and note that values are equal to or less than one till about $W = 200$ km). Thus a 75 km thick channel with 4×10^{20} poise could account for the sea level observations off the east coast of the United States.

A 75 km thick 4×10^{20} p channel (i.e., a channel with a diffusion constant of 3.6 km^2/yr—see next section) would not substantially affect the uplift in central Fennoscandia, however (i.e., $391/25 \gg 1$ in

Channel Thickness	$\mathcal{L}$ Bonneville	Equivalent Half-Space Viscosity if $\eta_{channel} = 4 \times 10^{20}$ poise
50	9.6	3.8×10^{21}
75	4.3	1.7×10^{21}
100	2.7	1.1×10^{21}

Table IV-9 The channel response of Lake Bonneville. $\mathcal{L}$ for Lake Bonneville computed from equation IV-11 for $\bar{\lambda} = 5.2\ R_0$, $R_0 = 95$ km. Lake Bonneville will uplift as if it overlay a 1.7×10^{21} poise half-space as required by equation IV-16, if it overlies a 4×10^{20} poise channel 75 km thick.

Table IV-8). If the channel were thicker, the central Fennoscandian uplift would be affected by the low viscosity channel (i.e., 153/25 or 24.3/25 ~ 1). The Fennoscandian uplift can thus place limits on the permissible thickness of the low viscosity channel. We consider this point further when we analyze the Fennoscandian uplift in Section IV.D.

4. On the Physical Origins of a Thin Worldwide Low Viscosity Channel and Its Relation to Non-Adiabatic Density Gradients in the Mantle

Morgan (1972) has proposed that deep mantle upwelling occurs at some 20 plumes, each of which is about 150 km in diameter. Elsewhere Morgan assumes the mantle is sinking at a rate about one two-thousandths the rate of upwelling in the plumes (.1 cm/yr). According to Morgan's hypothesis, both the low viscosity and low velocity channel might be the hot outspreading limbs of the plumes. Such limbs would be about 75 km thick at the location of the plume but would decrease in thickness away from the plume due to cooling. There appears to be a correlation between the location of Morgan's plumes and areas that have pronounced low velocity (and perhaps low viscosity) channels.

Morgan's plume hypothesis is of interest to us for two reasons: (1) It provides a genetic explanation of the low viscosity/velocity channel. (2) It implies that there are no non-adiabatic density gradients in the mantle. The second conclusion follows from the sinking rate of 1 mm/ year. This is almost sufficient to maintain an adiabatic temperature gradient through the phase transitions between 350 and 650 km depth (even if they are univariant, see Section III.A.3.b), and, in fact, throughout the mantle as a whole.

Morgan's hypothesis is not the whole story, however. It accounts for neither the plume-like nature of the mantle upwelling nor subduc-

tion. R. H. Dicke suggested a way in which these two phenomena may
be related: He pointed out that, if there is convection in the earth's
mantle, it can exist only for the purpose of transporting heat from the
interior of the earth to its surface. Near the surface, the mantle material will cool. In the cool, near-surface areas, convection is unlikely.
Heat must be conducted to the surface, where it is radiated into space.
Conduction is not effective through too thick a layer, however. The
thickness of the upper spreading limbs of the convection "cells" cannot be much thicker than about 75 km if they are to cool significantly
by conduction in the 200 million years they have before they are subducted. Since the limbs are thin, so must the upwelling plumes (or
rolls) be thin (in at least one dimension). Further, since viscosity is a
function of temperature, the plumes will heat up and narrow until an
equilibrium heat balance is attained. In effect, Dicke says the Rayleigh
number must be considerably larger than the critical Rayleigh number
if convection is to be effective in dissipating heat at the earth's surface.

The Rayleigh number, Ra, for the mantle is indeed vastly larger than
the critical Rayleigh number, if, as we shall see in the next section, the
viscosity of the mantle is around 10^{22} poise. Taking rough estimates of
the needed parameter values for the mantle from Turcotte and Oxburgh (1962):

α $\equiv$ Coefficient of thermal expansion 2×10^{-5}°C^{-1}

ρ $\equiv$ Density of mantle 4 gm/cc

g $\equiv$ Gravitational acceleration 10^3 cm/sec^2

d $\equiv$ Depth extent of mantle 3000 km

ΔT $\equiv$ Non-adiabatic temperature difference

k $\equiv$ Thermal diffusivity 10^{-2} cm^2/sec

and

μ $\equiv$ Viscosity of mantle 10^{22} poise

$$Ra = \frac{\alpha \rho g d^3 \Delta T}{k \mu} = 21{,}600 \, \Delta T.$$

The critical Rayleigh number for fixed-fixed boundary conditions
(which give the largest critical Rayleigh number) is 1700. For $\Delta T =$
100°C, the Rayleigh number of the mantle is 1271 times greater than
critical. Even for $T = 1$°C $Ra = 12.7 \, Ra_c$. From experiment, convective upwelling becomes narrower and more concentrated as the
Rayleigh number progressively exceeds critical. Thus, if the viscosity
of the lower mantle is around 10^{22} poise we can expect narrow convective upwelling in the mantle which could take the form of plumes with
cylindrical cross-section, or rolls with sheet-like zones of upwelling.

Dicke's way of viewing convection might suggest that subduction is more important than Morgan's uniform mantle-wide sinking in returning material to the mantle. This might seem to be a consequence of a thin surface thermal-flow boundary layer. Both mechanisms are probably in fact operative. The plume upwelling injects extra material into the uppermost mantle. The weight of this material will tend to cause a downward flow throughout the whole rest of the mantle. At the same time the cold skin, pushed outward by the spreading plume limbs, cools and becomes increasingly gravitationally unstable. Finally it turns under and is subducted.

Turcotte and Oxburgh (1969) have put some of these concepts in quantitative form by applying boundary layer theory to mantle convection. They do not emphasize the genetic necessity of thin convection cells for effective mantle heat transport, but it is implicit in their treatment. A low viscosity channel, an adiabatic mantle, and mantle convection are related phenomena. The low viscosity channel we observe from uplift phenomena has a natural explanation in terms of plausible mantle dynamics.

D. The Upper Mantle

The viscosity of the upper mantle is probed most directly by the Fennoscandian uplift. Information is also provided by data on the post-glacial adjustment of the east coast of the United States. We defer discussion of these data to the next section because it also critically bears on the viscosity of the lower mantle.

In this section we shall (1) review the extensive Fennoscandian literature, (2) infer mantle viscosity structure directly from the uplift data in a fashion suggested first by McConnell, (3) present several "forward" model calculations to highlight critical geological data that could be collected and further elucidate competing viscosity models, and (4) show that the uplift in central Fennoscandia is well fitted by a uniform 10^{22}p viscosity mantle and a gradual load removal.

1. A Review of the Fennoscandian Literature

Daly (1934) was one of the first to draw information on the viscosity structure of the mantle from the glacial uplift in Fennoscandia. He distinguished two extreme models that he named "bulge" and "punching" hypotheses, respectively. Discussion today continues to focus on these two hypotheses.

The bulge model presumed that flow is restricted to a thin channel.

Thus glacial loading produces a peripheral bulge and unloading a peripheral trough. This model has been developed and supported by Van Bemmelen and Berlage (1935), Lliboutry (1971), Artyushkov (1971), and Walcott (1972).

Daly could not find the geological evidence he expected in support of the bulge model. This perplexing lack of the expected evidence persists today as reported by Flint (1971, p. 346): "But the evidence on this [Daly's channel bulge] is wholly negative, even where shore features and stream terraces should reflect such movement and afford a basis for measuring it. Either the slopes are so gentle that they have not been recognized, or the displaced material is distributed in a different pattern."

Daly proposed the "punching" model to explain how the flow could be distributed in a pattern that would avoid peripheral bulges. He appealed to the crust to force flow at great depths. The glacial load would "punch" through the crust primarily along the periphery and isostatic adjustment would be attained through a piston-like depression of the glacially loaded area. Shearing at the margins would cause a slight elastic depression of the surrounding regions. Similarly, a slight elastic uplift of the surrounding regions would occur upon unloading. Eventually these peripheral areas would settle back down, but the initial response of the surroundings would be sympathetic (in the same direction) to the motion of the areas centrally located with respect to the load. The shear zone might reflect itself in the hinge lines (lines of abrupt increase in slope of ancient tilted shorelines) observed in Canada and Fennoscandia. The observed hinge lines, evidence of super-elevation of areas bordering the Fennoscandian ice cap long after the ice had retreated northward, and the local horizontality of strand lines in peripheral areas in Late Glacial time provided, in Daly's view, evidence supporting the punching hypothesis.

Shortly after the publication of Daly's book, Haskell (1935, 1936, 1937) found that the isostatic adjustment of a half-space of uniform viscosity to loading and unloading behaved much like Daly's punching model. In a uniform viscosity half-space, flow occurs at great depth naturally; no recourse need be made to a rigid crust. Haskell showed half the flow across a vertical plane under the edge of a load occurs at a depth equal to the width of the load (~ 1100 km for Fennoscandia). Furthermore, the areas peripheral to the load respond sympathetically with the central regions. Unloading produces an initial regional uplift; the peripheral areas later sink back to equilibrium. The friction of the punched through region on the surrounding elastic crust is not needed to produce this sympathetic movement. Haskell (1937, p. 25) recognized the parallelism between his calculations and Daly's punching hypothesis:

> These conclusions are in complete agreement with Daly's "down-punching" hypothesis, which pictures the isostatic compensation of a great ice sheet as taking place by dominantly vertical movement under the load with lateral displacement distributed through a great depth. The only way in which it would be possible for the lateral flow to be confined to a shallow layer beneath the crust with the consequent formation of a wave-like bulge in the peripheral zone is to suppose that at some depth there is a considerable increase in viscosity so that flow is limited in depth by an effectively rigid lower boundary. The evidence adduced by Daly in support of the "down-punching" hypothesis may therefore indicate that there is no such sharp increase in viscosity within the earth's silicate shell.

From the Fennoscandia data Haskell determined the viscosity of the earth's silicate shell (mantle) to be $.95 \times 10^{22}$ poise and showed that the viscosity did not change with the interval of time over which the analysis was made. Thus Haskell showed the viscosity of the mantle was not stress dependent and was Newtonian. The uplift predicted by Haskell's analysis fit the geological data well. It was not dependent on the details of the glacial retreat, because only data after the ice dissipation were analyzed. However a consequence of Haskell's analysis was the prediction that only about 20 m of uplift remain today in central Fennoscandia. Haskell initially assumed this was the case following a suggestion by Nansen, who had studied the uplift geologically (Gutenberg, 1941, p. 754). It was on this point that other workers were later to base objections.

Van Bemmelen and Berlage (1935) put Daly's bulge hypothesis in quantitative form and analyzed the uplift in Fennoscandia assuming flow occurs in a thin channel. Their model is not *a priori* a physically valid one since they assumed that all load harmonics decay inversely proportional to the wavelength squared. The shorter wavelengths might be expected to decay less rapidly as the channel begins to appear infinitely thick to them (i.e., like a half-space). However the elastic crust makes Van Bemmelen's and Berlage's approximation a good one. Van Bemmelen and Berlage showed that the uplift in central Fennoscandia could be accounted for by flow in a 100 km thick asthenosphere that had a viscosity of 1.3×10^{20} poise, and further that about 210 m of uplift remain to be realized in central Fennoscandia.

For reference, it is useful to record here that Van Bemmelen and Berlage solved the isostatic adjustment of depression whose initial geometry was that of a Gaussian trough of half width σ:

$$h = h_0 e^{-x^2/\sigma^2}:$$

They found

$$h = \frac{h_0}{\sqrt{1 + \dfrac{4Dt}{\sigma^2}}} \, e^{\frac{-x^2/\sigma^2}{1 + 4Dt/\sigma^2}} \tag{IV-18}$$

D is a diffusion constant with units of $L^2 T^{-1}$. It is defined:

$$D = \frac{\rho g H^3}{3\eta}, \tag{IV-19}$$

where H is the thickness of the viscous layer, ρ is the density of the fluid in that layer, η is its viscosity in poise, and g is the gravitational field strength.[9] A 100 km thick channel with viscosity 1.3×10^{20} poise has a diffusion constant $D = 26$ km^2/year.

Van Bemmelen's and Berlage's model found favor because, as we have seen, gravity anomalies of -25 mgal to -30 mgal observed in central Fennoscandia were thought to suggest 180–200 meters if uplift remained there (see Vening Meinesz, 1937, and discussion in Section IV.B). The gravity data and their interpretation led to a series of analyses which tried to reconcile the observed uplift with ~ 200 m of remaining uplift.

One of the simplest analyses was made by Vening Meinesz (1937). He assumed each harmonic decayed exponentially, $h = h_0 e^{-t/\tau}$, and added 180 m to Nansen's uplift curve for central Fennoscandia. The decay time, τ, was changed from the 4100 years found by Haskell to $\sim 13,000$ years. Thus Vening Meinesz concluded the viscosity of the mantle under Fennoscandia was around 3×10^{22} poise rather than 10^{22} poise determined by Haskell. Vening Meinesz also pointed out that if the viscosity of the mantle decreased with depth the upper mantle viscosity could be larger, and he quantified this relation to some extent, assuming various exponential decreases in viscosity.

Niskannen (1939) derived an expression identical to Van Bemmelen's and Berlage's (equation IV-18) but re-labeled $D/\sigma^2 = K$. In the central areas of uplift (IV-18) then becomes

---

[9]These equations are Van Bemmelen's and Berlage's (1935) equations (11), (12), (39). For convenience we have regrouped and relabeled the parameters. Equations IV-18 and IV-19 are derived in Appendix VI.

$$h = \frac{h_0}{\sqrt{1 + 4Kt}}, \tag{IV-20}$$

and the central rate of uplift may be expressed:

$$\frac{\partial h}{\partial t} = 2Kh_0[1 + 4Kt]^{-3/2} = v = v_0[1 + 4Kt]^{-3/2}. \tag{IV-21}$$

Note that h_0 is the total amount of uplift that will eventually be achieved, h is the uplift remaining at any time, v_0 is the initial rate of uplift. Niskannen used equation IV-21, the rates of uplift of 1.2 cm/year at present (8800 years after unloading) and a rate of uplift of 13 cm/year at $t = 0$ (8800 BP, the assumed time of unloading), which he deduced from Liden's uplift curve for the Angerman River, to deduce that

$$K = 1.07 \times 10^{-4} \, \text{year}^{-1}.$$

Then, assuming 250 m of uplift had occurred to the present time, Niskannen used (IV-20) to show $h_0 = 460$ m and that 210 m of uplift remain at present. Following a suggestion by Lamb and Wien, Niskannen identified K:

$$K = \frac{g\sigma}{2\eta}. \tag{IV-22}$$

Taking $\sigma = 750$ km, he then deduced that the viscosity of the mantle under Fennoscandia was about 3.6×10^{22} poise. This is a somewhat strange conclusion when it is remembered the equations used by Niskannen are appropriate for flow in a thin channel, and K is more appropriately interpreted as $\frac{D}{\sigma^2}$. This suggests D, the channel diffusion constant, is 60 km^2/year.

Gutenberg (1941), in reviewing Niskannen's results, made the interesting observations that

$$\frac{h_0^3}{v_0} = \frac{h^3}{v} = \text{const} = \frac{h_0^2}{2K}, \tag{IV-23}$$

a result which follows by substituting (IV-20) into (IV-21). Further, if v is derived from an expression like (IV-18),[10] it can be shown that the zero uplift contour increases its distance from the center of uplift:

[10]In Table App. VI-1 it is shown uplift for a cylindrically symmetric Gaussian depression is described by an expression like (IV-18) with x replaced by r and $\sqrt{1 + 4Kt}$ replaced by $1 + 4Kt$.

$$r^2_{\substack{\text{zero}\\\text{uplift}\\\text{contour}}} = \frac{\sigma^2(1 + 4Kt)}{2} \tag{IV-24}$$

Further, Gutenberg pointed out the decay time, T (i.e., the time for remaining uplift to be cut in half) for Niskannen's model is strongly dependent on time: $T = 4000$ years at the beginning of uplift, $T = 16,000$ years at 3000 B.C. and $T = 30,000$ years today. These again are all observations appropriate for a channel flow model. Paradoxically, Gutenberg followed Niskannen in deducing from an expression like (IV-22) that $\eta_{\text{Fennoscandia}} = 2 \times 10^{22}$ poise and concluded that Daly was correct in his down-punching hypothesis and that flow is induced well below the 1000 km level. He did find, however, that if 200 m is added to Nansen's curve, the viscosity predicted by Haskell's theory is no longer constant with time but suggests that uplift is dependent on the driving force in a greater than linear fashion. Viscosity appears to increase with time.

Lliboutry (1971) analyzed the uplift in central Fennoscandia in a fashion similar to Niskannen and Gutenberg but interpreted the results consistently with the channel flow model from which the theory derived. He showed, using Liden's data, that, if about 185 m of uplift remain at present in central Fennoscandia, $v^{1/3}$ is a linear function of h, the amount of uplift remaining at any time. Indeed:

$$\frac{h^3}{v} = 730 \text{ km}^2/\text{year} = \frac{h_0^2}{2K}. \tag{IV-23}$$

Lliboutry assumed h_0, the total ultimate uplift, to be 604 m and so found $K = 2.5 \times 10^{-4}$ year^{-1}. This is about two and a half times the value Niskannen derived. (Lliboutry calculated another case which gave a $K = 3.2 \times 10^{-4}$ year^{-1}.) From $K = 2.5 \times 10^{-4}$ year^{-1}, and a value $\sigma = 374$ km (obtained from h_0 and an estimated ice volume at the salpausselkä stage), Lliboutry then found a value for D:

$$D = K\sigma^2 = 35 \text{ km}^2/\text{year} \tag{IV-25}$$

From (IV-19) this D can be interpreted as a 100 km thick low viscosity channel with a viscosity of 0.97×10^{20} poise, a result in good agreement with Van Bemmelen's and Berlage's findings of a 100 km 1.3×10^{20} p channel. Lliboutry, however, obtained a different result, since he defined $D = \frac{\rho g H^3}{12\eta}$ rather than $\frac{\rho g H^3}{3\eta}$. As shown in equations III-23, 25,

Van Bemmelen's and Berlage's expression is the correct (physically meaningful) one.   We have also verified it by propagator matrix (layered viscosity) calculations.   Lliboutry would also have obtained a different result had he treated the Fennoscandian load as a cylindrical Gaussian depression rather than a Gaussian trough.

Post and Griggs (1973) show a constant $\dfrac{h}{\nu^{1/n}}$ results for a non-Newtonian flow law where the strain rate is proportional to the nth power of the stress.   Analyzing the Fennoscandian uplift data as Lliboutry did (i.e., assuming ~ 180 m of uplift remain in central Fennoscandia at present), Post and Griggs found $n = 3.21$ and noted this was in good agreement with laboratory creep tests on presumed mantle rocks. Their conclusions depend entirely on the amount of uplift assumed to remain at present.   The similarity between channel and non-Newtonian uplift is to be expected (see pages 2, 272).

A serious flaw in all channel flow or non-Newtonian flow analyses presented to date is that complete isostatic equilibrium is assumed at the time of the removal of the ice load.   No load cycle is taken into account.   Since the duration of ice loading is quite comparable to the duration of unloading ($\sim 20{,}000$ years, see Section IV.A.2), it is apparent that the load cycle must be accounted for.   Gutenberg pointed out that the time to reduce the remaining isostatic adjustment by a factor of two increases very rapidly as equilibrium is approached for a channel flow model, attaining, in the case of Fennoscandia, a value of 30,000 years after $\sim 10{,}000$ years of unloading.   Thus an important consequence of a load cycle will be (in most cases) a dramatic reduction in the amount of predicted uplift remaining.

McConnell (1968b) presented a most interesting analysis of the Fennoscandian uplift.   He transformed the sequence of strandlines determined by Sauramo, and deduced the dependence of decay time on harmonic wavelength directly, with no intervening model assumptions. He showed that these data were consistent with an elastic layer 120 km thick and rigidity 6.5×10^{11} dyne cm^{-2} (i.e., a flexural rigidity of 250×10^{23} N-m), and a low viscosity channel whose viscosity decreased from 4.1×10^{21} poise just beneath the lithosphere to 2.7×10^{21} between 220 and 400 km depth, and then increased rapidly to more than 6.8×10^{22} poise below 1200 km.   McConnell felt that much higher viscosities of $\sim 10^{25}$ poise could be attained below 800 km, and favored this for reasons other than the analysis of the Fennoscandian uplift (non-hydrostatic equatorial bulges and gradual changes in the length of day).

McConnell assumed that no uplift occurred past about 800 km from the center of uplift (i.e., all strandlines were adjusted so they were at a common elevation at this distance). Since this may not be the case, as McConnell points out, the long wavelength data are not strictly reliable, although the indication of a low viscosity channel (or elastic crust) is.

Parsons (1972, Chapter 9) has shown, using Gilbert-Backus inversion theory, that if the elastic crust assumed by McConnell is retained, a low viscosity channel beneath the crust is only barely resolvable from the data McConnell used, if indeed it is resolvable at all.

2. Direct Inference of Mantle Viscosity From Strandline Uplift

When the ice dissipated from Fennoscandia and the land began to uplift, a series of seas and lakes, ancestors of the present day Baltic, was formed. The shorelines of these lakes and seas, called strandlines, record the earth's adjustment. McConnell (1968) and Hankel transformed the strandline data collected by Sauramo (1958) to obtain the relaxation time for a series of load wavelengths from ~ 6000 km to 900 km. McConnell's decay time spectrum (which is smeared to reflect differences between different pairs of strandlines) is compared to the spectra calculated for a 10^{22} poise mantle, a 75 km low viscosity channel of various viscosities, and an overlying lithosphere of various flexural rigidities in Figure IV-26. That figure also shows the uplift data we have investigated in the previous section and data from central Canada that we shall discuss in the next section. The wave number associated with each load is assigned according to (App. VI-10); the decay time is deduced from the uplift data.

The short load wavelength (order number > 15) components were computed using (III-20) and (III-21), modified by α to account for the elastic crust as discussed in Section III.A.2.d.[11] It can be seen that the decay time for a uniform viscosity half space increases proportionately with k (wave number) or n (order number)[12] as is indicated by (III-17). The effect of the lithosphere is to decrease the decay time of the higher order number harmonic loads. The effect of an uppermost mantle low viscosity channel is similar, although if it were not for the lithosphere the decay curve would bend around to parallel its lower order portion

---

[11]That is, the decay time τ is augmented by α, and the amplitude at isostatic equilibrium is reduced by $\dfrac{1}{\alpha}$, where α is given by (IV-4).

[12]As shown in equation App. V-13, $k = \dfrac{n + \frac{1}{2}}{r}$, where r = radius of the earth and n is large.

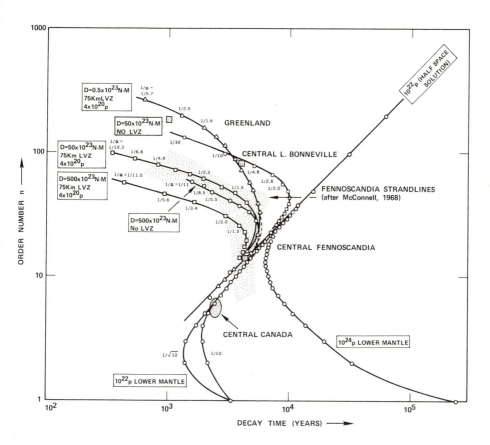

Figure IV-26. The decay spectra for various earth models compared to geological data. The order number of a load harmonic is plotted versus its decay time for each model. A uniform viscous half-space would be an approximately straight line with a slope of +1 (see equation III-17). A thin channel has a spectrum which also plots on an approximately straight line but with a slope of −0.5. (see equation III-25). Lines are approximately straight because decay time is plotted versus order number rather than wave number. The low-order number values were computed using the full spherical self-gravitating viscoelastic earth model. Below $n = 7$ self-gravitation and sphericity become important. $\frac{1}{\alpha}$ is the factor by which isostatic adjustment is reduced by the lithosphere's rigidity. The most important points to be drawn from this diagram are: (1) McConnell's Fennoscandian data require either a very thick lithosphere or an upper mantle low viscosity channel (probably require some low viscosity channel); (2) The Fennoscandian data require a lower mantle of $\leq 10^{23}$ poise viscosity, and (3) Greenland and Lake Bonneville require a lithosphere with flexural rigidity less than about $.5 \times 10^{23}$ N-m. A low viscosity channel 4×10^{20} p and 75 km thick would have a diffusion constant of 3.6 km^2/year (see equation IV-19).

at high order numbers, as the low viscosity channel begins to appear infinitely thick to the short wavelength harmonic load components (see Figure IV-27). $\frac{1}{\alpha}$ is also given along the curves to indicate the degree of filtering of the short wavelength components of the load by the lithosphere.

The low order number segments of the curves ($n < 15$) are taken from the viscous self-gravitating decay curves summarized in Figure III-16 (Model #1) and Figure III-18 (Model #4). In the case of Model #1 (10^{22} p mantle) two curves are shown for the lowest order numbers. These curves reflect the influence of the build up of buoyant forces at the core-mantle boundary. These forces slow the adjustment of low order number harmonics after a time so that the decay time calculated from the time to attain 90% equilibrium is greater than the decay time calculated from the time to attain 60%, $1 - \frac{1}{\sqrt{10}}$, equilibrium. Note the elastic response of the fluid has not been taken into account in this figure. Doing so would have the effect of increasing the apparent decay time somewhat, since the immediate elastic uplift following unloading is gradually recovered as isostatic equilibrium is approached.

The following conclusions can be drawn from Figure IV-26:

(1) The strandline data of Sauramo as analyzed by McConnell is well fit by a 10^{22} p mantle overlain by a 75 km low viscosity channel of 4×10^{20} p viscosity and a lithosphere with flexural rigidity 50×10^{23} N-m. Such a channel would have a diffusion constant of 3.6 km²/yr. The data could be fit as well by a less viscous or thinner low viscosity channel if the lithosphere had a greater flexural rigidity.[13] It seems unlikely the low viscosity channel could be eliminated entirely since the short wavelength load harmonics would then be substantially reduced in amplitude by lithospheric filtering $\frac{1}{\alpha}$.

(2) The uplift of Greenland and Lake Bonneville require a lithosphere of substantially lower flexural rigidity. The value suggested by Walcott of $.5 \times 10^{23}$ N-m appears appropriate.

(3) A lower mantle (below 1000 km) viscosity as high as 10^{24} p appears ruled out by the Fennoscandian strandline data, although 10^{23} p would be acceptable. This is not indicated directly by McConnell's

---

[13]Parsons (1972, Chapter 9) came to a similar conclusion for different reasons. He assumed a lithosphere with flexural rigidity of 250×10^{23} N-m. Using numerical inversion techniques and McConnell's data, he found an upper mantle low viscosity channel was just resolvable if a 10% error in the data was assumed.

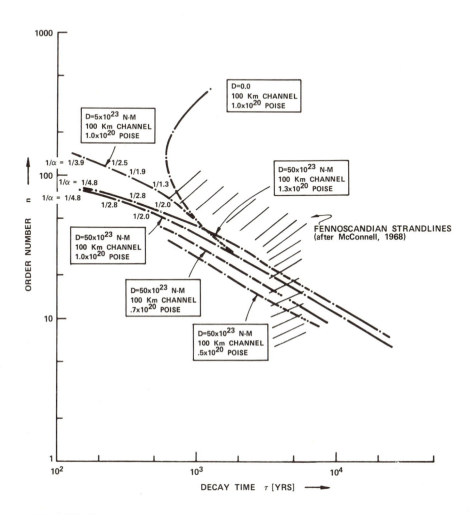

Figure IV-27. Channel flow models compared to McConnell's Fennoscandia data. Notice that the lithosphere eliminates the return to half-space behavior at large order numbers and so makes Van Bemmelen's and Berlage's approximation (which is the same as the heat conduction analogy, i.e., the rate of adjustment is proportional to $1/k^2$, see equation III-25), a good one. A 100 km thick channel of 1.3×10^{20} poise (Van Bemmelen's and Berlage's model) was found to match the Fennoscandian uplift data the best of any of the channel models. See Figures IV-29, IV-30, and IV-32. Channels 100 km thick of .013, .007, .004, .001 $\times 10^{22}$ p viscosity would have diffusion constants of 26, 49, 92, and 341 km^2/year, respectively (equation IV-19).

data since the lower order number portion of his data is distorted by the fact all strandlines were assumed to be at a constant, unchanging elevation 800 km from the center of uplift.

We conclude that the Fennoscandian strandlines data can be well matched by a 10^{22} p lower mantle provided that the effects of the lithosphere are taken into account and a thin, uppermost mantle low viscosity channel is introduced. There is trade-off between the flexural rigidity of the lithosphere and the channel. A substantial lithosphere would require no channel.

Can a channel flow model fit the strandline data as well? Figure IV-27, calculated with the aid of (III-26) and (III-27), shows a straight channel flow model apparently cannot accommodate the Fennoscandian strandline data as well as a deep flow model can. Figure IV-27 does show that lithosphere effects make the heat conduction analogy ($\tau \propto 1/k^2$ for all values of k) quite a good one (this is Van Bemmelen's and Berlage's model). For a 50×10^{23} N-m lithosphere, elastic effects take over just as the channel begins to appear deep to the shorter wavelength load components.

3. Forward Calculations of the Fennoscandian Adjustment Assuming Deep Flow and Channel Flow Models

Although powerful, the approach of the previous section tends to obscure the geological impact of differences between the deep flow and channel models. Figure IV-28 shows isostatic adjustment of the earth model that best fits the Fennoscandian strandline data (10^{22} p mantle with 75 km, 4×10^{20} p LVZ and lithosphere with $D = 50 \times 10^{23}$ N-m) following the removal, at 10,000 BP, of a cylindrical square-edged load of radius 550 km applied for 20,000 years.[14] The sympathetic uplift of the surrounding regions immediately after load removal is evident. These regions are now sinking. Figure IV-30 shows that the model predicts an uplift history close to that observed for the central load area if the initial depression is assumed about 300 m deep. (This would imply equilibrium under an ice load ~1100 m thick. Figures IV-31 and IV-32 show that the model predicts a peripheral sinking about the magnitude observed and a general uplift configuration similar to that observed.

Figure IV-29 shows the isostatic adjustment to the unloading of a similar load if flow is restricted to a channel. Several channels were investigated, but the one suggested first by Van Bemmelen and Berlage

---

[14]See Appendix VI for details of model computation and other examples.

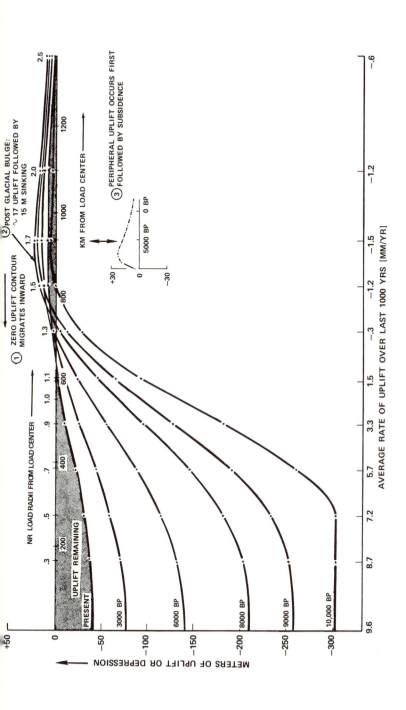

Figure IV-28. Cross-sectional portrayal of the isostatic adjustment of a 10^{22} p mantle overlain by a 75 km, 4×10^{20} p low viscosity channel and a lithosphere with flexural rigidity 50×10^{23} N-m after removal of a Fennoscandian sized load. The flexural rigidity is suggested as a minimum by gravity data (see Section IV.B). A square-edged cylindrical load corresponding to an ice thickness of 1100 m and 550 km in radius was applied for 20,000 years and then removed. Near-complete isostatic equilibrium was attained in 20,000 years, so the load cycle has little influence on the uplift. The low viscosity channel reduces but does not eliminate the sympathetic response of the surrounding area to the load removal (see Appendix VI for other examples and details of computation). Average rates of uplift over the last 1,000 years are given. The central uplift is compared to that observed in Figure IV-30. The present rate of uplift is compared to that observed in Figure IV-32. Significant points of difference with the channel flow model of Figure IV-29 are specifically enumerated on both diagrams.

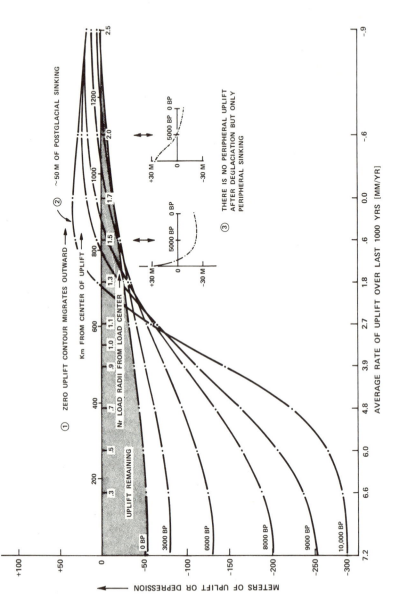

Figure IV-29. Cross-sectional portrayal of the isostatic adjustment of a 100 km thick channel of viscosity 1.3×10^{19} poise (i.e., $D = 26$ km²/year, equation IV-19) underlain by a rigid mantle, and overlain by a lithosphere with flexural rigidity 50×10^{23} N-m, after removal of a Fennoscandian-sized load applied for 20,000 years. A 550 km radius, cylindrically symmetric, square-edged load was applied for 20,000 years and then suddenly removed 10,000 years BP. The peripheral bulge produced by the squeezed out channel material is evident from the first post-glacial peripheral sinking is predicted by the model. The surrounding areas later uplift. Over the last 1,000 years most of the peripheral regions have been uplifting, not sinking. The central uplift is compared to that observed in Figure IV-30. The present rate of uplift is compared to that observed in Figure IV-32. This is the best fitting channel flow model and is identical to that proposed by Van Bemmelen and Berlage (1935). Significant points of difference with the half-space flow model of Figure IV-28 are specifically enumerated on both dia-

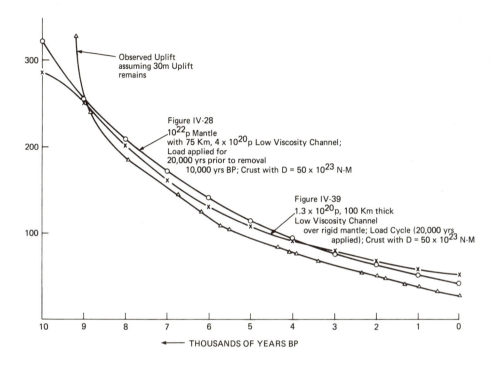

Figure IV-30. Comparison of the central uplift of the models given in Figures IV-28 and IV-29 to the uplift observed in central Fennoscandia (Liden, 1938, as corrected for eustatic sea level in Lliboutry, 1971). For the sake of comparison it is assumed 30 m of uplift remain at present in central Fennoscandia. The "observed" curve could be shifted up or down any amount to achieve a better fit by assuming greater or lesser amounts of uplift remain. The uplift curve was dated by varve techniques (annual sedimentary fluctuations that can be counted) and is probably reasonably accurate (± 200 years) to 8000 BP. Past 8000 BP there is some suggestion 800 varves (years) may have been lost (see discussion in Section IV.E.1).

(1935) was found to fit best the observed data in Fennoscandia. The quality of the match to the uplift in central Fennoscandia and to the present rate of uplift and peripheral sinking is shown in Figures IV-30 and IV-32. Agreement between model and observed uplift is at least as good as in the previous half-space case. As in the previous case, the load was applied for 20,000 years and then removed 10,000 years BP. The ice was taken to be 1100 m thick (i.e., such as would produce an equilibrium depression of 300 m) since this was found to provide the best match to the geological data in this case also.

Several points can be made from Figure IV-29:

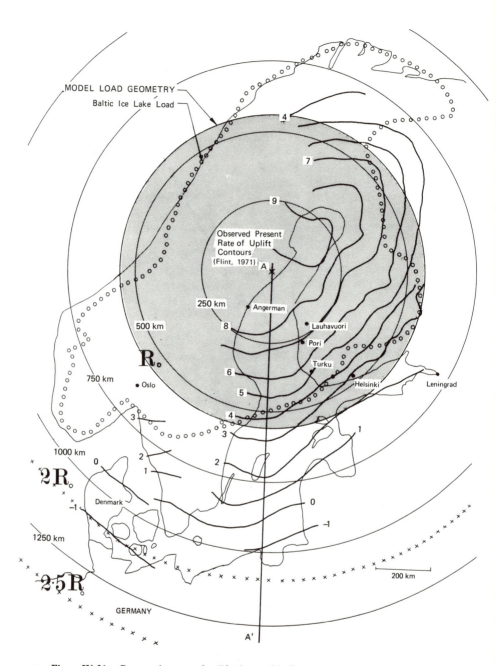

Figure IV-31. Present-day rate of uplift observed in Fennoscandia. Superimposed on a present-day map of the Fennoscandian coastline are (1) the ice configuration of the Baltic Ice Late Stage (Flint, 1971, p. 351), (2) the cylindrical ice load (shaded) assumed in the models presented in Figures IV-28 and IV-29 (R_0 = 550 cm), (3) contours indicating distance from the model load center, (4) present-day rate of uplift contours from Flint (1971, Fig. 13.2). The rate of uplift contours are deduced both from tide gauge measurements along the Baltic and from precise leveling surveys (see Kääriäinen, 1963, 1966). The section AA′ is chosen so that the dimension of the model and actual ice load agree well and so that at the same time the section goes through well-determined rate of uplift contours. The data along AA′ are compared to the model predictions in Figure IV-32.

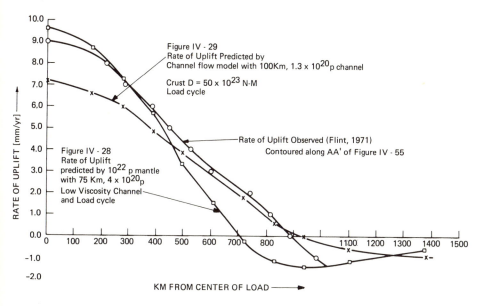

Figure IV-32.   The average rate of uplift over the last 1000 years predicted by the models in Figures IV-28 and IV-29 along a radius from the load center compared to the uplift observed in Fennoscandia today taken from Figure IV-31 and along AA' of that figure. It can be seen that both models are in good agreement with the observed uplift.

(1) At the close of 20,000 years of loading, a peripheral bulge ~35 m high surrounded the ice-loaded area. This bulge sloped steeply from its crest toward the ice edge, but gradually from its crest away from the ice load.

(2) Upon deglaciation, this peripheral bulge quickly collapsed. The collapse started near the edge of the former ice load and migrated outward. A peripheral sinking of about 50 m occurred between deglaciation and present time. A significant collapse of the outer portions of the bulge was delayed several thousand years after the ice dissipation. No post-glacial uplift preceded the sinking (as in the 10^{22} p half-space case of Figure IV-28).

(3) Zero uplift contours in the channel flow case (Figure IV-29), defined by the crossing of older cross-sectional lines with that marked 0BP, migrate *outward* as the strandlines get younger, whereas in the half-space case (Figure IV-28), the zero uplift contours migrate *inward* with time. That is, in the channel flow case, the distance from the center of load that a given age strandline will appear, at present, neither uplifted nor depressed increases as the strandlines become younger, whereas in the half-space case it decreases.

(4) Only about 50 m of uplift are predicted to remain in central Fennoscandia by the channel flow model. This is a consequence of the load cycle. The half-space model (Figure IV-28) predict 40 m of remaining uplift.

Some of these points have been made, or could have been made, before. For example, Gutenberg pointed out that the channel flow model implies an outward migration of the zero uplift contour (see equation IV-24). It can easily be seen from equation IV-20 and the principle of superposition that a load cycle implies a substantial reduction in the uplift remaining at the present time. If the load is applied for T years and then removed, the central uplift remaining t years after the load removal is:

$$h = \frac{h_0}{\sqrt{1 + 4K(t + T)}} - \frac{h_0}{\sqrt{1 + 4Kt}}. \qquad \text{(IV-26)}$$

Taking Lliboutry's (1972) values for $K = \dfrac{D}{\sigma^2} = \dfrac{35 \text{ km}^2/\text{yr}}{(374 \text{ km})^2}$, $T = 22{,}000$ years, and $t = 10{,}000$ years, it can be easily shown that (IV-26) suggests only 38 m of uplift remain if $h_0 = 300$ m. If h_0 is taken to be 600 m, the remaining uplift would be 76 m. (If T were taken to be infinity, these values would be 80 m and 178 m, respectively.)

Forty to fifty meters of remaining uplift is consistent with gravity anomalies in Fennoscandia if the effects of the lithosphere are taken into account. As we saw in Section IV.B.2.c, the residual 3.5 milligal negative gravity anomaly suggests 25 m of remaining uplift.

In light of the size of the channel bulge (Figure IV-29) and particularly its steep slope toward the ice edge, Daly's and Flint's warning about the lack of geological evidence for such a bulge appears to tell against a channel viscosity model. With the better gravity maps now available and the equal success of a half-space model, there appear to be no compelling reasons for favoring the channel flow model. If eustatic sea level changes and hydro-isostatic adjustment to the ocean meltwater load are subtracted from the peripheral sinking geologically observed, 50 m post-glacial peripheral sinking is probably not indicated (Anderson, 1960, Pl. 8). There is an indication of uplift followed by subsidence in some interpretations of sea level from Holland (Mörner, 1969, Figure 153). There is perhaps some indication of an inward migration of the zero uplift isobase with time. Kääriäinen (1953, p. 70) states that, although full agreement on the movement of the zero isobase has not been reached, "probably ... the region of uplift [in Fennoscandia] is in the process of decreasing, i.e., the zero isobase moves toward the center [of uplift]." Kääriäinen reports that Model has estimated the zero isobase moves toward the center of uplift at a rate of between ten and twenty meters per year. This is considerably faster than would be suggested by Figure IV-28, but would be in reasonable agreement with half-space flow if no upper mantle low viscosity channel were present (see Table App. VI-2, and Figure App. VI-4). Inward migration of the zero uplift isobase would suggest a uniform viscosity deep flow model.

The difference between the channel and deep flow models (Daly's bulge and punching models) are significant enough that an unequivocal geological distinction between the two should be possible. Both appear able to fit the undisputed geological evidence equally well, however.

4. Determination of Mantle Viscosity from the Fennoscandian Uplift if the Deep Flow (Relatively Uniform Viscosity) Model Is Adopted

The filtering of the models of Table IV-1 was not severe enough to invalidate comparison of those models to the uplift in central Fennoscandia, although it is too severe for comparison of peripheral model uplift to peripheral geological uplift. A feeling for the smearing of the ice load can be obtained from Figure IV-34, which shows the uplift of a suddenly unloaded Fennoscandia-sized continent. The geometry of that continent is shown in Figure IV-33.

Figure IV-36 shows the history of uplift of central Fennoscandia with the gradual load removal shown in Figure IV-13 for the various earth models of Table IV-1. Figure IV-35 shows the rate of central uplift as a function of model time. The relatively large effects of a 335 km, 10^{21} poise low viscosity channel (Model #3) and of a high viscosity lower mantle (10^{24} p mantle beneath 1000 km depth—Model #4) should be noticed particularly. The models take into account the gradual removal of the ice load, so on the basis of central uplift alone a substantial (335 km thick) low viscosity (10^{21} p) channel superimposed on $\sim 10^{22}$ p mantle (Model #3) can be ruled out. A high viscosity (10^{24} p) mantle beneath 1000 km depth becomes suspect.[15] The influence of lower mantle viscosity should not be surprising in light of the findings in Appendix VI that 20% of the horizontal flow occurs at a depth equal

---

[15] The fact central Fennoscandia is within 125 m of equilibrium after 10,000 years of unloading indicates a 20,000 year load cycle probably will not decrease the predicted present rate of central uplift enough to make this model acceptable.

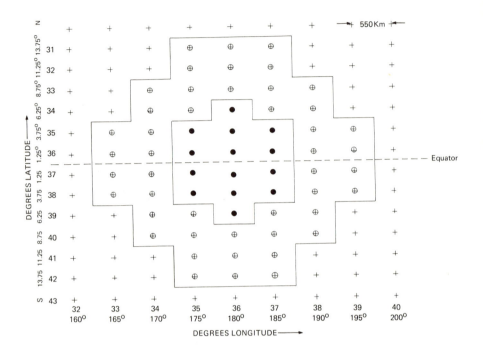

Figure IV-33.   Outline of approximately circular load distributions used in heuristic calculations. Glaciers are assumed to exist initially, in isostatic equilibrium, on a single continent that is just glacier-sized. The meltwater is returned to an ocean that constitutes the rest of the globe. The melting is assumed to occur instantaneously.

to two and a half times the load radius. This translates in the case of Fennoscandia to a depth of about 1400 km.

Figure IV-35 shows that the central rate of uplift for Model #1 at present model time agrees with that geologically observed. Figure IV-37 shows that Model #1 and Model #2 (uniform 10^{22} p mantle viscosity without and with a non-adiabatic density gradient between 335 and 635 km depth) bracket the observed history of uplift of central Fennoscandia. We can therefore conclude that, if the mantle has low viscosity to depths ~ 1000 km, the uplift of central Fennoscandia indicates a substantially uniform mantle of about 1.0×10^{22} poise viscosity to at least 1200 km depth with at most a thin uppermost mantle low viscosity channel and a small non-adiabatic density gradient between 335 and 635 km depth (into the mantle).

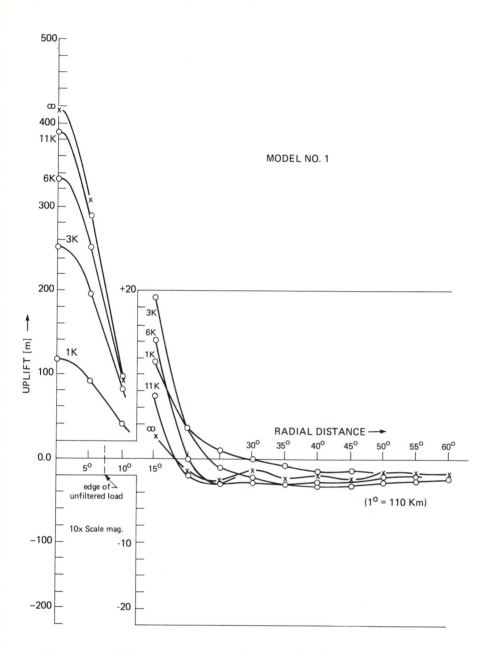

Figure IV-34. Uplift cross-sections at various times after the sudden removal of a Fennoscandian-sized glacial load that had attained isostatic equilibrium on a uniform 10^{22} poise mantle (Model #1). The ice load was 1.76 km thick, but filtering cuts this maximum thickness 15% to 1.49 km. Hydro-isostatic adjustment of the ocean surrounding the glaciated continent (and constituting the rest of the heuristic model globe) can be seen from the ~2 m ultimate peripheral sinking. The load is shown in Figure IV-33.

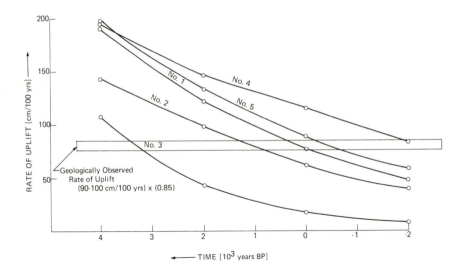

Figure IV-35. The rate of uplift in the northern gulf of Bothnia at various model times for the earth models of Table IV-1. A bracket is given for the geologically observed rate of uplift there. If the viscosity structure of the earth agrees with a given model, the curve for that model should cross the present geological rate of uplift bracketed near T_{model} = 0BP. Thus Model #3 (low viscosity channel) is contradicted by geologic observation, and doubt is cast on Model #4.

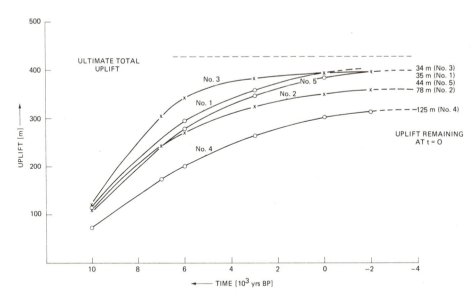

Figure IV-36. History of uplift in the northern gulf of Bothnia calculated on the basis of the earth models of Table IV-1. Full isostatic adjustment for all models is assumed to have been attained before the glaciers began to melt. The uplift remaining at present model time is noted for all models although it is probably only meaningful for Models #1, 3, 5. For the others, the load cycle should have been taken into account.

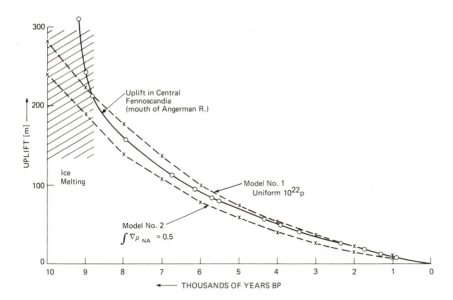

Figure IV-37. The uplift calculated in central Fennoscandia for Models #1 and #2 compared to the uplift geologically observed.

It is instructive to note that the same mantle viscosity can be inferred from the uplift data directly if 30 m uplift remaining is assumed. The uplift data from the mouth of the Angerman River (Table IV-10) then plot as a straight line on semi log paper (Figure IV-38) indicating that the isostatic uplift can be described as an exponential decay with a time constant of 4400 years. No attempt was made to optimize the exponential fit of Figure IV-38 by choosing values other than 30 m of remaining uplift. From the principles of Appendix VI, choosing $R_0 = 550$ km, $\rho = 3.313$ gm/cc, $\bar{k} = 2.18 \times 10^{-8}$ cm^{-1},

$$\eta = \frac{\rho g \tau}{2\bar{k}} = 1.03 \times 10^{22} \text{ poise.} \qquad (IV-27)$$

Viewed in this way the determination of a 4400 year decay constant for Fennoscandia and the resulting determination of a mantle viscosity of 10^{22} poise fundamentally depends on the recognition that only 30 m of uplift remain. Accounting for elastic uplift, the effects of neighboring loads and the gradual removal of the ice are relatively minor garnishments. Also the prediction of 30 m remaining uplift can be viewed as only a minor extrapolation of a curve whose slope is well determined over 8800 years of uplift (Figure IV-38).

Years before 2000 AD	Emergence curve (elevation of shore-lines above present sea level)	Land uplift relative to present (Emergence curve corrected by Morner's sea level curve)	Meters from isostatic equilibrium (uplift curve plus 30 m)
9200	280	310	340
9042	219	244	274
8839	194	212	242
7944	138.9	157.9	187.9
6741	104.1	113.1	143.1
6178	90.4	95.6	125.6
5713	80.2	84.0	114.0
5535	76.2	79.7	109.7
4354	54.4	55.9	85.9
4094	51.1	52.1	82.1
3918	48.2	48.2	78.2
3408	40.7	40.7	70.7
2365	26.3	26.3	56.3
1779	18.0	18.0	48.0
1317	12.2	12.2	42.2
939	8	8	38
539	2	2	32
87	0	0	30

Table IV-10 Uplift at mouth of Angerman River, Sweden.   Data are taken from Lliboutry (1972, Table 1), but are originally from Liden (1938).   The last column is plotted versus first in Figure IV-38.

E. The Viscosity of the Lower Mantle

Comparison of computed and observed eustatic sea level curves (Section IV.A.5), a suggested systematic difference between continental and oceanic sea level minima (Section IV.A.6), the apparent lack of a Holocene sea level high (Section IV.A.7), and eclipse data and analysis of the rotation of the earth together with the lack of a correlation between gravity anomalies and the post Wisconsin load redistribution (Chapter I), all suggest a fluid lower mantle with a viscosity $\sim 10^{22}$ poise.   The viscosity of the lower mantle is perhaps most directly probed by the North American load redistribution and most convincingly revealed by the associated uplift there.   The uplift history of

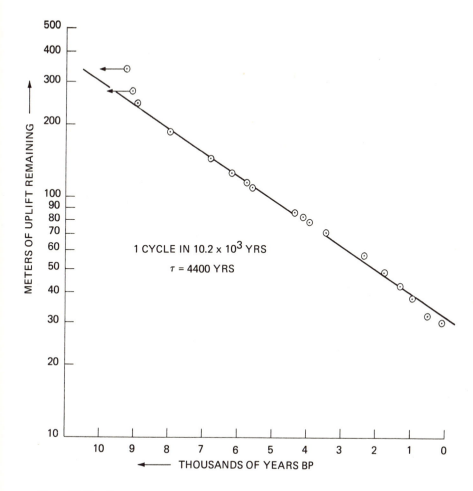

Figure IV-38. Demonstration that the uplift in central Fennoscandia can be well-characterized by an exponential function with a decay constant of 4400 years providing about 30 m of uplift are assumed to remain at present. Data are from Table IV-10. Chronology past 8000 BP may be in error due to missing varves in varve chronology (see Section IV.E.1, Figure IV-40). Arrows show suggested correction of 800 years (Stuiver, 1971).

the east coast of the United States appears to rule out the possibility of accounting for the Canadian uplift by a channel flow model and to require deep flow in a Newtonian lower mantle of ~ 10^{22} p viscosity.

In this section, we first review the literature bearing on the Canadian uplift, discussing in detail channel flow models that have been proposed. We then discuss in order: History of uplift of the United States' east coast; model predictions of Canadian uplift history and present rate of uplift; the worldwide pattern of isostatic adjustment; and information available from the earth's gravity field bearing on the viscosity of the deep mantle.

1. Review of the Canadian Uplift Literature

The earliest studies of glacial uplift in Canada mapped the marine limit—i.e., the elevation of the highest marine strandline in an area regardless of the age of that strandline (see Figure IV-39). Strandlines are identified by accumulations of marine shells, preferably in sand or silt (which indicates the shells are in place), or by the limit of wave action determined by detailed ground observation and air photography. In this way, it can be shown that the maximum recorded uplift in Canada increases in a regular fashion from the limit of glaciation toward the interior of the glaciated region. Farrand and Gajda (1962) found a maximum uplift of 270 m over Hudson Bay. This prominent central dome is outlined by isobases that are younger than those nearer the periphery. This suggests that more glacial rebound has occurred in the central load regions than near the load periphery. Troughs centered over Fox Basin and the District of Keewatin appear to coincide, in Farrand's and Gajda's view, with centers of late deglaciation, suggesting that considerable uplift occurred in an area prior to complete ice dissipation. A prominent ridge of high marine limit elevations extends from Queen Maud Gulf to the Bothnia Peninsula and over Devon Island to the northern part of Ellesmere Island, suggesting a substantial ice load existed in this part of the arctic archipelago. With the exception of the trough over Fox Basin these features are confirmed by more recent studies (Andrews, 1970a, Figure 5-5).

As useful and significant as the marine limit is, the history of uplift is more significant for geophysical interpretation. The history of uplift can be deduced if debris at identified marine strandlines (such as shells, driftwood, bone, or peat) is radiocarbon dated. A sequence of such strandlines at a given locality determines an emergence curve which can be converted to an uplift curve by subtracting eustatic sea level. Past eustatic sea levels are taken negative so uplift is greater than emergence. Farrand (1962) did this for a number of localities in Canada and found the pattern of uplift to be similar at the points he studied. Very rapid initial uplift was followed by a strongly exponential decrease in the uplift rate. The uplift curves seemed simply displaced in time according to the time of deglaciation.

Broecker (1966b) and Brotchie and Silvester (1969) proposed models

Farrand and Gajda (1962, Fig. 1)

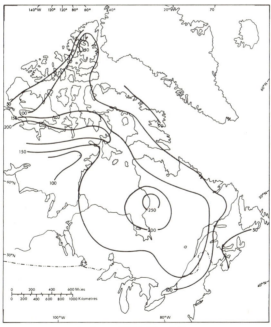

Andrews (1970, Fig. 5-5).

Figure IV-39. Present elevation of the marine limit according to Farrand and Gajda (1962, Fig. 1, p. 8) and Andrews (1970, Fig. 5-5, p. 62). Isobases of Farrand and Gajda are in feet. *Note* that Andrews omits the trough over Fox Basin and dome over Baffin Island shown by Farrand and Gajda. The figures ©, respectively, in 1972 by the Department of Energy, Mines, and Resources (Canada) and in 1970 by the Institute of British Geographers, are reproduced by permission of the authors.

for the Canadian uplift based on simple exponential uplift following (incremental) load removal. Their models were inspired by Farrand's observations or observations similar to his (Washburn and Stuiver, 1962).

Andrews (1968c, 1970a, b) compiled the Canadian uplift in the particularly usable form of contour maps showing the uplift over certain intervals of time. Andrews assumed all uplift could be described by an exponential decay with a time constant of 2560 years. This approximation was suggested by sites studied by Andrews as well as by Farrand's study. In Appendix VII, Andrews' interesting method is reviewed extensively and its implications for uplift over the last 8, 7, 6, 4 and 2 thousand radiocarbon years, the present rate of uplift, and the amount of uplift remaining at present, are shown. We interpret Andrews' data differently from the way he interpreted them regarding the present rate of uplift and the amount of uplift remaining, but the history of uplift to present is interpreted similarly. The reasons for doing this are discussed in Appendix VII. Andrews' simple method enjoys remarkable success in describing the uplift in Canada. This success has significant implications. It is also significant that Andrews' method suggests directly that very little uplift should remain at present. At least 7500 years, or three decay times 2500 years long, have elapsed since effective deglaciation. Thus about 5% of the total glacial uplift should remain at present. Assuming a total isostatic uplift of 1000 m, 50 m of uplift should remain today. As shown in Appendix VII, Andrews' analysis as we have modified it suggests 15 to 20 m.

Walcott (1972a) has recently compiled emergence contour maps similar to Andrews' directly from the primary data without assuming any particular form for the uplift. There is generally good agreement between studies of the historical uplift in Canada. In the second part of Appendix VII we compile uplift contour maps for 2, 4, 6, 7 and 8 thousand radiocarbon years before present using data from Walcott, Farrand, and other sources. Andrews' data are shown but given less weight. This compilation shows the general agreement between the various studies and the success of Andrews' simple method. The various calculated uplift histories will be compared to these data.

In Appendix VII, we also compile a present rate of uplift map. Direct evidence on rate of uplift is provided by a tide gauge station at Churchill, Manitoba, and a recent precise leveling survey in Quebec Province. For reasons that will be indicated in our later analysis, 2.4 mm/year is added to the 3.8 ± 1.2 mm/year rate of uplift indicated by the Churchill Tide Gauge Station (Barnett, 1970) to account for a recent rise in eustatic sea level. This correction brings the rate of uplift measured by the tide gauge stations into good agreement with that which would be inferred from the present slope of the uplift curve. The rest of the data used to compile the rate of uplift map is obtained from estimates of the present slopes of the uplift curves.

Differences between the compilation of data in Appendix VII and those of other authors are discussed in Appendix VII. Our compilation falls within the error estimates of other compilations and can be briefly summarized: (1) The general central present rate of uplift is probably centered over James Bay and is probably at least 10 mm/year in magnitude. (2) The central uplift over the last 7, 6, 4, and 2 thousand radiocarbon years as recorded by strandlines and corrected for changes in eustatic sea level is about 200 ± 25, 120 ± 10, 50 ± 5, and 20 ± 2.5 m, respectively. The error estimates are Walcott's (1972a).

The maximum present rate of uplift over the last 2000 years is estimated at Cape Henrietta Maria by Walcott to be 17 ± 5 mm/year. By extrapolation, Walcott estimates that the present maximum rate of uplift in Canada is near 20 mm/year. Walcott suggests that the maximum uplift over the last 2000 years was about 40 m. Since we compare the uplift history and rate of uplift of a relatively broad central region (because of model filtering), narrow maxima are not of primary concern. For this reason we have cited uplift and rate of uplift appropriate for the central load regions and not maxima. Such data do not require extrapolation.

Two corrections to the uplift data are required, however: (1) There is evidence that the radiocarbon time scale does not coincide with the solar time scale past about 2500 years BP. (2) The deformation of the earth's geoid due to the past isostatic disequilibrium of Canada must be accounted for.

As mentioned previously the uplift data for Canada are derived from sequences of radiocarbon dated, elevated marine strandlines.[16] This is important since recent studies (Stuiver, 1971) suggest that a shift between radiocarbon and real time occurred between 2500 and 5500 BP. This shift, as shown in Figure IV-40, is such as to make the older (>5000 BP) observed strandlines somewhat (~1300 years) older than their radiocarbon age dates would indicate. There is some uncertainty as to the time period over which this correction should be applied.

---

[16]The Fennoscandian data do not share this dependence on radiocarbon dating since varve dating was extensively used. It appears that the varve chronology of Liden (1938) used in Section IV.D is accurate, with perhaps 200 years added to correct for recent varves missing, to about 8000 BP. Comparison with varve data from the Lake of the Clouds then suggests about 800 varves may be missing from Liden's sequence (Stuiver, 1971).

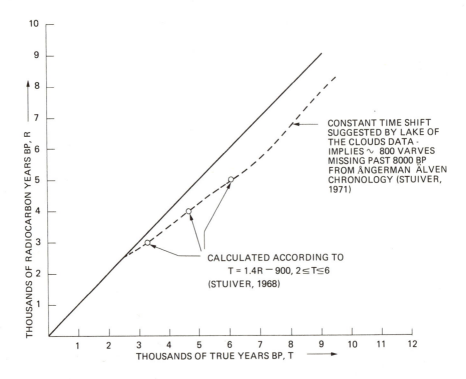

Figure IV-40. Real age plotted on a function of radiocarbon age.   Tree ring, historical, and varve data indicate a shift between radiocarbon and solar chronology (Stuiver,
1971) between 2000 and 6000 BP. Past about 8000 BP, Fennoscandian varve data indicate a return to near coincidence between radiocarbon and solar time, but data from
Lake of the Clouds varves indicate the shift is maintained. The Lake of the Clouds data
appear to have better control than the Fennoscandian data, so some (Stuiver, 1971) feel
that 800 varves may be missing from the Fennoscandian data past about 8000 BP. We
adopt that interpretation here.

Since the ice removal data for North America come from radiocarbon
end morain data (see Section IV.A.4), corrections must be applied to
both the theoretical model of ice removal and the geological uplift data.
Since these corrections will be self-compensating to some extent, we
shall generally assume that model time is the same as radiocarbon time.
In Section IV.E.6, we examine the implications of this assumption.
Corrections are minor and in the nature of second order refinements.

The retreat of the Wisconsin ice in Canada left a negative gravity
anomaly as well as a land depressed and in isostatic disequilibrium.
The two phenomena are of course related. Sea level, which follows or
defines an equipotential gravity plane (geoid), was thus anomalously

low in North America in the years immediately following the melting
of the Wisconsin ice. As Canada uplifted, the degree of isostatic disequilibrium and the negative gravity anomalies associated with it were
reduced. As the uplift occurred, perturbations in the geoid (and sea
level) flattened out. If you like, the return of high density material
flowing up under Canada attracted water toward Canada. This local
increase in sea level over the period of historical land uplift in Canada
must be accounted for if we are to infer the magnitude of glacial uplift
in Canada from strandline data.

It is shown in Section IV.E.7 that the depression of the geoid in
Canada was about 16% of the isostatic depression of the land there.
Thus the uplift inferred from emergence curves and corrected for
eustatic sea level variations must be augmented by about 16% to correct
for the changes in the local geoid elevation that attended the glacial
uplift. The maximum uplift in central Canada over the last 7, 6, 4, and
2 thousand radiocarbon years becomes 230 ± 25 m, 140 ± 10 m,
58 ± 5 m, and 23 ± 2.5 m. The geoid corrections are less important
relative to other uncertainties for the (smaller) uplifts that have occurred over the last 5 thousand radiocarbon years.

Gravity has played an influential role in the study of the Canadian
uplift, just as it did in the study of the Fennoscandian uplift. As shown
in Figure IV-41, a large negative free air gravity anomaly about -30
mgals in magnitude is centered over the region of Canada that was
heavily glaciated and is now uplifting (Fisher, 1959a, b; Innes and
Weston, 1966; Walcott, 1970b, 1972a). The ellipticity of the gravity
anomaly as a whole appears to coincide with the ellipticity of the Wisconsin glacial system. Within the broad anomaly, local lows correlate
with areas last deglaciated (i.e., Fox Basin, Keewatin Province, and
Quebec-Newfoundland area). Present topography correlates well with
the gravity anomalies (Walcott, 1970b, pp. 718ff.). This coincidence
has been taken to indicate association, and it has been generally concluded on the basis of the observed gravity anomalies that about 200 m
of uplift remains in Canada.

There has been some recognition that such a large amount of remaining uplift conflicts with the short (2560 years) relaxation time observed for the Canadian uplift. Walcott (1970b) proposed a double
relaxation time to circumvent this difficulty:

remaining uplift at time t years $= 150 \, e^{-t/1000} + 450 \, e^{-t/50,000}$

$$(IV-28)$$

He suggested such a double relaxation time could be explained by

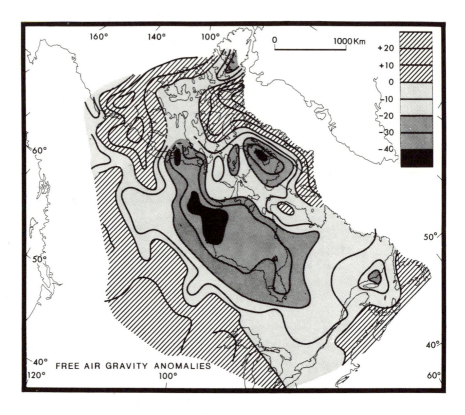

Figure IV-41. Smoothed free-air gravity anomaly map of Canada, showing the broad area of "undercompensation" over the Laurentide and Innuitian uplifts. The average anomaly over the central dark shaded region is 35 mgal. The figure, reproduced with the author's permission, is from R. I. Walcott, *Reviews of Geophysics, 10,* p. 877, ⊚ 1972 by the American Geophysical Union.

elastic effects from the crust, a layered viscosity structure, or a higher order dependence of strain rate on stress. In light of previous discussions, it is unlikely that it can be accounted for by crustal effects or phase transitions. A layered viscosity structure will work only if the high viscosity layer overlies the lower viscosity layer. Otherwise equilibrium will be established in the fluid layer before significant deformation of the lower, more viscous, layer is achieved. We shall discuss the possibilities of higher order flow laws later.

Both Walcott (1972b) and Artyushkov (1971) have proposed channel flow models for the Canadian uplift. Walcott suggested a layer with a diffusion constant $D \sim 50$ km^2/year. Artyushkov suggested a layer diffusion constant between 19 km^2/year and 95 km^2/year. He correctly

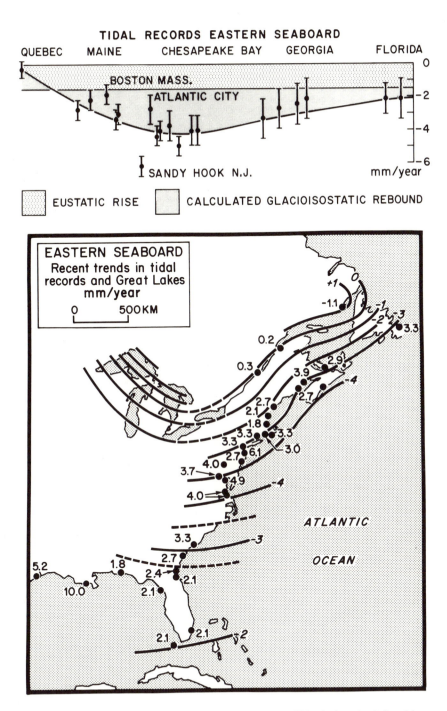

Figure IV-42. Present rate of submergence of east coast of North America inferred by Walcott from data of Hicks and Shofnos (1965) and Dohler and Ku (1970). Data are tied into uplift data for Great Lakes, and the rate of submergence profile is drawn by inference from the history of sinking. A recent eustatic rise of sea level of 1.5 mm/year is assumed. The figures, reproduced with the author's permission, are from R. I. Walcott, *Reviews of Geophysics, 10,* p. 874, © 1972 by the American Geophysical Union.

showed that, if 200 m of uplift remains, the viscosity of the mantle beneath the asthenosphere channel must be at least 10^{23} poise if rates of uplift greater than those observed are to be avoided. Walcott showed that the channel flow model might account for the present peripheral sinking observed around the Canadian, Fennoscandian, and other glacial systems (Walcott, 1972a, b). The data Walcott (1972a) compiled on the post-glacial isostatic behavior of the United States east coast are particularly significant. We shall review the history of uplift in this area later.

Figure IV-42 shows the present sinking of the United States east coast determined by tide gauge data and compiled by Walcott (1972a). Neither Walcott nor Artyushkov took the load cycle into account in their modeling. Both assumed the glacial load had been applied long enough for complete isostatic equilibrium to have been attained prior to removal.

2. Channel Flow Models for the Canadian Uplift

Figures IV-44–IV-48 show the uplift in Canada that would follow the sudden removal of a cylindrical square edged glacier as shown in Figure IV-43 for the channel earth models listed in Table IV-11. The load was assumed to have been applied for 20,000 years and removed suddenly 10,000 years BP. An elastic crust with a flexural rigidity of 50×10^{23} N-m smears the ice load somewhat. A cylindrical ice load 2160 m thick[17] and 1650 km in radius with square edges was assumed. The thickness was chosen because it produced model central uplift histories most like those observed. If a larger ice load is desired, the uplifts and rate of uplift calculated can simply be scaled up in proportion to the load increase. The radius was chosen to match the size of the Wisconsin ice load in an approximate fashion (see Figure IV-43 and IV-9 or 10). The uplift following the sudden removal of the ice load was calculated as discussed in Appendix VI.

The effects of the load cycle are clearly apparent in Figure IV-44. At the time of ice removal, a peripheral bulge 140 m high surrounded the model ice load. The bulge sloped steeply toward and less steeply away from the ice margin. It quickly disappeared upon the removal of the ice load. The post-glacial peripheral sinking was greater in magnitude than the central uplift.

Model C1, having a diffusion constant of 49 km^2/year is quite close

---

[17]At isostatic equilibrium such a load will produce a 600 m depression (ρ_{mantle} = 3.313, ρ_{ice} = .92).

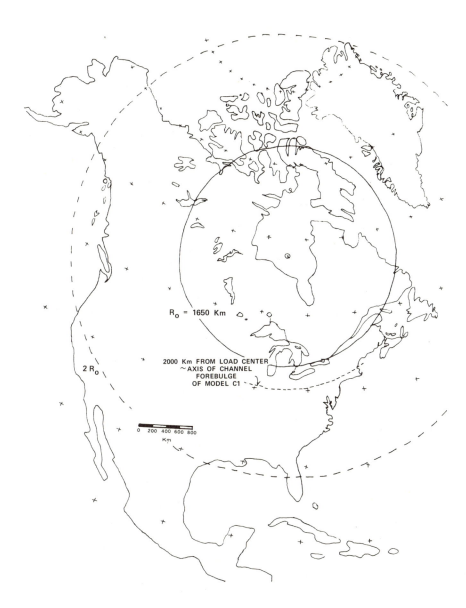

Figure IV-43. Equal area map of North America showing the cylindrical square-edged ice load used to calculate the channel models of Table IV-11. Lithosphere with flexural rigidity of 50×10^{23} N-m smears the load about 200 km from the ice edge (see Table IV-7).

$$\text{Model C1} \quad \eta = 7.0 \times 10^{19} \, p$$
$$D = 49 \, \text{km}^2/\text{year}$$
$$\text{Model C2} \quad \eta = 4.0 \times 10^{19} \, p$$
$$D = 85 \, \text{km}^2/\text{year}$$
$$\text{Model C3} \quad \eta = 1.0 \times 10^{19} \, p$$
$$D = 341 \, \text{km}^2/\text{year}$$

Table IV-11 Channel flow models for North America used to calculate Figures IV-44–46. Cylindrical square-edged ice load 1650 km in radius (as shown in Figure IV-43) and a lithosphere with flexural rigidity of 50×10^{23} N-m are assumed in all models. All models assume the ice load was applied for 20,000 years and then suddenly removed 10,000 years BP. The discussion in Section A of this chapter indicates 20,000 years of load application is about the maximum duration that would be reasonable. An ice thickness sufficient to cause 600 m of central depression at isostatic equilibrium (~ 2160 m ice load) is assumed. This figure can be taken larger by simply multiplying the calculated uplifts by the appropriate factor (i.e., 2 if 4320 m ice load assumed). D is the diffusion constant approximately characterizing the layer, computed from (IV-19).

to models proposed by Walcott (1973) for the mantle viscosity structure under Canada (~ 52 km²/year). With the load cycle taken into account, there are several reasons it is not a good model for the Canadian uplift:

(1) The central uplift is not of large enough magnitude to accommodate the marine limit data of Figure IV-39. Furthermore the model predicts the marine limit should be greater near the edges of the loaded region than toward the center. This conflicts with distinct domal pattern of the marine limit where the maximum uplift is greatest at the center even though the strandlines are younger there than nearer the edge.

(2) The peripheral bulge during glacial times of the magnitude predicted (~ 140 m) is not observed.

(3) A post-glacial peripheral subsidence of the magnitude predicted (~ 160 m) is not observed.

(4) Although the present rate of subsidence along the east coast of the United States matches that observed fairly well (see Figure IV-48), the history of central uplift, known geologically, conflicts with the central uplift predicted by the model (Figure IV-47).

Part of the first problem could be removed by increasing the ice thickness. This might bring the central uplift since deglaciation into better agreement with the observed marine limit. Some elastic uplift might also be added, but not more than about 50 m without encountering problems explaining how the ice could hold a region down elas-

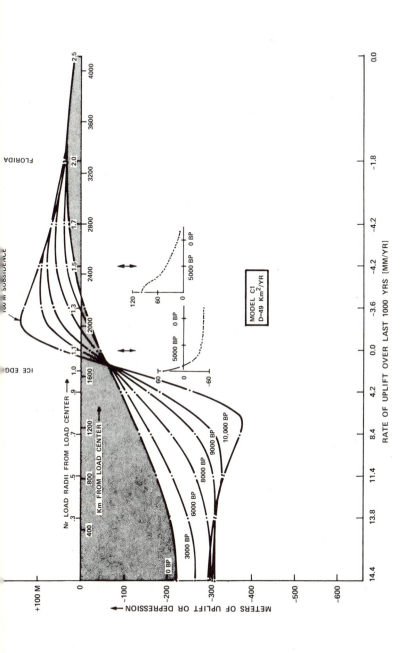

Figure IV-44. Cross-section of the post-glacial adjustment of Canadian channel flow Model C1 of Table IV-11 (100 km viscous channel with viscosity 7×10^{19} p, load as shown in Figure IV-43). The channel model is close to that proposed by Walcott (1970, 1973), and Artyushkov (1971). Notice that with the load cycle taken into account the post-glacial central uplift is less than the post-glacial peripheral depression, which is extremely large (160 m). The post-glacial central uplift is not large enough to account for the observed marine limit (Figure IV-39). This could be remedied by increasing the thickness of the ice load at the cost of a greater peripheral subsidence and a poorer near present agreement in the history of central uplift (Figure IV-47). The present peripheral sinking matches the sinking observed well (Figure IV-48), although the magnitude of the rate of sinking may be a bit large. Increasing the ice thickness would make this match worse.

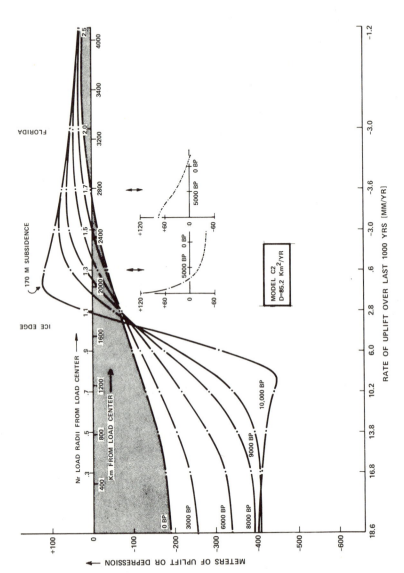

Figure IV-45. Cross-section of the post-glacial adjustment of Canadian channel flow Model C2 of Table IV-11 (100 km viscous channel with 4×10^{19} p viscosity, load as shown in Figure IV-43). The central uplift is now large enough to agree with the observed marine limit (Figure IV-39), but post-glacial peripheral sinking is still nearly comparable in magnitude to post-glacial central uplift. The present rate of peripheral sinking agrees less well with that observed than the more fluid channel calculated in Figure IV-44 (see Figure IV-48), although the history of central uplift agrees better.

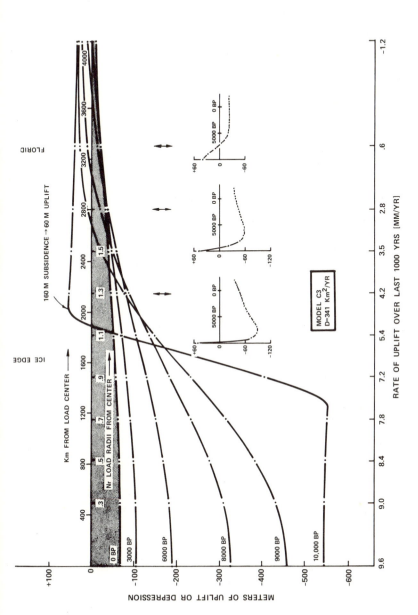

Figure IV-46. Cross-section of the post-glacial adjustment of Canadian channel flow Model C3 of Table IV-11 (100 km viscous channel with 1.0×10^{19} poise viscosity, load shown in Figure IV-43). The glacial peripheral bulge is smaller, but the post-glacial subsidence is practically as great as in the previous two cases. The central uplift history matches that observed quite well (Figure IV-47), but the present uplift of the peripheral areas does not agree at all with that observed (Figure IV-48). Peripheral depression has changed to peripheral uplift. We reasoned this should be the case by analogy to heat conduction in Section IV.C.2.

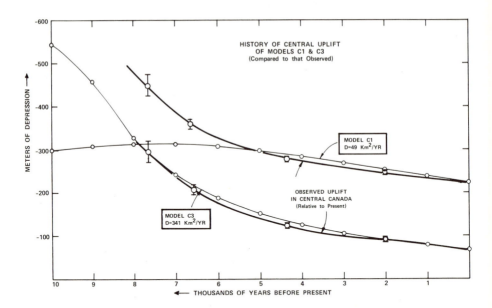

Figure IV-47. Central uplift for Canadian channel flow Models C1 and C2 of Table IV-11 compared to the central uplift observed. Radiocarbon time scale is corrected to solar time scale by using Figure IV-40. Uplift data are for the central region of Canada, not the maximum uplift that has occurred in the area. A load such as would produce 600 m of depression at equilibrium was chosen because the uplift history in near recent times is best matched by this magnitude load. It could be chosen larger at some cost to the quality of this match. It can be seen that the more fluid channel (C3) fits the central uplift history very well and that the less fluid channel contradicts the observed data between 7.7 and 6 thousand years BP (solar time).

tically yet still permit strandlines to be formed in uncovered pockets. It is difficult to see how the trend to greater post-glacial uplift toward the load margins could be eliminated. Increasing the ice thickness would make agreement between the model and observed history of uplift in recent times poorer (Figure IV-47).

Points (2) and (3) may be somewhat controversial, but I do not believe they can be brought seriously into question. Walcott (1973) cites Trowbridge (1921), Frye (1963), Frye and Leonard (1952), and McGinnis (1968) as providing sparse evidence for a zone of submergence surrounding the Canadian glacial system. McGinnis searched for a glacial flexural bulge with a maximum ~66 km from the ice edge and amplitude 80–185 m. From Section IV.B we know that more recent studies indicate McGinnis' estimate of circumferential flexural uplift was too large and probably too close to the ice margin. Nevertheless,

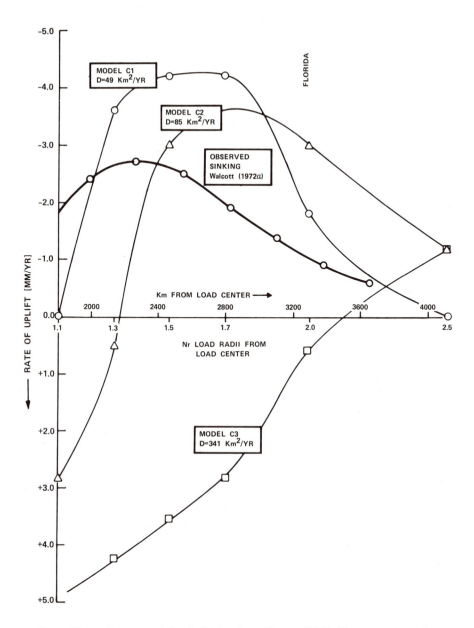

Figure IV-48. Present peripheral sinking from Figures IV-44–46 compared to the present sinking observed along the North American east coast (Figure IV-42). Note that we assume a present 1.5 mm/year rise in eustatic sea level as Walcott did. It can be seen that the present behavior of peripheral areas agrees well with that observed for Model C1, less well for Model C2, and very poorly for the most fluid channel model, Model C3. It can be seen from Figure IV-46 that this is because the peripheral areas of Model C3 have ceased sinking and begun uplifting. Comparison of this figure and Figure IV-47 indicates that it is difficult to find a channel flow model that will simultaneously accommodate the observed uplift in central Canada and the present observed peripheral bulge behavior if the load cycle is taken into account.

McGinnis found no forebulge could be distinguished in the geological data. He commented that, since increased discharge could reduce a river's slope, only reversed gradients could provide clear evidence of a forebulge. This he took to undermine Frye's and Leonard's (1952) and Frye's (1963) earlier arguments that low Pleistocene stream gradients were evidence of a forebulge. Trowbridge (1921), according to McGinnis, believed the Mississippi River bedrock valley had subsided about 180 feet since Wisconsin time. The temporal coincidence with uplift in Canada suggested to McGinnis a causal relation. However, 55 m is nowhere near the 160 m of subsidence suggested by Model C1 as a minimum (thin ice load). It is more in accord with the collapse of the post-glacial uplift appropriate for a deep flow model, as we shall see. As quoted previously in Section IV.D.1, Flint has recently characterized the evidence of a glacial forebulge of the sort Daly proposed and calculated in Model C1 as being "wholly negative" (Flint, 1971, p. 346). It seems a forebulge the size indicated by Model C1 in Figure IV-44 is not indicated by geological observations.

The history of central uplift predicted by Model C1 is compared to that observed in Figure IV-47. Central Canada (Model C1) continues to sink for a time after the load is removed. Channel material squeezed out from near the load edge is refilled from both the central and peripheral regions for a time. After about 3000 years, central uplift begins. The match between observed and model uplift is quite good since about 5000 BP. Prior to this, the match is bad. There seems to be no acceptable way to achieve agreement in early post-glacial times. Increasing the ice load will cause disagreement in near recent times. Although the match with peripheral subsidence (Figure IV-46) is quite good, the history of central uplift conflicts with observation.

Models C2 and C3, shown in Figures IV-45, 46 and analyzed in Figures IV-47, 48 show that as the channel is made more fluid (i.e., as the diffusion constant is increased above that suggested by Walcott and others) the match between model and observed central uplift histories is improved, but agreement with currently observed peripheral subsidence quickly deteriorates. The model that best fits the observed central uplift (C3) shows peripheral subsidence only south of Florida (see Figure IV-48). Further, although the glacial forebulge decreases to perhaps acceptable amplitudes as channel fluidity is increased, the amount of post-glacial peripheral sinking remains large (~ 160 m). Thus it appears that channel models are not promising candidates for explaining the isostatic adjustment that followed the North American load redistribution.

The most diagnostic observation of the Canadian uplift is the history of peripheral bulge behavior. We treat this matter immediately in the next section. Notice that for all the channel models (C1–C3; Figures IV-44–46) peripheral depression occurred immediately following load removal in the region between the ice edge and Florida. In the case of the more fluid channels (C2, C3), that immediate post-glacial sinking was later followed by uplift.

3. The Diagnostic Uplift History of Areas Peripheral to the Wisconsin Glacial Load—The East Coast of North America

Figure IV-50 shows the uplift history of the approximately circular heuristic Wisconsin-sized glacial load shown in Figure IV-49. The full self-gravitating viscoelastic formalism has been used to compute this simple case for earth Model #1 (i.e., realistic elastic parameters, inviscid core, 10^{22} p mantle, see Table IV-1). The earth has but a single continent which has reached isostatic equilibrium under 2.5 km thick ice load. The smearing of the square-edged ice load by the filtering required to eliminate the Gibbs phenomena can be seen, as can the sinking of the ocean basins under the meltwater load. A broad trough, resulting from restriction of flow to the channel between the core-mantle boundary and lithosphere can be seen. The most important thing to note for our purposes here, however, is the peripheral *uplift* that followed deglaciation. From Figures IV-50 and 51, it can be seen that uplift attained an amplitude of about 20 m, 1000 years after load removal, and extended well past the tip of Florida.

Post-glacial *uplift* of areas peripheral to the load is a direct consequence of deep flow. It deep flow is permitted, the mantle material moving in to eliminate the isostatic anomaly caused by the load removal is not particular, at first, about the precise edges of the anomaly. Peripheral areas are therefore uplifted. As time goes on, the fluid has time to notice the higher frequency components of the load removal (the edges of the load), and the peripheral areas sink back toward isostatic equilibrium. This behavior has been pointed out by Haskell (1935, 1936, 1937), Burgers and Collette (1958), and has been verified experimentally by Burgers and Collette and Cathles (1965). The phenomenon is often a source of confusion.

The significance of the deep flow peripheral behavior is that it is exactly opposite that of channel flow. This is shown schematically in Figure IV-52. In the channel flow case, peripheral areas first subside and then uplift. In the deep flow case, peripheral areas first uplift and then subside. The history of uplift of areas peripheral to a load

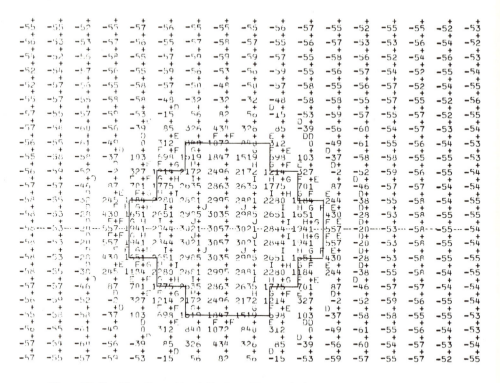

Figure IV-49. The effect of load filtering. The total ultimate isostatic uplift expected for a 1 km square-edged water load with initial borders as outlined but filtered with a .5% Gaussian filter. Ultimate uplift = −.306 (water load). Values of uplift are printed in tenths of meters and apply to the crosses above the listings. Letters may be connected to produce 50 m contours. The dotted line is the Earth's equator. In the unshown portions of the Globe "Oceanic" uplift ranged from −5.5 to −5.9 meters. As can be seen the initial square edged load is smeared out about (1½) (5°) in longitude and (3) (2.5°) in latitude or about 825 km in all directions.

redistribution thus provides a diagnostic test of mantle viscosity structure.

The geological data show a late glacial and early post-glacial uplift of the east coast of the United States, followed by a late glacial subsidence. This was pointed out by Kaye and Barghoorn (1964) as shown in Figure IV-53 and was later reiterated and extended to other locations by Mörner (1969, pp. 442–443). Figure IV-53 shows that when the elevations of radiocarbon dated peat (which indicates past sea level) are corrected for known eustatic changes in sea level, a relative sea level curve results that would put the Boston area under water about 13,000 radiocarbon years BP. The data indicate the Boston area was about

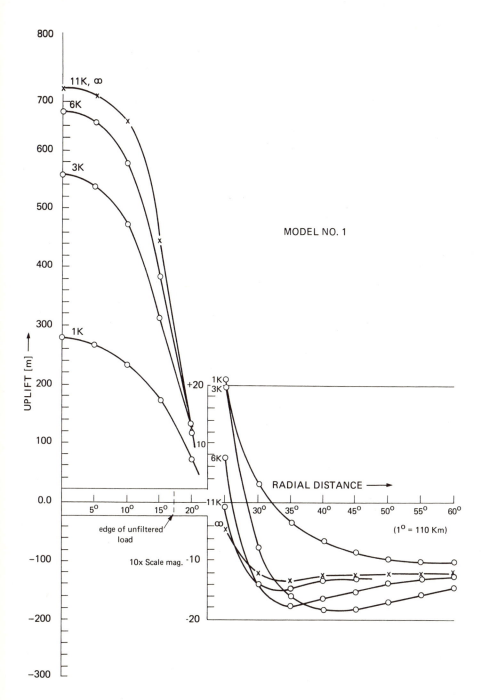

Figure IV-50. Uplift cross-sections at various times after the sudden removal of a heuristic Canadian-sized glacial load that has attained isostatic equilibrium on a uniform 10^{22} poise mantle (Model #1). A trough associated with the restriction of flow to the mantle by buoyant forces at the core-mantle boundary is apparent outside the positive peripheral uplift. The initial viscous uplift is not as rapid as might be thought from the figure since there is about 140 m of immediate elastic uplift following load removal. This initial elastic uplift is recovered as uplift proceeds and acts to slow the rate of uplift observed.

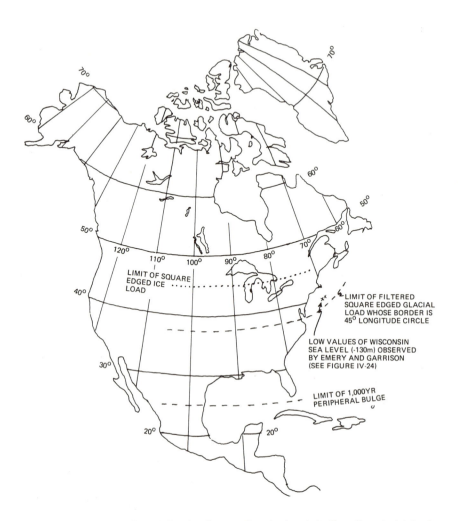

Figure IV-51. The effect of filtering in smearing the heuristic Canadian glacial load and the extent of positive uplift 1000 years after the sudden removal of the heuristic Canadian ice load of Figure IV-50 are shown on a map of North America.

280 feet lower than present between 14,000 radiocarbon years before present and about ten feet higher than present 5.5 radiocarbon years BP. Strandline data are supported by geological evidence. Kaye and Barghoorn (1964) report that the Lexington glaciation was able to override a soft clay strata (Clay III) without appreciably disturbing it because the area was under water at the time. This undisturbed clay had puzzled geologists and, for a time, had delayed recognition of the

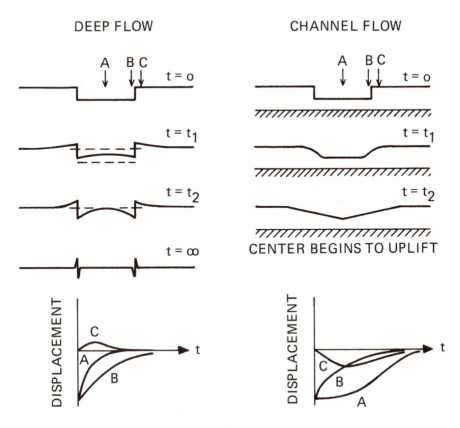

Figure IV-52. Schematic portrayal of uplift behavior of deep flow (uniform viscosity) and channel flow models. Note that the behavior of peripheral areas (curves marked C) are diametrically opposite to one another.

Lexington glaciation in the Boston basin. As the land uplifted, the full weight of the ice compacted the clay as observed in places by civil engineers through soil mechanics tests.

Uplift followed by subsidence is shown by the emergence data from Maine, Prince Edward Island, and the Bay of Fundy directly (see Figure 9 of Walcott, 1972a). If the emergence data for Massachusetts, Connecticut, or New Jersey are corrected for eustatic sea level changes, late and early post-glacial uplift with late post-glacial subsidence is again indicated (see Figure IV-55-a-f). This behavior is documented beyond reasonable doubt.

The natural next question is: Can a uniform Newtonian mantle of $\sim 10^{22}$ poise generate peripheral uplift histories in good agreement with

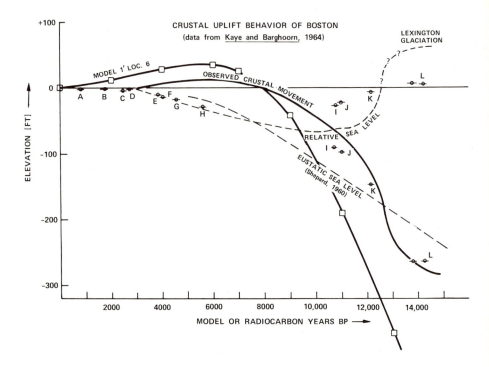

Figure IV-53.   Past elevations of Boston inferred by Kaye and Barghoorn (1964) from radiocarbon dated peat samples. Data give a clear indication of a late glacial and early post-glacial uplift (∼280 feet) followed by a more recent subsidence of about 10 feet. Figure IV-55 a–g shows this behavior is not unique to Boston. Relative sea level was higher at Boston during the Lexington glaciation than at present. This explains in Kaye's and Barghoorn's view how the thin Lexington glacier was able to override Clay III without disturbing it on its way into the Boston basin (i.e., the land was under water and buoyant forces relieved ice load). Late glacial peripheral uplift followed by subsidence is a phenomenon unique to Newtonian deep flow models. It can be seen that the calculated uplift for Model #1' (10^{22} p mantle—see Table IV-1), location #6 (see Figure IV-54) agrees well with that observed. As discussed in the text, since radiocarbon corrections are involved in both the model ice removal and geological uplift data, model and radiocarbon time are taken to correspond as a first approximation. The consequences of this are analyzed in Section IV.E.6. Present geological time is taken to correspond to −2000 BP model time (i.e., 2000 years future). The figure, reproduced with the permission of the first author (the computed uplift curve is superimposed), is from C. A. Kaye and E. S. Barghoorn, *Bull. Geological Society of America, 75,* p. 76, © 1964 by the Geological Society of America.

those observed? What limits can be placed on non-adiabatic density gradients or viscosity variation with depth? Figures IV-53, 54, and 55 a–g answer these questions.

Figure IV-54 shows the model continental outline of the United States' east coast, the points at which model uplift was calculated, and the location of geological uplift data collected by Scholl, Craighead and Stuiver (1969), Redfield (1967), Bloom and Stuiver (1963), Stuiver and Daddario (1963), and others, and compiled by Walcott (1972a). Uplift following deglaciation was computed for the models shown in Table IV-1. Account was taken of the gradual removal of the load from

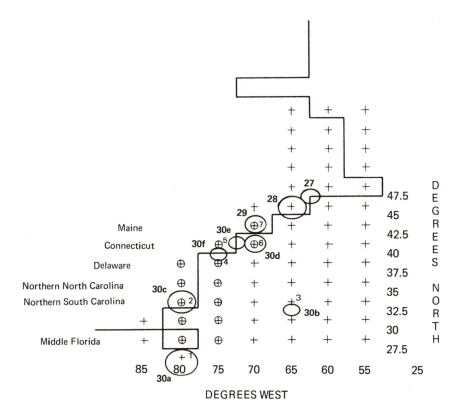

Figure IV-54. The model outline of the east coast of North America is shown by the heavy lines. Areas for which there is emergence data are circled and labeled with the codes Walcott (1972a) gave them. For example, 30a is the tip of Florida, 30b Bermuda, etc. Small crosses indicate the points at which model uplifts were calculated. Those points used in Figure IV-56 a–e are circled. The points used in Figure IV-55 a–g are indicated by small numbers above the crosses.

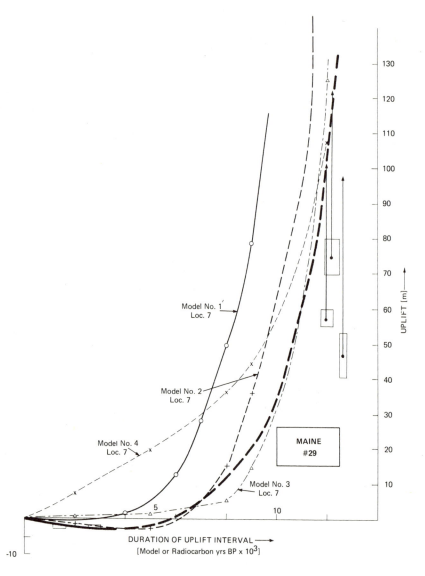

Figure IV-55a. The uplift of the Maine coast over time intervals of increasing duration. The start of the time interval is present geological time and $T_{\text{Model}} = -2\,\text{KBP}$. The geological uplift is determined by correcting emergence curves presented by Walcott (1972a, Fig. 9, 10), for changes in eustatic sea level using the particular values shown in Figure IV-15 and a smooth interpolation between. This correction is shown by the vertical arrows. The geological uplift curve is then drawn with a heavy line through the arrow tips. The next heaviest lines (solid and dashed) are used for Models #1′ and #2, and lighter lines for Models #3, #4. Location numbers refer to Figure IV-54.

The geological uplift shows a late glacial uplift followed by a post-glacial subsidence. This feature is mimicked by Models #1′, #2, and #3 but not by Model #4. Remembering the smearing of glacial load caused by filtering, the low viscosity lower mantle models provide adequate agreement. Model #4, the high viscosity lower mantle model, conflicts with the geological data in near recent times by an unacceptable amount.

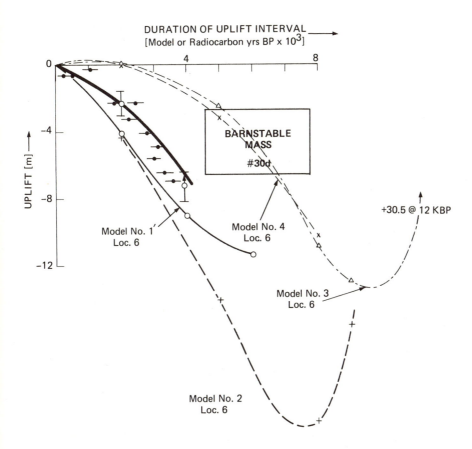

Figure IV-55b. The observed uplift of the Massachusetts coast near Barnstable is compared to that predicted by various earth models over time intervals of increasing duration from present. Conventions are the same as in Figure IV-55a. Agreement between model and geological uplift is satisfactory for Models #1', #2 but not for #4. Some low viscosity channel in the upper mantle may be indicated by need to shift the curve for Model #1' upward toward Model #3 a slight amount. Model #3 would be more acceptable if T_{Model} = 0BP were taken to correspond to present geological time. This would shift the Model #3 curve 2,000 years to the left into fair agreement with geological observations. The same is true of Model #4.

Canada and of the hydro-isostatic adjustment of the ocean basins under the meltwater load. This was discussed in Section IV.A.

Figure IV-53 shows that the Boston uplift is well matched by the uplift calculated for a model earth with a uniform mantle viscosity of 10^{22} p (Model #1) at a point corresponding to the location of Boston (Location #6 Figure IV-54), provided present geological time is taken

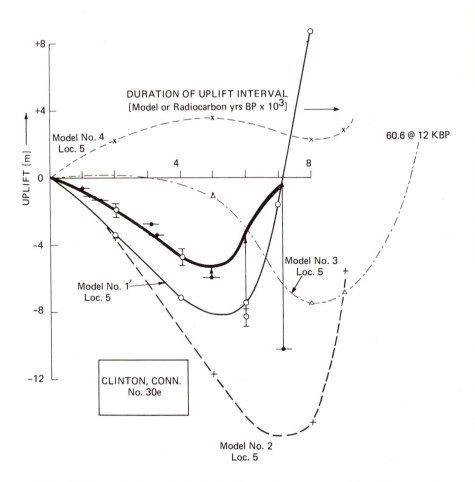

Figure IV-55c. The observed uplift at the Connecticut coast near Clinton is compared to that predicted by various earth models over time intervals of increasing duration from present. Conventions are the same as in Figure IV-55a. Geological uplift was followed by subsidence. This feature is mimicked by the low viscosity lower mantle Models (#1', #2) but not by Model #4 (10^{24} poise lower mantle). The best agreement is for Model #1'. The agreement is bad for Model #2 and unacceptably bad for Model #4. Model #3 would be acceptable if $T_{\text{Model}} = 0\text{BP}$ were taken to correspond to present geological time. This would shift the Model #3 curve 2000 years to the left.

to be 2000 years future model time. Figure IV-55 a–g shows this is also true for other studied locations down the east coast of the United States. These figures compare the observed data to the predictions of other models (see Table IV-1). The relative agreement between the observed uplift and the different models is discussed in the figure cap-

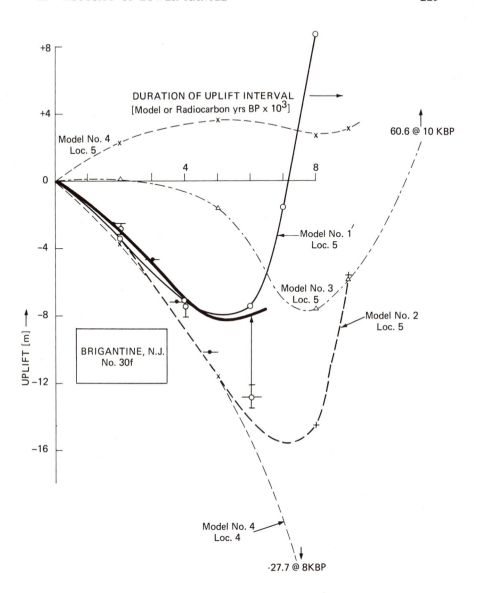

Figure IV-55d. The observed uplift of the New Jersey coast near Brigantine is compared to that predicted by various earth models over time intervals of increasing duration from present. Conventions are the same as in Figure IV-55a. The geological uplift is again noticeably U-shaped. Agreement is best for Model #1′. Agreement is bad for Model #2. Model #4 shows a wide variation between location #4 and #5, and thus the possibility of a fit cannot be ruled out. Agreement for Model #3 could be improved by taking $T_{\text{Model }\#3} = 0\text{BP}$ to correspond to present geological time and shifting the curve 2000 years to the left.

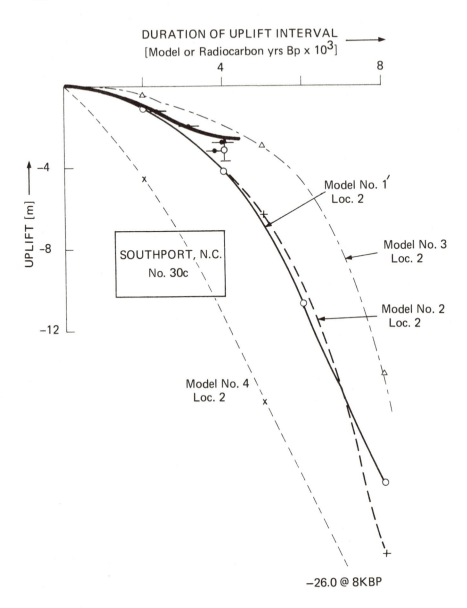

Figure IV-55e. The observed uplift of the North Carolina coast near Southport compared to the uplift predicted by various earth models over time intervals of increasing duration. Conventions are the same as in Figure IV-55a. Agreement is good for Models #1', #2 and #3. Model #4 contracts the uplift geologically observed. Model #3 is best but would not be so good if $T_{Model\ \#3}$ = 0BP were taken to correspond to geological present as has been suggested by other locations.

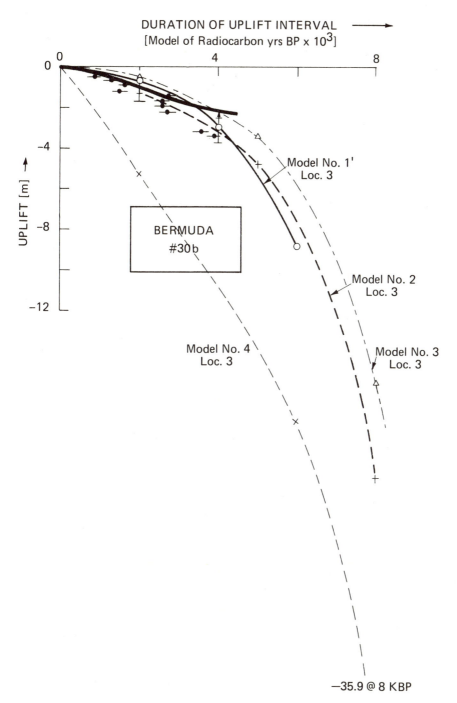

Figure IV-55f. The observed uplift of Bermuda compared to that predicted by various earth models over time intervals of increasing duration. Conventions are the same as in Figure IV-55a. Agreement is good for the low viscosity lower mantle models (#1', #2, #3), but the uplift predicted by Model #4 (high viscosity lower mantle) contradicts the uplift geologically observed. Model #3 is again best but would not be so if $T_{\text{Model } \#3}$ = 0BP were taken to correspond to geological present.

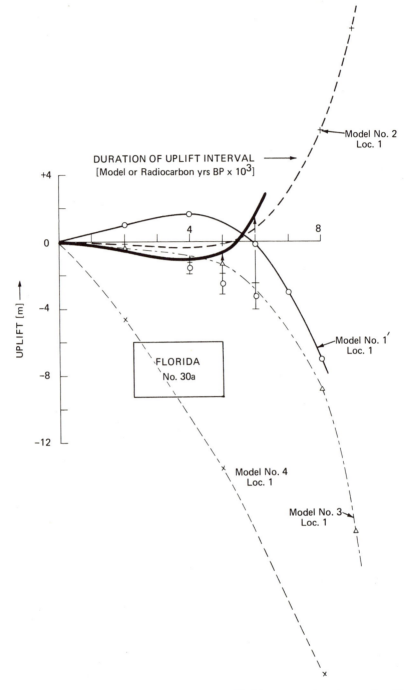

Figure IV-55g. The observed uplift of southern Florida compared to that predicted by the various earth models over intervals of increasing duration from present. Conventions are the same as in Figure IV-55a. Agreement for Model #2 is best. Model #1′ and #3 are probably acceptable. Model #4 contradicts the uplift geologically observed by an unacceptable amount. Model #3 would contradict the uplift observations if $T_{\text{Model }\#3}$ = 0BP were taken to correspond to present.

tions and summarized in Table IV-12. Agreement is the best for Model #1. It is clearly less satisfactory for Models #2 and #3. Model #4 is unacceptable. The data thus indicate a reasonably uniform Newtonian mantle of 10^{22} poise with minimal non-adiabatic density gradients and at most a thin uppermost mantle low viscosity channel. The rheology is Newtonian since, as discussed in Chapter I, a non-Newtonian mantle rheology would produce peripheral uplift phenomena similar to those of channel flow.

A general picture of the computed behavior of the Canadian peripheral bulge for various earth models can be gained from Figure IV-56, a–e. These figures show the uplift of the circled crosses of Figure IV-54 as a function of model time since the ice began dissipating 17,000 model years BP. The uplifts of inshore and offshore locations at a given latitude are connected by a vertical line. The resulting joined pairs form envelopes which are labeled according to their approximate latitude (i.e., Portland, Maine). The change in elevation over the last 3,500 years is also determined and compared to that geologically observed (as interpreted by Mörner (1969, pp. 442–443). The smearing of the ice load due to limited resolution ($n \leq 36$) and filtering are evident from the fact that positive ultimate uplift indicated by all models for locations as far south as Delaware and northern North Carolina, where there was never ice. This uplift is not physically meaningful and is not predicted by the modeling. It is an unavoidable mathematical artifact deriving from the necessity of filtering the load description. The smearing of the load undoubtedly overemphasizes the uplift expected in Maine and Connecticut. Although seemingly big, the load distortion probably does not significantly affect the peripheral uplift behavior. The peripheral uplift is described by long wavelength harmonics and these can carry the shorter (smeared by filtering) harmonics piggyback without difficulty.

Figure IV-56, a–e, shows that, if $T_{model} = -2$ KBP corresponds to present, the maximum peripheral sinking since deglaciation is about 30 m for Model #1. This is an offshore and therefore a maximum estimate. The amount of sinking over the last 6,000 years has an inshore maximum near Delaware of about 15 m, in good agreement with Walcott's (1972a, Figure 16) emergence maps. Figure IV-56c clearly shows the dramatic effect a 335 km 10^{21} poise uppermost mantle low viscosity channel would have on the uplift behavior of the peripheral regions. Such a thick channel can be ruled out on this basis. Figure IV-56d shows that even a modest change in the lower mantle viscosity structure will have clearly noticeable effects on the character of uplift of the peripheral areas.

Location	Model #1' Best	Model #2 Best	Model #3 Best	Model #1' Contradicted	Model #2 Contradicted	Model #3 Contradicted	Model #4 Contradicted
Maine #7			$\checkmark$				$\checkmark$
Mass #6	$\checkmark$					$(\checkmark)_{-2}$	$\checkmark$
Conn #5	$\checkmark$				$\checkmark$	$\checkmark$	$\checkmark$
New Jersey #5	$\checkmark$				$\checkmark$	$(\checkmark)_{-2}$	
North Carolina #2		$(\checkmark)_{-2}$	$(\checkmark)_{-2}$				$\checkmark$
Bermuda #3		$(\checkmark)_{-2}$	$(\checkmark)_{-2}$				$\checkmark$
Florida #1		$\checkmark$		$\sim$		$(\checkmark)_0$	$\checkmark$

Table IV-12 Analysis of agreement between the peripheral bulge behavior predicted by various earth models and that geologically observed as shown in Figures IV-55 a–g. Location numbers refer to Figure 54. It can be seen that the historical behavior of the peripheral bulge predicted by Model #1' is in best agreement with that geologically observed. Model #4 (10^{24} poise lower mantle) does not predict an acceptable peripheral bulge behavior. Parentheses indicate the checked statement applies only if the model time indicated is taken as present geological time. It can be seen that Model #3's credits and debits cancel. Hence it is not a very satisfactory model.

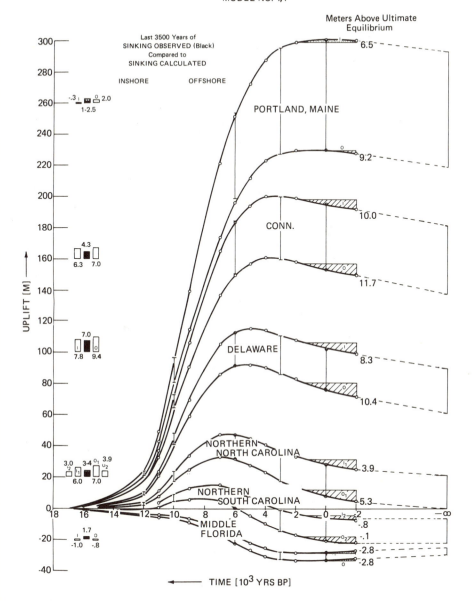

Figure IV-56a. The peripheral uplift history predicted by Model #1′. *In*shore and *Off*shore locations are connected by vertical lines to form envelopes which are labeled according to approximate geographic latitude. The precise location of points plotted can be seen in Figure IV-54, where the inshore and offshore locations are circled. The elevation of the various localities above ultimate equilibrium is shown, and the net change in elevation over the last 3,500 years compared to that deduced geologically by Mörner (bar graphs). The ultimate equilibrium elevation is shown by horizontal lines from the $T = -\infty$ datum. The uplift of the peripheral bulge agrees well with Mörner's values and peaks in Delaware as his values do. It can be seen a maximum subsidence of about 30 m is suggested between Delaware and North Carolina. About 15 m of subsidence is indicated on the last 6000 model or radiocarbon years, which agrees well with the values contoured by Walcott (1972a, Fig. 16).

MODEL NO. 2

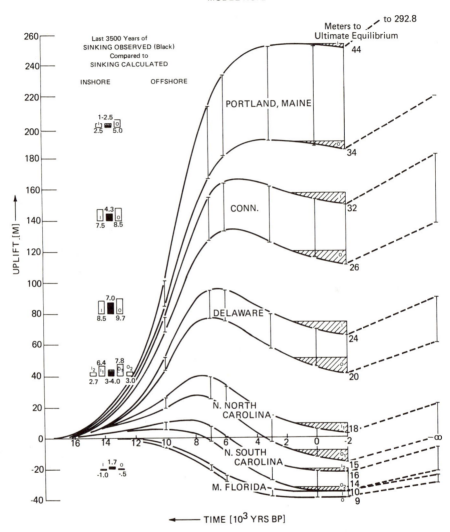

Figure IV-56b. The peripheral uplift history predicted by Model #2. The same conventions on Figure IV-56a are used. The bulge behavior is more peaked than in Model #1, 1'. The bulge depression over the last 3,500 years is actually greater in most areas than it was for Model #1, #1' (10^{22} poise without the non-adiabatic density gradient). For this reason peripheral bulge agreement is not as good as that for Model #1, #1', although still quite acceptable.

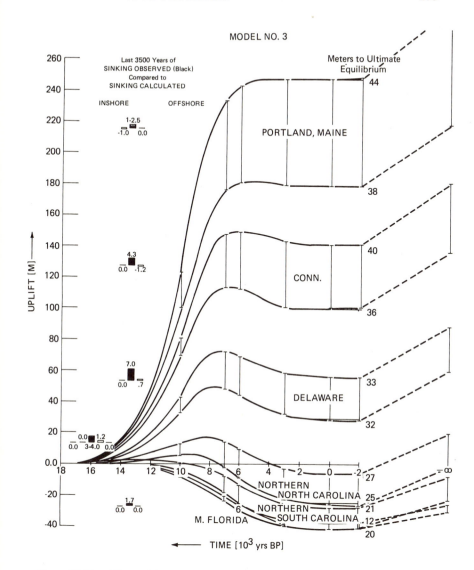

Figure IV-56c. The peripheral uplift history predicted by Model #3. Conventions are the same as in Figure IV-56a. It can be seen that the 335 km thick 10^{21} poise low viscosity channel brings adjustment of the peripheral areas virtually to a halt by present time (− 2 KBP or 0 KBP). Depression over the last 3500 years does not agree with that observed at all. The uplift is broader (less peaked at Delaware) than it was without the low viscosity channel. The peripheral area is generally depleted of material, especially under Connecticut, Delaware, and N. Carolina. This depletion acts to iron out differences in the uplift of neighboring areas on the peripheral bulge. The broadening of depressions and early cessation of adjustment mediate against the existence of a thick low viscosity channel under the east coast of the United States.

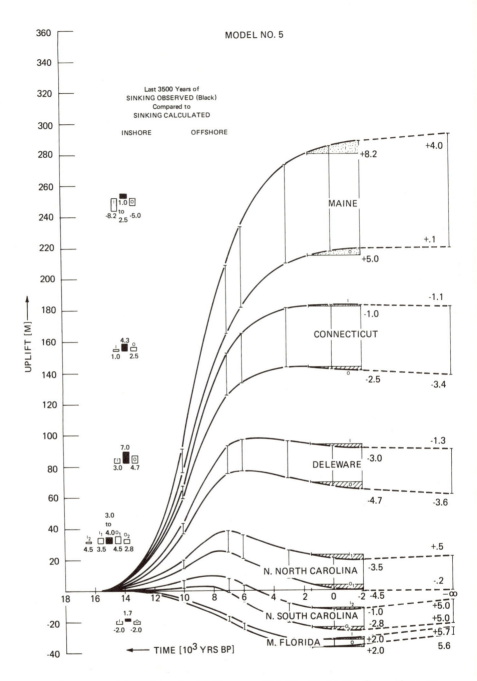

Figure IV-56d. The peripheral uplift history predicted by Model #5. Conventions are the same as in Figure IV-56a. Comparision to Figure IV-56a shows even minor modifications in lower mantle viscosity have marked effects on the peripheral bulge behavior. Agreement with the uplift geologically observed is not as good for this model as it is for Models #1 or #1'.

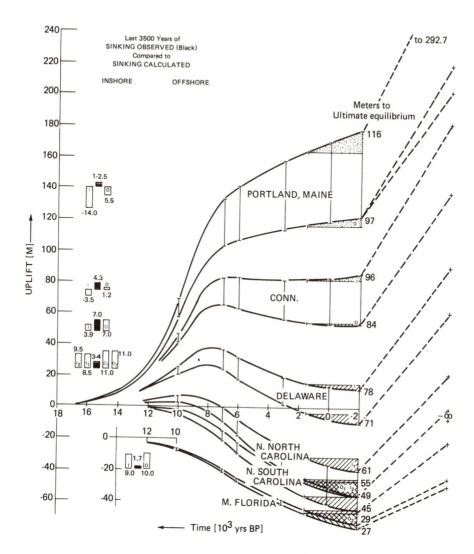

Figure IV-56e. The peripheral uplift history predicted by Model #4. Conventions are the same as in Figure IV-56a. The bulge behavior for this earth model with a high lower mantle viscosity is markedly different from the previous models whose lower mantle viscosities were near 10^{22} p. Of greatest note is the formation of a large peripheral depression and a southward increase in sinking rate to the Carolinas with Florida sinking almost as rapidly. As we saw in Section IV.E.2, a load cycle will increase the magnitude of post-glacial peripheral depression (since a positive bulge would be converted to a negative one). Even without such a factor, which would be necessary if agreement were attempted in central Canada and which is indicated by the failure of Model #4 to attain near complete equilibrium since deglaciation, the geologically observed bulge contradicts that predicted by Model #4.

Figure IV-57, a–e, compares the computed present rate of sinking of the United States east coast to that observed by tide gauge measurements (see Figure IV-42). The present observed peripheral rate of subsidence is almost perfectly matched by that predicted by Models #1 or #2 if eustatic sea level is rising at about 2.4 mm/year, and $T_{model} = -2$ KBP is taken to correspond to present geological time. Using Donn et al.'s (1962) estimate for the present volume of the world's minor glaciers of 4.5×10^5 km^3, a 2.4 mm/year rise in eustatic sea level would correspond to a 0.2% per year wasting of the world's minor glaciers or a 10% reduction in their size over the last 50 years. Such a melting rate is not unreasonable in light of the known, rapid retreat of small glaciers over the last few decades. A 2.4 mm/year

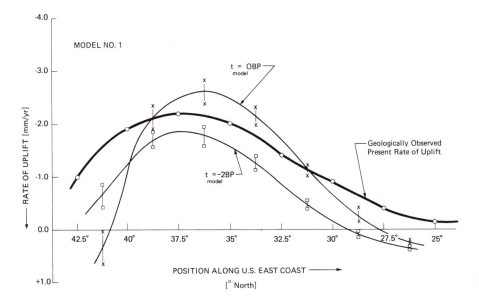

Figure IV-57a. The rate of sinking inferred from present tidal trends along the east coast of the United States is compared to the rate of uplift predicted by Model #1' at $T_{Model} = 0$ and -2 KBP. The present rise of eustatic sea level is assumed to be 2 mm/ year. The observed rate of uplift curve is from Dohler and Ku (1970) and Hicks and Shofnos (1965) via Walcott (1970) and is shown in Figure IV-42. The calculated curves are plotted for locations shown in Figure IV-54. Inshore and offshore locations are plotted and connected by a line. It can be seen that agreement between the predicted rate of uplift (particularly at $T_{Model} = -2$ KBP) and that observed is strikingly good. The agreement both in magnitude and shape is nearly perfect if the geological curve is shifted down .4 mm/year suggesting the present rate of rise of eustatic sea level is about 2.4 mm/year. Conventions for the figures that follow are the same as for this figure.

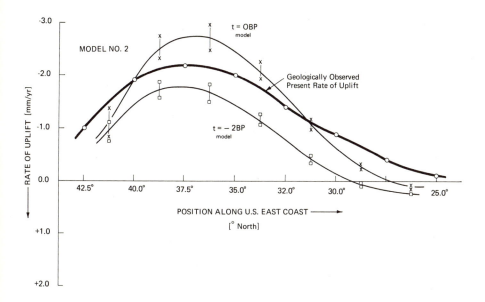

Figure IV-57b. Peripheral rates of sinking predicted by Model #2 ($\nabla \rho_{NA} \neq 0$) agree well with those observed. A present rise of eustatic sea level of 2.5 mm/year is suggested.

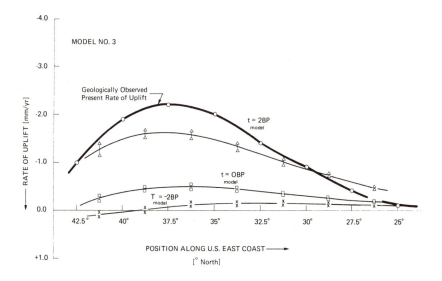

Figure IV-57c. Model #3 ($\nabla \rho_{NA} \neq 0$, low viscosity channel) does not agree with the rate of sinking observed. A low viscosity channel as thick as 335 km with a viscosity as low as 10^{21} p (diffusion constant 128 km²/year) cannot exist above a 10^{22} p mantle.

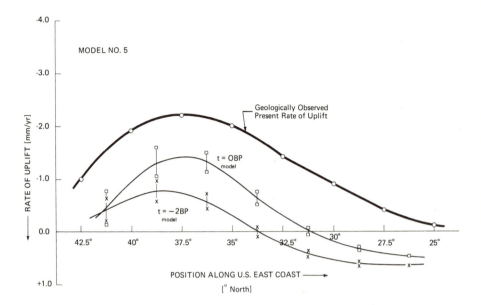

Figure IV-57d.   The rate of sinking predicted by Model #5 (η = 1,2,3, × 10^{22} p) is too small at present (-2 KBP) to agree with that geologically observed.   Agreement is not as good as for Model #1. The lower mantle must be close to 10^{22} poise.

present rise in eustatic sea level has the added advantage of bringing the tide gauge measurement at Churchill, Manitoba (3.8 ± 1.2 mm/ year; Barnett, 1970) into good agreement with the rate of uplift inferred from emergence curves near that locality (7.5 mm/year).   This advantage was mentioned in Section IV.E.1 and Appendix VII.B.

Figure IV-57, a–e, indicates that the observed present rate of uplift rules out an upper mantle low viscosity channel as thick as 335 km and as fluid as 10^{21} poise (equivalent diffusion constant 128 km²/year). Model #5 is clearly inferior to Model #1 in peripheral history and uplift rate; even quite minor increases in lower mantle viscosity are not indicated.   Figure IV-57e shows the geological data will decidedly not tolerate a high viscosity lower mantle (10^{24} poise beneath 1000 km depth).

We can conclude that diagnostic geological data from the east coast of the United States indicate a Newtonian mantle viscosity quite close to 10^{22} poise throughout.   A low viscosity channel in the uppermost mantle cannot be too thick or too fluid (i.e., must be less than D = 128 km²/year).   Non-adiabatic density gradients between 335 km and 650 km with an integrated magnitude of .5 gm/cc might be tolerated,

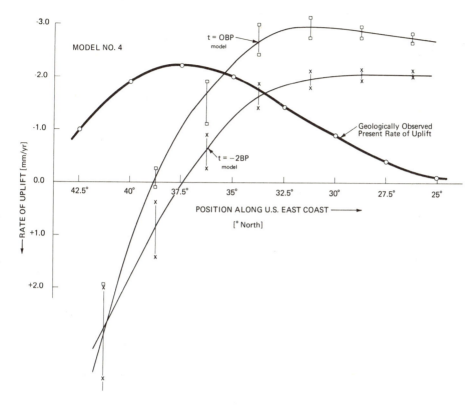

Figure IV-57e.   The present rate of sinking of the peripheral bulge predicted by Model #4 (10^{24} poise lower mantle) clearly contradicts the rate of uplift observed.

although analysis suggests that the gradients have an integrated value less than this if they exist at all.   The next section places stricter limits on the viscosity of the lower mantle.   Section IV.E.6 interprets the meaning of $T_{model} = -2$ KBP corresponding to the geological present.

4.   The History of Uplift and Present Rate of Uplift in Canada; Stricter Limits on an ~ 10^{22} p Lower Mantle

a.   The History of Land Uplift in Canada

Figure IV-58 compares the observed uplift in central Canada to the uplift calculated for a point near the mouth of James Bay (56.25°N, 80°W).   As discussed in Section IV.E.1, we compare the calculated uplift to the uplift in an area about 800 km in diameter in the central

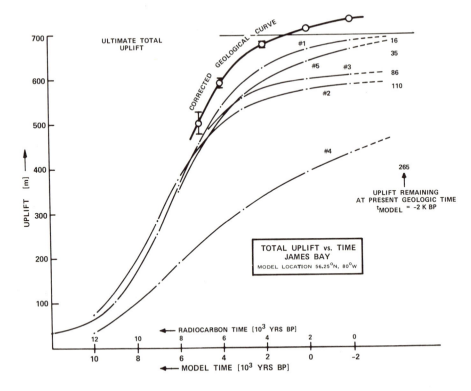

Figure IV-58. Computed and observed uplift histories for the mouth of James Bay (model location 56.25°N, 80°W). The geological curve is corrected as described in the text for the deformation of the geoid that attended the surface deformation. Notice that the geological curve is noticeably curved and thus will fit only earth models with lower mantle viscosities around 10^{22} poise. The observed uplift curve matches the uplift curve of Model #1' virtually exactly. (*Note*, the geologic curve is plotted on a different but equivalent uplift scale and so can be shifted up or down parallel to the *y*-axis to achieve a match with the computed curves.)

region rather than the absolute maxima since model filtering precludes comparison of narrow maxima. The geological uplift in the central region was previously concluded to have been 230 ± 25 m, 140 ± 10 m, 58 ± 5 m, and 23 ± 2.5 m over the last 7, 6, 4, and 2 thousand radiocarbon years before present, respectively. Figure IV-58 shows that the uplift predicted by Model #1 matches that observed almost exactly, providing $T_{model} = -2$ KBP is taken to correspond to the geologic present. Even a slight shift along the time axis (± 500 years) would seriously detract from the fit.

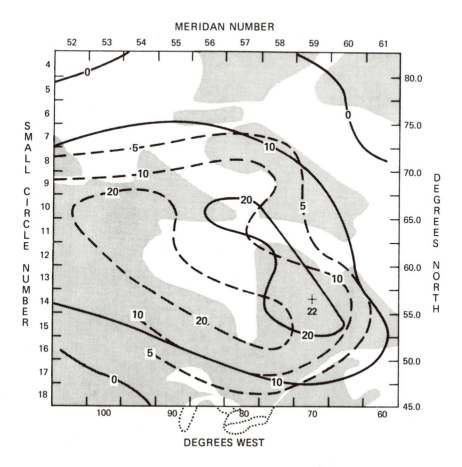

Figure IV-59a. The uplift (solid curves) calculated to have occurred between -2 KBP and 0 KBP for Model #1' compared to the uplift geologically observed over the last 2000 radiocarbon years taken from Figure App. VII-9. The geological uplift should be augmented about 16% to account for deformation of the geoid. The central 20 m contour then becomes 23 m which is comparable to the calculated 20 m contour, only shifted to the west.

Model #4 in Figure IV-58 is again distinctive. Nowhere has it enough curvature to effect a match with the uplift geologically observed. This observation is fundamental. The "uplift curvature" a given model exhibits may be associated loosely with a "response time." This "response time" is an intrinsic property of the model and is not altered substantially by changing boundary conditions (load cycles or the like). There is no effective way to introduce curvature into Model

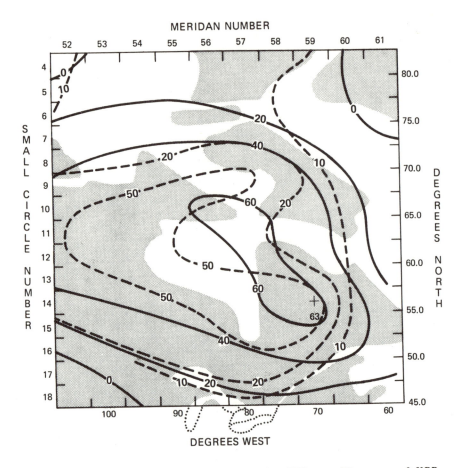

Figure IV-59b. Uplift calculated (solid) over the last 4000 years ($T_{\text{Model \#1'}} = 2$ KBP to -2 KBP) compared to the geological uplift (dotted) over the last 4000 radiocarbon years. 16% augmentation of the 50 m geological contour brings it to 58 m, a value comparable to the calculated 60 m uplift. Geological data taken from Figure App. VII-10.

#4 other than changing the model (viscosity profile) itself. Model #4 conflicts with the geological data.

The computed curves for models that do not contain an upper mantle non-adiabatic density gradient (#1, #5) show that there is remarkably little uplift remaining in central Canada at "present." The larger amount of remaining uplift indicated by models with a non-adiabatic density gradient (#2, #3) is misleading, since the effects of the load *cycle* were not taken into account. Were this done, these models would

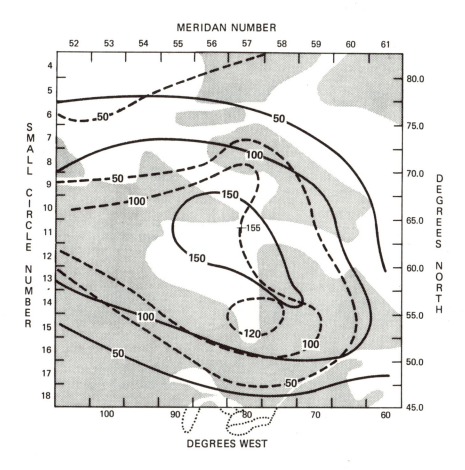

Figure IV-59c. Uplift calculated (solid) over the last 6000 years ($T_{\text{Model }\#1'} = 4$ KBP to -2 KBP) compared to the geological uplift observed over the last 6000 radiocarbon years (dotted, from Figure App. VII-11). Augmentation of the geologic uplift by 16% (to account for geoid deformation) would bring the 120 m geological contour to 140 m, a value not far from the computed 150 m.

indicate a remaining uplift of about the same magnitude as Models #1 and #5. Thus the modeling suggests that little uplift remains in Canada.

The history of uplift match is not limited to just one location. Figure IV-59, a–e, compares the uplift calculated for Model #1' (uniform 10^{22} poise mantle with no non-adiabatic density gradients) and the geologically observed uplifts over various time intervals (taken from Figures App. VII-9–13. The geological uplift should be augmented about 16% to correct for the deformation of the earth's geoid. Par-

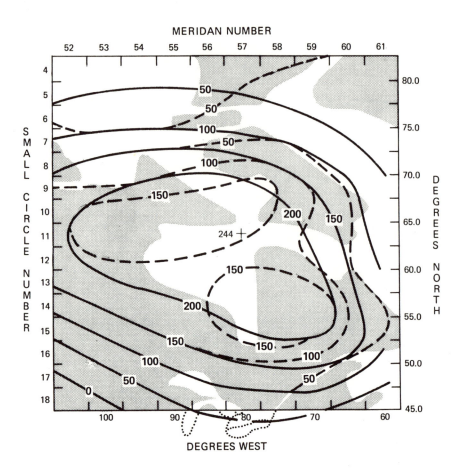

Figure IV-59d. Calculated uplift over the last 7000 years ($T_{\text{Model \#1'}} = 5$ KBP to -2 KBP) is compared to the geological uplift observed over the last 7000 radiocarbon years (see Figure App. VII-12). Sixteen percent augmentation of the 150 m geological contour would bring it up to 175 m, a value not too far from the comparable 200 m calculated contour.

ticularly if this is done, agreement between observed and calculated uplift in Canada is quite good. The general match of observed and calculated uplift in the central regions of the load over the last 7000 years is extremely good: Over the last 2000 years a maximum of about 23 m is inferred geologically (inner uplift contours) and 20 m calculated. Over the last 4000 years comparable corrected central uplift for observed and calculated uplift are 58 and 60 m; over the last 6000 years one might pick 140 and 150 m, over the last 7000 years 175 and 200 m,

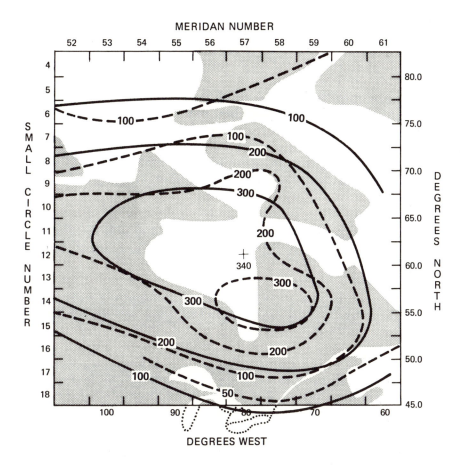

Figure IV-59e. Calculated uplift over the last 8000 years ($T_{\text{Model } \#1'}$ = 6 KBP to -2 KBP) is compared to the uplift inferred to have occurred over the equivalent radio-carbon period by extrapolation of the geologic data (see Figure App. VII-13). (1.16) × (300) = 350 which is quite close to the computed maximum of 340.

and over the last 8000 years 350 and 340 m. (The geologic data for 8000 radiocarbon years BP are extrapolated.) The general configura-tion of computed uplift over the various intervals suggests the model load chosen can be criticized for having assumed too much ice in the Arctic archipelago, for having the late dissipating ice load in the eastern highlands rather than western lowland areas, and for being a bit too smeared out as a result of filtering and limited resolution ($n \leq 36$). These are minor points in the nature of details. The general quality of the match is excellent.

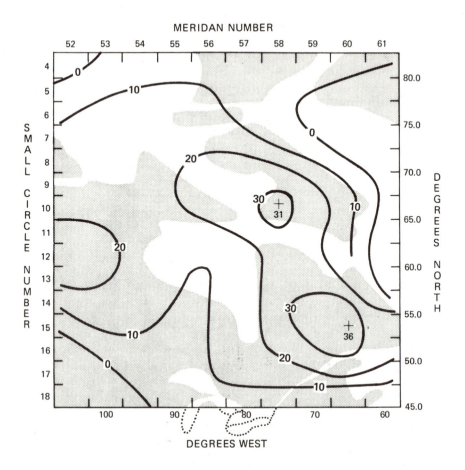

Figure IV-60. The amount of uplift remaining for Model #1' after $T_{\text{Model } \#1'} = -2$ KBP. Because of the evidence implicit in Figures IV-59 a,b and 64 a–c, the maximum remaining uplift isopores should probably be shifted about $7\frac{1}{2}$ degrees west. Modeling suggests that the ice load dissipated last from basin and not upland areas.

Figure IV-60 shows the amount of uplift remaining at "present" for Model #1' (after $T_{\text{model}} = -2$ KBP) is at maximum only about 35 meters. Most uplift remains under areas last deglaciated (see Figure IV-12). Maximum calculated remaining uplift contours correspond closely to the maximum calculated present rate of uplift contours (Figure IV-64, a–c) and to contours of maximum calculated uplift of the past few thousand years (Figure IV-59a, b). This, of course, should be expected. Thus the magnitude of the remaining uplift in Figure IV-60 is probably about right, but the remaining uplift isopores (maxima) should be shifted about $7\frac{1}{2}$ degrees to the west to be centered over lowland areas and provide better agreement with the geologically observed uplift over the last few thousand years and the present geologically observed rate of uplift.

The small amount of uplift remaining conflicts with the large amount of remaining uplift suggested by the large negative gravity anomaly centered over Canada. As previously noted (Section IV.E.2), it might be difficult to find any model that preserves substantial remaining uplift, that takes into account the ice load cycles, and that fits all the available geologic data. We shall return to the question of gravity in Section IV.E.7.

In reviewing the Canadian uplift literature (Section IV.E.1), we commented that Andrews' use of a single exponential decay function with a decay time of 2560 years to characterize the post-glacial uplift in all of Canada enjoyed remarkable success considering its simplicity. To

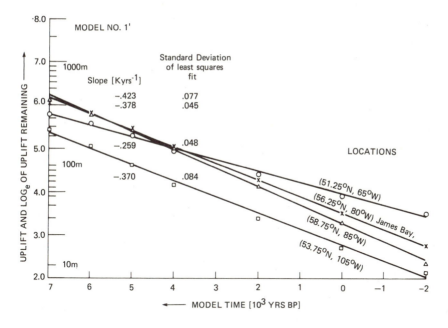

Figure IV-61. The logarithm of the uplift remaining over the interval 7000 to −2000 years BP (model time) for Model #1' in Canada fit to a straight line. The worst fit (in central Canada) was obtained for location 53.75°N, 105°W, yet even this fit is good. The last 9000 years of glacial uplift can thus be well-characterized as a simple exponential decay with a spatially varying decay constant, provided, perhaps, that the viscosity of the mantle is reasonably uniform. The standard deviations of 75% of the locations fit in Figure IV-62 was less than 0.050.

assess the theoretical support Andrews' method might have, a least squares fit was made to the logarithm of the calculated uplift remaining over the last 9000 model years of uplift for Model #1'. It was found that the post-glacial uplift could be very well characterized by an exponential decay, but the decay constant varied somewhat as a function of geographical location. Peripheral areas, where sinking followed initial uplift, could not be treated by this method and were excluded. Figure IV-61 shows the least square fit for a selected set of locations. The location 53.75°N, 105°W was the worst fit in all of central Canada. Even this fit is quite acceptable.

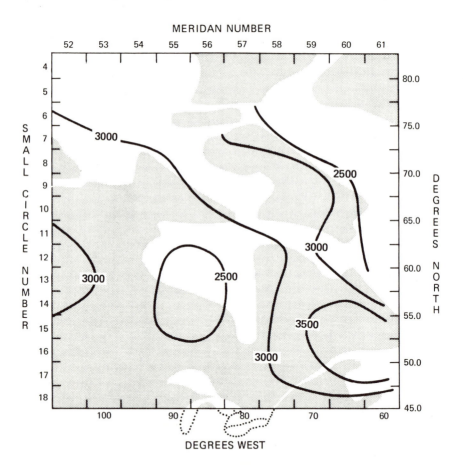

Figure IV-62. The spatial variation of decay constants characterizing the uplift of Model #1' in Canada between 7000 and −2000 years BP (model time). The least squares fit was made to seven uplift values at each location ($T = 7, 6, 5, 4, 2, 0, −2$ thousand years BP).

The decay times of individual locations are contoured in Figure IV-62. The drop in decay time near the northeast edge of uplift is due to the influence of peripheral bulge behavior. The spatial variation (other than the northeast edge) of decay time in Figure IV-62 is attributable to several factors. If the ice were removed instantaneously and the mantle infinitely thick, there would be a radial increase in decay time away from the center of uplift due to an increase in high frequency harmonic components. The core-mantle boundary complicates this slightly. In addition, the gradual retreat of the ice superimposes smaller load removals, the adjustment to which is slower because of their smaller dimensions. Consequently the decay contours are not simple. The *general* agreement with Andrews' value of 2560 years is quite striking. Andrews' method and observations therefore enjoy theoretical support if the earth's mantle has a viscosity similar to that of Model #1' (10^{22} poise). Andrews' method in turn supports the conclusion that only about 20 or 30 m of uplift remain in Canada; in fact it leads directly to this conclusion, as we noted in the introduction to this section.

In conclusion: The history of uplift in Canada supports a $\sim 10^{22}$ poise lower mantle and places quite stringent restrictions on the precise value. Present model time must be taken to be -2 KBP $\pm$ 500 years. Full isostatic equilibrium was obtained under the ice load (10,000–20,000 years duration) and has been nearly reattained since deglaciation. Very little uplift remains in Canada today. This observation is supported directly by the form of the observed uplift curves and the 2500–3000-year time constant that describes them.

b. The Present Rate of Land Uplift in Canada

Figure IV-63 compares the observed rate of central uplift to the rate of uplift computed at the mouth of James Bay, for model times near "present," for the various earth models of Table IV-1. If it is remembered that the model calculations can only be compared to the uplift in a central region (approximately 800 km in diameter) because of filtering, it can be seen the $\sim 10^{22}$ p lower mantle viscosity models all indicate a central uplift rate quite near that observed at $T_{\text{Model}} = 0$BP and a bit low at $T_{\text{Model}} = -2$ KBP. Model #4 is distinctive in predicting high central rates of uplift for all model times near "present" and in having an almost linear decrease in uplift velocity with time. The low viscosity channel of Model #3 has little effect on the uplift of central Canada (i.e., it is quite similar to Model #2).

The configuration of model rate of uplift at model times 0BP and -2 KBP (2000 years future) for Model #1 and at -2 KBP for Model

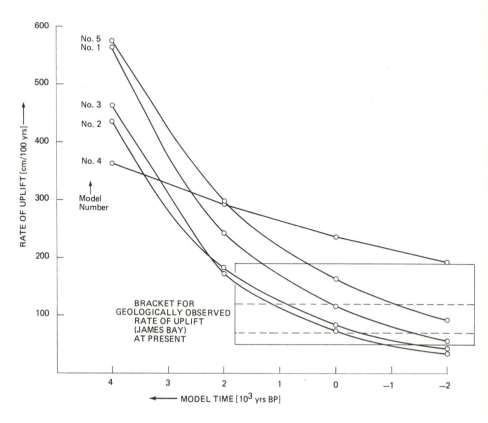

Figure IV-63. The rate of uplift at the mouth of James Bay, computed as a function of model time, for various earth viscosity models (Table IV-I), compared to the rate of uplift geologically observed. The range of geologically observed rates of uplift from Andrews .5m/100 years to Walcott's 1.9m/100 years is shown (see Appendix VII). Because of model filtering, the geological rate of uplift for the central area (~800 km in diameter) is also shown, bracketed between .7 m/100 years to 1.2/100 years (shown dotted). Note that low viscosity channel (Model #3) has little influence on the uplift of central locations (i.e., is similar to Model #2) but that minor modification of deep mantle viscosity has a substantial effect (difference between #5 and #1). Model #4 (10^{24} poise lower mantle) is distinguished from the other models that have lower mantle viscosities near 10^{22} poise by having a nearly constantly decelerating uplift.

#1' (10^{22} poise lower mantle) are shown in Figure IV-64a, b, c. Shown in dotted contours for comparison is the geologically observed present rate of uplift from Figure App. VII-9. The configuration of computed rate of uplift matches the geologically observed configuration quite well, although it is a bit too broad (smeared out). This is in part due to resolution limitation and filtering in the computed models. It may also suggest (as did the eustatic sea level curves) that the ice was a bit

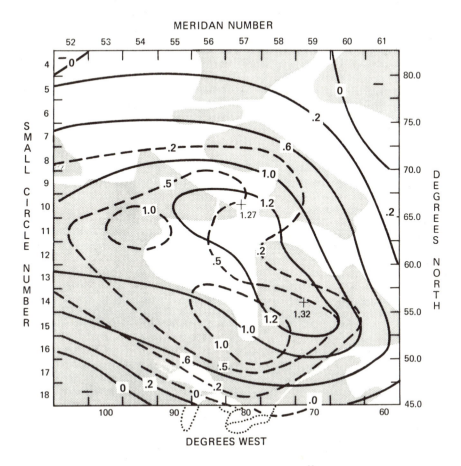

Figure IV-64a. The rate of uplift of Model #1 (uniform 10^{22} p) at present model time compared to the geologically observed rate of uplift in Canada (dashed). Note particularly the differences in Quebec province, Keewatin, and in the Arctic archipelago. Notice also calculated contours are a bit too smeared out.

more gradually removed or was a bit thicker in the last stages of deglaciation than was assumed in Models #1 and #1'. This would tend to make the present uplift sharper by increasing the present maximum rate of uplift under the last deglaciated areas. Comparison of Models #1' and #1 (Figures IV-64a, b) shows that even a slightly more gradual final removal has a substantial effect on the rate of uplift under the last ice dissipated.[18] The maximum rate of uplift for Model #1' (.94 m/ 100 years) is about 22% greater than that of Model #1 at the same time

[18]The only difference between Model #1 and #1' (see Table IV-1) is that the last 2000 years of ice removal has been shifted 500 years toward the present.

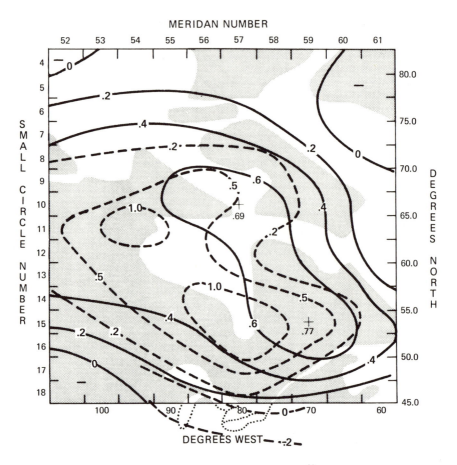

Figure IV-64b.   The rate of uplift of Model #1 (uniform 10^{22} p) at 2000 years future model time compared to the geologically observed rate of uplift in Canada (dashed).

(.77 m/100 years). The outer configuration of uplift is almost identical, but the uplift under the last ice remnants is greater and so the uplift is more peaked on the whole. The maximum rate of uplift of Model #1', .94 m/100 years, is quite close to rate of uplift geologically observed in central Canada. Thus the optimum match for the present rate of uplift in Canada may be for Model #1' at 2000 years future model time.

Apart from the general sharpness of the uplift contours, the model rate of uplift in the Arctic archipelago is clearly too large, and there is a definite relative shift between the present maximum rate of uplift geologically observed and that predicted by the models. The geological

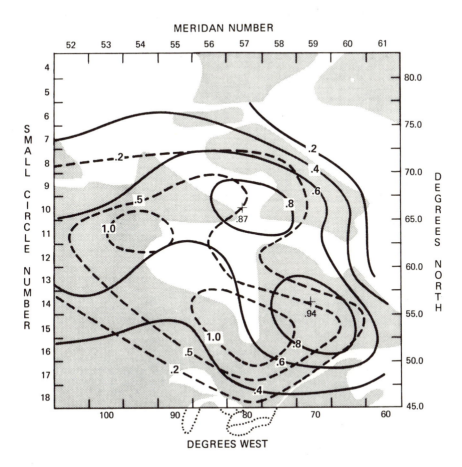

Figure IV-64c. The rate of uplift for Model #1′ at 2000 years future model time. Model #1′ is exactly the same as Model #1 (see Table I) except that the load removal at $T = 7000$ BP and 5000 BP were shifted to 6500 BP and 4500 BP, respectively. Comparison with Figure 64b shows even such a slight shift in the last stages of load removal has a significant ($\sim 22\%$) effect on the rate of uplift under the last dissipated ice.

rate of uplift maxima appears to be more centrally located than those predicted and not strictly under what was apparently the last ice dissipated. Part of the difference may be due to errors in the geological uplift contours. Particularly in Quebec province, the data control is very poor (see Figures App. VII. 1, 6, 7). However, it does appear that the uplift in James Bay is more rapid than the models would predict, and more rapid relative to areas for which the models predict greater rates of uplift. The same could be said for the geologically indicated rate of uplift isopore in Keewatin province.

The most likely explanation for these discrepancies is that we have modeled the ice removal slightly wrong. We probably should have tapered the ice thickness in the archipelago as we did on the southern (continental) edge of the glacial system (see Figure IV-11). The modeling also suggests that the effective load removal proceeded in a more uniform fashion, with late ice of substantial weight being centered over Fox Basin, Keewatin, and the James Bay area. We assumed in the modeling that the last ice load was closely associated with the last

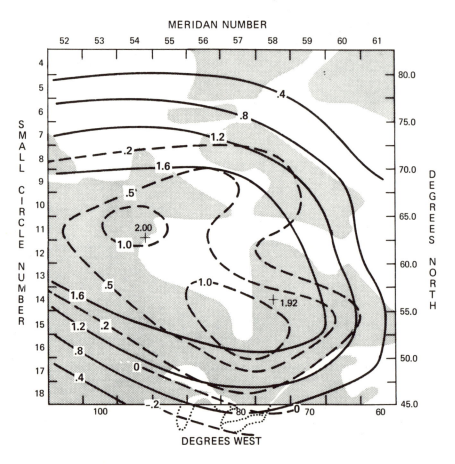

Figure IV-65. The rate of uplift at T_{model} = −2 KBP for Model #4 compared to the rate of uplift observed geologically (dotted contours). By comparison to Figure IV-64 a–c, it can be seen that the high viscosity lower mantle model has a broader central region of high rate of uplift, and steeper gradients on the edges of this region. The rate of uplift configuration for this model is not as close to that observed as are the rates of uplift configuration of the ~ 10^{22} p lower mantle models.

dissipated ice and thus centered in Quebec province, Baffin Island and to a lesser degree in Keewatin (see Figures IV-9, 12). It is not logically necessary to assume this. The last ice in mountain areas might have been quite a bit lighter and thinner than the last ice in basins. For whatever reasons, there is a general indication that the major ice load was removed last from lowland rather than mountain or upland areas. The same indication was given by the recent history of uplift (Figures IV-59, a, b).

Figure IV-65 (and Figure IV-66a, b) shows that the configuration of central rate of uplift predicted by Model #4 is not as similar to that observed as the configurations of the $\sim 10^{22}$ p lower mantle models are. Large gradients in the rate of uplift occur at the edges of a broad rate of uplift central plateau. Figure IV-14 shows that this piston-like uplift is also reflected in the total uplift to $T_{Model} = 0BP$.

In conclusion, we see the observed present rate of uplift in Canada is well modeled by Model #1 or #1'. Model #2 (not shown) would have modeled the rate of uplift about equally as well, but #3 not so well (see Figure IV-66a, b). The rate of uplift is sensitive to the last details of load removal. Although the magnitude of the central rate of present uplift is perhaps not, in itself, a particularly useful guide to the details of the lower mantle viscosity structure, a low ($\sim 10^{22}$ p) viscosity lower mantle is certainly permitted, and a lower mantle viscosity as high as 10^{24} p is discouraged by the piston-like character of its predicted uplift.

5. The Worldwide Pattern of Uplift

Some feeling for the overall pattern of uplift has already been gained through Figures IV-1 and IV-14. Figure IV-1, an oblique perspective view of the worldwide uplift computed for Model #1', shows that the adjustment of ocean basins (evident from the increasing "relief") is rapid for a 10^{22} poise lower mantle model. The figure also shows that, as one might expect, mantle material is at first pushed under the continents and only later finds its way to the glacially unloaded areas in the northern hemisphere. This is apparent from the zero uplift contours of Australia and Antarctica, which grow at first but shrink after 8000 BP. The uplift in Canada follows the retreat of the ice margin. It is quite cycle-shaped 11,000 years BP. This lends some support to Broecker's (1966b) modeling of the uplift attending a plate-like load pulled steadily backward. Later, at 8,000 years BP, the uplift is noticeably bidomal, even though the ultimate uplift is domal. This is a result of the shape and style of melting of the ice. Uplift in the Himalayas, Siberia, and Fennoscandia is prominent. The uplift in Europe at first

MODEL NO. 1
UNIFORM 10^{22}p
MANTLE

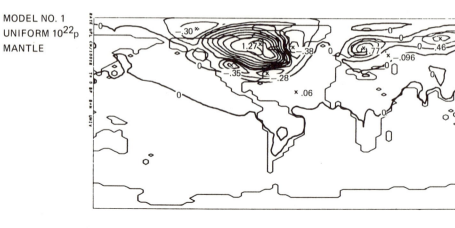

MODEL NO.2
VDG

Figure IV-66a. Contours of rate of uplift calculated for various earth models at T_{Model} = 0BP displayed on a "mercator" projection with undistorted latitude. Sinking contours are every .1 m/100 years from heavy zero contour; rising contours are every .2 m/100 years from heavy zero contours.

is quite broad (11,000 BP). Only later do the individual domes of Fennoscandia and Siberia distinguish themselves.

Figure IV-14 shows the total uplift contours for the various earth models at T_{Model} = 0BP. Differences in the models are highlighted by this figure and Figure IV-66 a, b, which shows the rate of uplift of the same models at T_{Model} = 0BP and -2 KBP. In particular, these figures show that a relatively uniform depression of the ocean basins can be expected only if non-adiabatic density gradients are small and the viscosity of the mantle uniform (Figure IV-14). Otherwise, troughs

MODEL NO. 3
VDG
LVZ

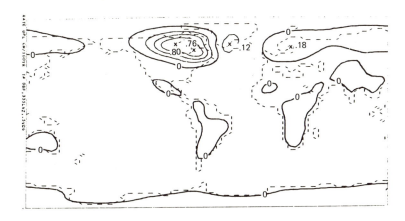

MODEL NO. 5
1,2,3 x 10^{22}p
LOWER MANTLE

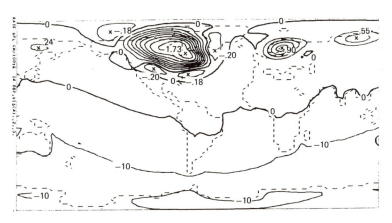

MODEL NO. 4
10^{24}p
LOWER MANTLE

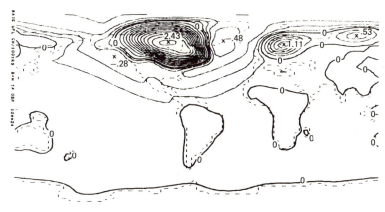

Figure IV-66a. *cont'd.*

MODEL NO. 1
UNIFORM $10^{22}p$
MANTLE

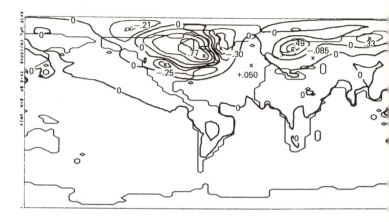

MODEL NO. 2
NON-ADIABATIC
DENSITY
GRADIENT

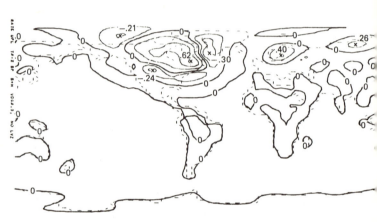

Figure IV-66b. Contours of rate of uplift calculated for various earth models at $T_{Model} = -2$ KBP (2000 years future model time) displayed on a "mercator" projection with undistorted latitude. Sinking contours are every .1 m/100 years from heavy zero contour; rising contours are every .2 m/100 years from heavy zero contour.

peripheral to the glaciated areas (particularly Canada) develop. Increasing lower mantle viscosity even a little manifests itself in dramatic consequences. Notice, for example, that virtually the entire southern hemisphere should be still sinking under the meltwater load if the mantle viscosity is just slightly larger than 10^{22} poise (Model #5, Figure IV-66 a, b). The lack of adjustment for Model #5 is clear from Figure IV-14. Channel troughs have developed in Model #2. Once the non-adiabatic gradient between 355 and 635 km depth is uplifted, buoyant forces induce channel flow. Thus a peripheral trough is de-

MODEL NO. 3
VDG
LVZ

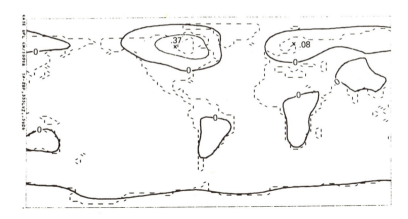

MODEL NO. 5
1, 2, 3 x 10^{22}p
LOWER MANTLE

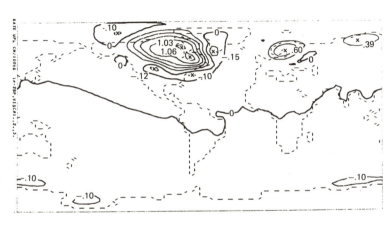

MODEL NO. 4
10^{24}p
LOWER MANTLE

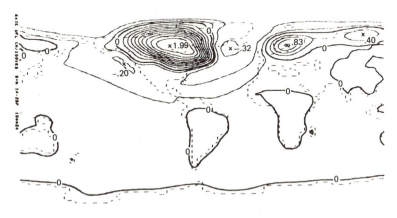

Figure IV-66b *cont'd.*

veloped (Figure IV-14), and the marginal rate of sinking is augmented (Figure IV-66 a, b).

The dramatic effect of a thick upper mantle low viscosity channel in eliminating the peripheral response is evident from Figure IV-66 a, b. The trough associated with the non-adiabatic density gradient is not affected, however, as indeed it should not be (Figure IV-14).

A comparison between Figure IV-66a and Figure IV-66b indicates that the rate of uplift of the low viscosity lower mantle models is decelerating rapidly. This should be expected since Figure IV-14 indicates most of the uplift has been accomplished for these models. The maximum total model uplift is 698 m in Canada and 415 m in Fennoscandia. For the models with non-adiabatic density gradients between 335 and 635 km depth (integrated values .5 gm/cc) the maximum ultimate uplift would be 3.313/3.813 times the above, or 606 m and 360 m, if no flow occurred above the non-adiabatic density gradients.

Unfortunately the data on recent global adjustment are not in general well enough known (to the writer at least) to be of much use in evaluating lower mantle viscosity directly. Some of the phenomena associated with some of the models seem large enough to measure. One interesting phenomenon that may be of significance can be mentioned.

Return for a moment to the simple heuristic example of unloading a single continent overlying a uniform 10^{22} p mantle, shown in Figure IV-50. Notice that, outside the peripherally uplifted region, a trough is formed. This trough is associated with the restriction of flow to the channel between the core-mantle boundary and the earth's surface.[19] Its maximum magnitude is about -7 m and it uplifts back toward equilibrium as the peripheral uplift decays and as isostatic equilibrium is approached. The central zone of uplift (under the initial glacial load) is thus surrounded, in the final stages of adjustment, first by a zone comprising sinking peripheral uplift and then by a zone of rising peripheral depression.

This double zoning is evident around the Pleistocene load systems and is shown by Model #1 in Figure IV-66, a, b. The rising zone is

---

[19]Figure III-16 shows that after a few thousand years the decay of harmonic loads of order two and three is significantly slowed by the deformation of the core-mantle boundary. The decay of harmonics $n = 4, 5$ overtakes $n = 1, 2, 3$. The result is the broad outer peripheral depression:

The outer trough is thus just a consequence of limiting flow (after a time) to the "channel" between the earth's surface and the core-mantle boundary.

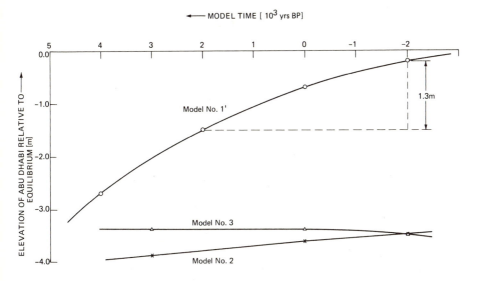

Figure IV-67. The uplift of Abu Dhabi predicted by various earth models is shown as a function of model time. The uplift is associated with the return to equilibrium of the outer trough formed after the removal of the glacial load by the restriction of flow to the channel between the surface and the core mantle boundary by buoyant forces built up at the deformed core-mantle boundary. Adjustment of the outer trough behavior is impeded by a non-adiabatic density gradient (see Figure IV-66a,b) and so the uplift behavior of Models #2 and #3 (models with non-adiabatic gradient) differs from that of #1'. The uplift of Model #1' over the last 4000 years (1.3 m) is almost exactly that inferred in a careful study by Kinsman, Patterson, and Park (personal communication) to have occurred in Abu Dhabi (1.25 m). This coincidence supports Model #1' and suggests non-adiabatic density gradients in the upper mantle may be everywhere near zero.

quite broad as it should be for order number n = 4, 5. Notice also that an upper mantle non-adiabatic density gradient (Model #2) breaks the outer trough response into chaotic pieces.

The potential significance of this phenomenon comes from a single measurement of a 1.25 m fall of sea level in Abu Dhabi over the last 4000 years. Figure IV-67 compares the sea level regression in Abu Dhabi to that predicted by various earth models. Of course, Abu Dhabi could be uplifting for any number of reasons completely unassociated with global Pleistocene adjustments. It is certainly not possible to take the coincidence of theory and a single measurement as indicated by Figure IV-67 as evidence in support of a 10^{22} p mantle viscosity or against non-adiabatic density gradient. We mention it here primarily to draw attention to the phenomenon, which is of academic interest, to emphasize the fact isostatic adjustment to the Wisconsin load redistribution was and is a global phenomenon, and to illustrate the use of the global data.

6. The Meaning of Future Model Time Corresponding to Present Geological Time and Radiocarbon Time to Model Time

In the modeling we have done so far, we have made two assumptions: We have identified model time directly with radiocarbon time. We have associated 2000 years future model time with the geological present. The first approximation was made because the model removal of the ice load was specified in radiocarbon time. The second was required by the data in order to achieve a good match between predicted and observed isostatic adjustment.

Correction of the radiocarbon time scale would make the load removal more gradual. It would last over a period about 1,300 years longer in duration than the load removal we modeled. Correction would also shift the time of effective load removal back almost 1,300 years. Since the earth would have more time to adjust to the load redistribution, a higher model mantle viscosity would be permitted for any given model match. During the load removal, the calculated eustatic sea level curves for Model #1 would agree better with those observed. (We concluded in Section IV.A.5 that eustatic sea level suggested a lower mantle slightly less viscous than 10^{22} poise.)

By associating 2000 years future model time with present geologic time, we effectively allowed 2000 more years for isostatic adjustment to take place. This is approximately also what correction of the radiocarbon dates of load removal would do. Thus the fact that we were required to pick a future model time to correspond to present geologic time may simply be an indication that the time scale by which the model and geological load removal was measured needed a slight correction (i.e., an extension of the period of ice removal by about 1300–2000 years and a shift backward in time by this amount). Alternatively, or in addition, a present geological time that corresponds to 2000 model years in the future may indicate that the viscosity of the lower mantle is slightly less than 10^{22} poise. Since 1300 years may not be a sufficient time shift, we shall adopt this latter viewpoint to some extent.[20]

---

[20]In Cathles (1971) I was not aware of the necessity of radiocarbon time scale correction and so interpreted the entire time shift in these terms. This gave a lower mantle viscosity of $\sim.8 \times 10^{22}$ poise.

The results of previous sections may now be summarized as suggesting an average upper mantle viscosity of $1.0 \pm .1 \times 10^{22}$ poise and an average lower mantle viscosity of $.9 \pm .2 \times 10^{22}$ poise.

$$\eta_{\substack{\text{upper} \\ \text{mantle}}} = 1.0 \pm 0.1 \times 10^{22} \text{ poise} \qquad (\text{IV-29})$$

$$\eta_{\substack{\text{lower} \\ \text{mantle}}} = 0.9 \pm 0.2 \times 10^{22} \text{ poise} \qquad (\text{IV-30})$$

7. Data from the Earth's Gravity Field

If the earth's gravity field, with the hydrostatic rotational bulge subtracted, is expanded in spherical harmonics it may be expressed by a series of coefficients. These coefficients are commonly defined:

$$\Phi(r, \theta, \phi) = \frac{GM}{r} \left\{ 1 + \sum_{\ell=2}^{\infty} \sum_{m=0}^{\ell} \left(\frac{b}{r}\right)^{\ell} P_{\ell m}(\cos\theta) \right.$$
$$\left. \cdot [C_{\ell m} \cos m\phi + S_{\ell m} \sin m\phi] \right\}. \quad (\text{IV-31})$$

$P_{\ell m}$ are normalized such that

$$\int_{\theta=-90}^{+90} P_{\ell m}(\cos\theta) P_{\ell' m}(\cos\theta) \sin\theta \, d\theta = 2(2 - \delta_{0m}) \delta_{\ell \ell'}, \quad (\text{IV-32})$$

so that the integrated square of any surface harmonic over the unit sphere is 4π. M is the mass of the earth, r the distance from the earth's center, and G the gravitational constant. The coefficients $C_{\ell m}, S_{\ell m}$ may be determined from satellite data and conventional gravimetry (see Kaula, 1966; Goldreich and Toomre, 1969) and may be listed as shown in Table IV-13.

The expansion given in (IV-31) may be given some physical meaning if it is compared to the potential that would be produced by the harmonic redistribution of material on the earth's surface, assuming there is no elastic or viscous response to that redistribution:

$$\Phi(r > b, \theta, \phi) = \frac{GM}{r} \left\{ 1 + \frac{2\pi G\rho_0}{g_0} \left(\frac{b}{r}\right)^{\ell+1} P_{\ell m}(\cos\theta) \frac{2}{2\ell + 1} \right.$$
$$\left. \cdot [h_{\ell m}^c \cos m\phi + h_{\ell m}^s \sin m\phi] \right\}. \quad (\text{IV-33})$$

ℓ, m	Gravity Coefficients C_m	S_m	Power$^{1/2}$ σ_ℓ	Ice Load [m of water] h_m^c	h_m^s	Ice Load Gravity Anom. C_m^*	S_m^*	Power$^{1/2}$ σ_ℓ
1 0				126		3.58		3.71
1 1				17	-30	.48	$-.85$	
2 0	4.0			105		1.79		
2 1			4.8	13	-58	.22	$-.99$	2.10
2 2	2.42	-1.39		-26	2.7	$-.44$	.05	
3 0	0.97			54		.66		
3 1	1.90	0.11		3.9	-71	.05	$-.87$	
3 2	0.69	-0.78	2.75	-49	14	$-.60$	.17	1.26
3 3	0.55	1.29		-1.3	18	$-.02$	.15	
4 0	0.54			14		.13		
4 1	-0.59	-0.48		1.9	-66	.02	$-.63$	
4 2	0.28	0.69	1.53	-64	11	$-.61$	.10	0.92
4 3	0.89	0.19		1.6	21	.02	.20	
4 4	-0.32	0.00		3.7	11	.04	.10	

Table IV-13 Gravity coefficients observed (from Kaula, 1966) are compared to gravity coefficients that could be caused by the full Wisconsin glacial load redistribution (see discussion in text). Power$^{1/2}$, σ_ℓ, is calculated: $\sigma_\ell = (\Sigma_m (C_{\ell m})^2 + (S_{\ell m})^2)^{1/2}$. Caputo's hydrostatic earth is assumed (see Goldreich and Toomre, 1969). Legender polynomials are normalized as shown in (IV-32).

Here $h_{\ell m}^c$, $h_{\ell m}^s$ are the amplitudes of the harmonic redistribution of material, and the material is considered to have a density ρ_0.

If $h_{\ell m}^c$ and $h_{\ell m}^s$ were taken to be the coefficients that describe the amount of uplift remaining today from the Wisconsin load redistribution, (IV-33) could be used to estimate the gravity anomalies that should be observed.

Equation IV-33 may be used to obtain an upper bound on the size of the gravity anomalies that could be associated with the Wisconsin load redistribution, by taking $h_{\ell m}^c$ and $h_{\ell m}^s$ to be the coefficients of the load redistribution expressed in centimeters of water, and ρ_0 the density of water. These coefficients were determined in the course of our modeling and are listed in Table IV-13 together with the maximum gravity anomalies they suggest. The gravity anomalies are an upper bound on those to be expected, since we assume that there is no viscous or elastic response to the load redistribution. Both responses will tend to decrease the gravity anomalies produced by the redistribution.

It can be seen from Table IV-13 that, although the power in the gravitational perturbations caused by the load redistribution is comparable to that in the present non-hydrostatic geopotential field, there is no obvious correlation between the two. In fact the maximum possible P_{20} ice load anomaly is a factor of 2 less than the P_{20} anomaly observed. Thus Wang's (1966) suggestion that the non-hydrostatic bulge could be explained by Pleistocene load redistributions can be ruled out (see Chapter I).

O'Connell (1969, 1971) previously came to exactly these conclusions. He found the P_{20} ice bulge 2–3 times less than the non-hydrostatic P_{20} bulge observed. His load spectrum is nearly identical to ours. In addition O'Connell did a regression analysis on the ice and gravity potential and showed there was no correlation between the two. He concluded that the relaxation times for deformations of degree 2, 3, and 4 had to be less than 10 to 20 thousand years. From here O'Connell used restrictions imposed by length of day data and concluded the P_{20} relaxation time had to be 2000 years and the lower mantle had to have a viscosity near 10^{22} poise. From this it is clear caution must be exercised in inferring the amount of isostatic uplift remaining in Canada from gravity anomalies.

This last statement may be put another way. Figure IV-68 shows the global free air anomalies over 20 mgal in magnitude (Kaula, 1972). Large negative anomalies may be seen over Asia, Antarctica, parts of Africa, the Indian Ocean, in various parts of the Pacific, and over North America. In addition there are many positive anomalies. All these anomalies are too broad in dimension to be supported elastically by a lithosphere. All must find satisfactory geophysical explanations.[21] It is clear from Figure IV-68 that a great deal is going on in the world that manifests itself in non-hydrostatic gravity anomalies and yet is not related to the post-Wisconsin load redistribution or the isostatic adjustment remaining.

Because the present non-hydrostatic gravity field of the earth does not correlate with the Pleistocene load redistribution as a whole the finding that only ~30 m of uplift remains in Canada does not contradict the gravity anomalies there in a fundamental way. The position that isostatic adjustment is driven by small gravity perturbations superimposed on larger anomalies which are steadily maintained over periods much longer than 10,000 years (the period of isostatic adjustment) by unrelated processes (probably mantle correction) is permitted and

---

[21]The anomalies are probably related to mantle convection. It is difficult to find any other satisfactory answer for them. If the convection passes through the phase transition regions in the upper mantle, non-adiabatic gradients would be reduced or eliminated.

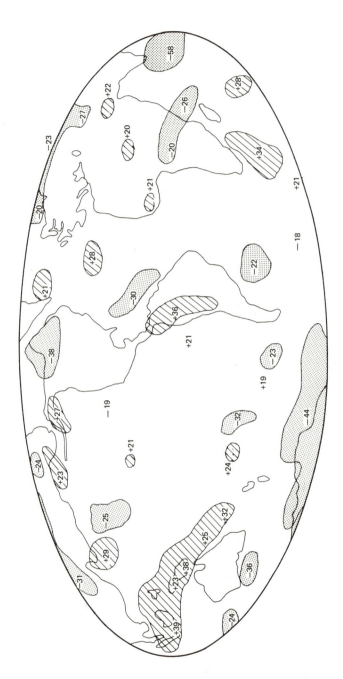

Figure IV-68. Free air gravity map of the world due to Kaula (1972, Fig. 1). Caution should be exercised in attaching special significance to the −38 m gal anomaly over Hudson Bay.

perhaps even suggested. Parsons (1972) has shown the time scale associated with convective changes should be large compared to the time
scale of glacial uplift phenomena.

For the sake of reference in other sections, a useful formula can be
derived from (IV-33). The elevation of the geoid (worldwide canal
system of constant gravitational potential) may be defined:

$$\Phi_1(\text{Geoid}, \theta) = 0 = \Phi_1(r = b, \theta)$$

$$+ \left[{}_g h^s_{\ell m} \sin m\phi + {}_g h^c_{\ell m} \cos m\phi\right] P_{\ell m} \frac{\partial \phi}{\partial r}.$$

ϕ_1 is the perturbation part of (IV-33).

To first order $\dfrac{\partial \phi}{\partial r} = g_0$, so

$${}_g h^c_{\ell m} \doteq \frac{2\pi G \rho_0 b}{g_0} \frac{2}{2\ell + 1} h^c_{\ell m}.$$

We have used the fact $g_0 = \dfrac{GM}{b^2}$. A similar relation holds for the sine
coefficients. If we drop the confusing zonal coefficients, we see that the
geoid will be raised an amount h_g under a harmonic load of order ℓ,
amplitude h, and density ρ_0:

$$h_g = \frac{2\pi G \rho_0 b}{g_0} \frac{2}{2\ell + 1} h. \tag{IV-34}$$

For the case of Canada, $\ell \cong 4.5$ and taking $\rho_0 = 3.3$,

$$h_g = .16\, h. \tag{IV-35}$$

Thus we see that the geoid will be depressed about 16% of the remaining uplift at any particular time. We used this relation in Section
IV.E.4.

F. Summary of Conclusions: Recommendations for Further Work

Figure IV-69 summarizes the viscosity and non-adiabatic density
profiles determined by this study of glacial uplift phenomena. The
profiles are for a strictly average radius through the earth. The upper

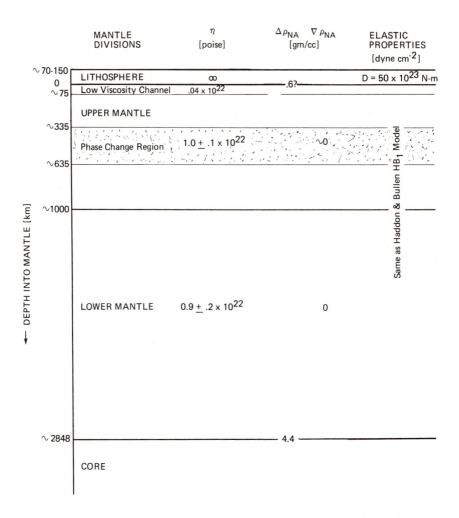

Figure IV-69.   Summary of earth parameters determined or used in this study.

mantle has a low viscosity channel about 75 km thick with a viscosity about 4×10^{20} poise. The rest of the upper mantle has a viscosity of very close to 1.0×10^{22} poise with non-adiabatic density gradients probably close to zero everywhere. The lower mantle may have a viscosity slightly less than the upper mantle. Additional conclusions concerning the rheology of the mantle (i.e., Newtonian or non-Newtonian viscosity), mantle convection, the amount of uplift remaining today, and the nature of the ice load were also reached and are summarized after Figure IV-69 is discussed in detail. Section references are given when the section mentioned is not the section of mantle discussed.

The Lithosphere

The ability of the earth to support small loads out of isostatic equilibrium for long periods of time indicates that the lithosphere may be thought of as an elastic shell with (for our purposes) infinite viscosity. A flexural rigidity of about 50×10^{23} N-m is suggested. This could correspond to a crustal thickness of 88 km with $\lambda = \mu = 3.34 \times 10^{11}$ dyne-cm^2, or to a thicker lithosphere of less rigidity. Lithosphere properties are variable geographically, but this "typical" lithosphere would enable density anomalies less than about 134 km in radius to be substantially elastically supported and never attain full isostatic equilibrium. The gravity anomalies associated with such density anomalies (i.e., in Fennoscandia) do not imply future glacial uplift.

The rigidity of the earth's lithosphere speeds up the isostatic adjustment of small scale loads but will not affect the adjustment of larger loads (III.A.2.e; IV.C.3).

The base of the crust may be a phase boundary even under continental land masses such as Antarctica or Greenland. If it is, phase boundary migration will not contribute significantly to glacial uplift phenomena unless the latent heat of the phase transition is less than 15 cal/gm. A value this low is unlikely. On this basis, we have assumed a subcrustal phase boundary will not contribute to glacial uplift.

The Low Viscosity Channel

Beneath the crust there is a low viscosity channel. Its existence is indicated by the uplift of Lake Bonneville, which requires a low viscosity channel of at least $\sim 10^{21}$ poise and 250 km thick but which can be accommodated by a channel 75 km thick and $\eta = 4 \times 10^{20}$ poise. Glacial uplift in Greenland requires a 75 km thick low viscosity channel with a viscosity of 4×10^{20} poise directly. A low viscosity channel is indicated by the style of uplift in the Arctic archipelago and late deglaciated areas in the Arctic such as Fox Basin. A low viscosity channel may be suggested by the details of old shorelines on the continental shelf of the United States and perhaps by the Fennoscandian uplift (IV.D.2). Thus, suggestions of a low viscosity channel are fairly widespread geographically.

Restrictions on the thickness of the low viscosity channel are indicated by the uplift in Fennoscandia and by the behavior of the peripheral bulge of the Canadian glaciation. A low viscosity channel as thick as 335 km and of 10^{21} poise viscosity (equivalent diffusion constant 128 km^2/year) would cause the uplift in Fennoscandia to be too rapid and so too near completion at the present time to agree with the observed uplift (IV.D.4). The present rate of adjustment of the Canadian peripheral bulge and its history indicate that a thick low viscosity channel does not exist under the east coast of the United States (IV.E.3). A 75 km thick low viscosity channel of 4×10^{20} poise (equivalent diffusion constant 3.6 km^2/year) would not affect these or other major uplifts perceptibly, however.

The average properties of the channel might be taken to be a thickness of about 75 km and a viscosity of about 4×10^{20} poise. It is quite likely that both the thickness and viscosity of the channel are markedly variable functions of latitude and longitude. These variations may be related to mantle convection, the channel being of lowest viscosity and thickest near areas of mantle upwelling (plumes).

The low viscosity channel seems related to the seismic low velocity zone.

The Upper Mantle

The viscosity of the mantle between 75 and 1000 km depth is most directly probed by the well-studied Fennoscandian uplift. That uplift is well-matched by a 10^{22} poise mantle. If such a model is adopted, the viscosity of the upper mantle is probably determined to 10% accuracy ($1.0 \pm .1 \times 10^{22}$ p), although it is something of an assumption to project this viscosity around the globe. The Fennoscandian uplift as analyzed by McConnell clearly shows the effects of a substantial lithosphere or a thin low viscosity channel (diffusion constant ~ 3.6 km^2/year) or both.

The Fennoscandian uplift could equally well be modeled by a more substantial low viscosity channel with a diffusion constant of 26.2 km^2/year (100 km thick, 1.3×10^{20} poise viscosity or the equivalent). Peripheral bulge behavior and the direction of migration of the zero uplift isobase can distinguish the deep flow and channel flow models. The geological data are not definitive at present. When the load cycle is taken into account, no more uplift remains for the best fitting channel model than for deep flow models. The channel model predicts ~ 50 m of remaining uplift, the deep flow model 30–40 m. A remaining uplift of 30–50 m agrees well with gravity data if the lithosphere is taken into account (IV.B.2.c).

Global deep flow modeling indicates that the Fennoscandian uplift is about equally well-matched whether or not there are non-adiabatic density gradients in upper mantle. The history of uplift of the Canadian peripheral bulge and perhaps the adjustment of Abu Dhabi suggest non-adiabatic density gradients in the upper mantle are near zero. This in turn suggests that there is a continual slow movement of material through the phase transition regions between 335 and 635 km depth (III.A.3.a, IV.C.4). The evidence is suggestive, not conclusive.

The Lower Mantle

The adjustment of the oceans basins, the uplift in central Canada, and the behavior of the Canadian peripheral bulges reflect the viscosity of the lower mantle.

Oceanic islands (and the edges of continental shelves) appear to register Pleistocene changes in sea level larger than coastal locations by the amount that would be predicted if the ocean basins adjusted rapidly to the meltwater load (IV.A.6; Bloom's Test of Isostacy). The amount of meltwater indicated by the "dip stick" oceanic islands agrees well with the amount that ice mechanics and end morain data predict.

If the lower mantle has a viscosity of 10^{22} poise or less, the rate of adjustment of the ocean basins is rapid enough to generate model eustatic sea level curves in close agreement with those observed. Eustatic sea level data suggest that a high viscosity lower mantle may be unlikely. A mantle of intermediate viscosity seems ruled out by the apparent lack of a broad Holocene high sea level (Wellman's Test of Isostacy, IV.A.7).

If the load cycle is taken into account, as it must be, channel flow models for Canada encounter difficulty accommodating both the known amount of uplift since deglaciation (marine limit) and the observed present sinking of the areas peripheral to glaciation. In addition, very large peripheral bulges would have been produced during glaciation and very large peripheral troughs produced upon deglaciation. Post-glacial peripheral subsidence would have been comparable to post-glacial central uplift (~ 160 m). Such subsidence is not observed. Deep flow is therefore indicated for the reasons first cited by Daly.

The most diagnostic information on mantle rheology is provided by data on the uplift history of the east coast of the United States. The data indicate a late glacial and early post-glacial uplift followed by subsidence. This behavior can *only* be generated by a Newtonian mantle of quite uniform viscosity and minimal non-adiabatic density gradients. For such a mantle of $\sim 10^{22}$ p viscosity, the behavior of the peripheral areas is remarkably well-accounted for both in the history of uplift and present rate of sinking.

The history of uplift and present rate of uplift in Canada is also well-matched by the predictions of a $\sim 10^{22}$ p Newtonian mantle. These data have high resolution once an approximate mantle viscosity is determined and are largely responsible for fixing the narrow error limits ascribed in Figure IV-69.

An $\sim 10^{22}$ poise lower mantle is required by Dicke's (1969) and O'Connell's (1969, 1971) analysis of eclipse data. The rotation of the earth requires rapid adjustment of the ocean basins (Chapter I).

Newtonian Mantle Viscosity

As mentioned in the first chapter, the assumption that the viscosity of the mantle was Newtonian was one of the fundamental assumptions of our model. Methodologically, we simply assumed this was the case.

If the viscosity of the mantle were non-linear, strain rate being proportional to some power greater than one (the only reasonable case) of stress, the isostatic adjustment to large scale light loads should be noticeably slower than to large scale heavy loads. However, we saw that the eustatic sea level curve would be better matched if the viscosity of the lower mantle were, if anything, even less than was suggested by the adjustment in Canada, being best for a completely fluid mantle. Since the Canadian load was far heavier than the ocean load, this indication is directly opposite that suggested by a non-linear mantle.

Further, if the Canadian uplift occurred more rapidly than material could be supplied from the loaded oceans, the mantle material for the uplift in Canada would necessarily have to be taken from the neighboring regions and a peripheral trough around Canada would develop.[22] Thus both the bulge behavior and the ocean adjustment for a non-Newtonian mantle should look something like that appropriate for channel flow (or a high viscosity lower mantle). We have found that this behavior is not indicated by the form of the past glacial uplift.

There is no indication that the rate of uplift in Canada slowed substantially more than predicted by the linear model as uplift proceeded and the driving stresses reduced.

The earth's mantle appears therefore to have a Newtonian viscosity.

[22]Another way to view this is: in a non-linear stress-strain rate relation, strain rate will be anomously large where stress differences are large. For a glacial load this is near the edges of the load. Hence peripheral troughs should develop upon unloading and bulges upon loading, much as in the channel flow case.

Mantle Convection

We have concluded that the mantle has a viscosity everywhere near 10^{22} poise. This permits both polar wandering and convection. The earth presently sports non-hydrostatic bulges ~ 90 m in amplitude (Chapter I). We have shown that these cannot be explained by the mass redistribution that followed the melting of the Wisconsin glacier even if no isostatic adjustment has occurred under that redistribution (IV.E.7). In fact, near-complete adjustment has occurred (only 30 m or so of uplift remain anywhere), and the earth's gravity figure shows no correlation with the Pleistocene load redistribution (O'Connell, 1969, 1971). Since no surface load redistribution even approaching the magnitude of those of the Pleistocene has occurred in the last 10,000 years, and the anomalies are too broad to be supported by the lithosphere, it is difficult to imagine any mechanism other than mantle convection to support the non-hydrostatic bulges observed.

Given that mantle convection is occurring, it must occur to transport heat from the earth's interior to space, and it can do this effectively only if the limbs of the convection cells (and therefore also areas of upwelling) are thin (IV.C.4). This physical reasoning is corroborated by calculation of a Rayleigh number for the mantle. For a viscosity of 10^{22} poise and even a very modest adiabatic temperature difference between the top and bottom of the mantle, the Rayleigh number is so supercritical that plumelike upwelling must be expected. The spreading limbs are probably associated with the subcrustal low viscosity channel.

The apparent lack of non-adiabatic conditions in the phase change region of the upper mantle suggests the convective return flow may be occurring, at least in part, uniformly through the mantle as a whole as Morgan has suggested (IV.C.4).

Uplift Remaining in Central Fennoscandia and Canada

The amount of uplift remaining follows directly from modeling. About 30 m of uplift remains in both Fennoscandia and Canada (Figures IV-14, IV-36, IV-60).

In both central Canada and Fennoscandia, the form of the postglacial uplift is convincingly exponential. The form of the uplift over the last 8000 years (and hundreds of meters) implies about 30 m of uplift remains in a direct fashion (Figures IV-38, IV-61, IV-62).

The Present Rate of Increase in Eustatic Sea Level

The present rate of adjustment of the Canadian peripheral bulge (Figure IV-57a) indicates a present rate of increase in eustatic sea level of about 2.4 mm per year. This rate translates to a 10% melting of the world's minor glaciers over the last 50 years and is supported at least qualitatively by the known recent rapid retreat of many of these glaciers. Such a rise in eustatic sea level would also bring tide gauge and emergence data for Churchill, Manitoba, into good agreement with the geologically observed recent uplift in central Canada (Appendix VII.B).

Where Ice Loads Dissipated Last in Canada

Isopores of present model rate of uplift and model uplift over the last few thousand years show close correlation to those model areas deglaciated last (Figures IV-12, IV-59 a–g, IV-60, IV-64 a–c). In the model calculations, we removed ice from upland areas last following isochron maps describing the glacial retreat. Isopores of the geologically observed present rate of uplift and uplift over the last few thousand years lie over lowland areas such as Fox Basin, Keewatin, and the James Bay area rather than over upland areas such as Quebec province and Baffin Island. This discrepancy suggests the significant ice load dissipated last from lowland rather than highland areas of Canada (IV.E.4).

Significant Analytical Details

In laying the physical and mathematical foundations needed to interpret glacial uplift phenomena, some analytical conclusions were reached that may warrant summary:

(1) The fluid and elastic response of the earth to loading may be considered decoupled without substantial loss of accuracy (at most a few percent) (III.A.2.g.2).

(2) P_1 loads are physically reasonable (III.B.4). The load deformation coefficients for P_1 elastic earth response are:

$h_1' = -.25 \pm .02$ (errors based on Figure App. IV-1),

$k_1' = 0.0.$

(Higher order load deformation coefficients are calculated in a conventional fashion and agree with values published in the literature; Appendix IV.)

(3) The decay of low order load harmonics is non-exponential. Buoyant forces build up at the core-mantle boundary and after a time, restrict flow to the mantle. The decay may be described as consisting of two connecting exponential sections (three if there are non-adiabatic density gradients in the upper mantle) (III.B.2). For a 10^{22} poise mantle, 60% decay is achieved in 1500 years for P_2 load harmonics and in 3800 years for P_1 load harmonics (Figure III-16).

Further Research

Computation of the global earth models could, of course, be carried out to higher order numbers than the $n = 36$ done here. This would be relatively easy as the higher harmonics could be computed using flat-space non-self-gravitating theory (see Figure IV-26). Transformation from a frequency to space description would not be much more time-consuming since the inverse transform could be done only in areas of particular interest. Going to higher orders would have the advantage of decreasing the model smearing of the ice load. It would probably pay off in significant added information about the earth's mantle only if combined with more detailed geologic information, however, and so may not be warranted for a few more years.

Calculated uplift in Canada could be refined by correcting for changes in elevation of the geoid. We applied an approximate correction to the geological data. An "exact" correction could be computed and applied to the computed uplift.

Geologic information defining the uplift in Quebec province would be of particular interest, especially in pinning down the mode of ice dissipation in Canada (see Appendix VII.B).

Core drilling in Antarctica and Greenland to determine the present elevation of localities formerly at sea level could delimit the extent to which the boundary between the crust and mantle may be considered a phase boundary under continents and/or place limits on the heat of transition (III.A.3, IV.B.3).

By far the most exciting and possibly productive area of future research appears to involve investigation of the upper mantle low viscosity channel and further delimitation of the degree of adiabaticity of the upper mantle. The low viscosity channel could perhaps be effectively studied on a *worldwide* basis by correlating the history of river stability profiles (infilling and erosion) with the change of elevation (particularly of river mouths) caused by the post-Pleistocene ocean loading and model-predicted channel bulge behavior near the oceanic margins.[23] Modeling could be done using non-self-gravitating half-space theory.

Confidence in the adiabatic character of the (upper) mantle would be substantially increased if localities riding on the outer (mantle) trough showed similar uplift histories to Abu Dhabi. (This trough is defined between the zero uplift contours enclosing the Persian Gulf in Figure IV-66 a, b.) In addition, it should be verified that a similar behavior cannot be produced with a non-adiabatic gradient in the upper mantle and a load cycle.

The "channel" above any non-adiabatic density gradient in the upper mantle is substantially thicker than the 75 km thick low viscosity channel, so it may be possible to identify channel phenomena associated with these gradients and to distinguish them from channel phenomena associated with the low viscosity channel.

Properly interpreted gravity studies may be of assistance in sorting out channel phenomena near presently glacially uplifting areas (III.B.3.b).

[23]This possibility was suggested to me by Dr. S. Judson.

Sketch Derivation of the Navier-Stokes Equation after Eringen

CONSERVATION of mass, momentum, and energy, and the general principles of thermodynamics apply to any material. Description of the physical properties of the material resides solely in the constitutive equations which express stress, heat flow, energy, and entropy in terms of strain, strain rate, temperature, rate of change of temperature, etc.

General restrictions can be placed on the form of these constitutive relations, however. For example, the present state of the body should be determined solely by its *past* history of temperature and motion and not by what its future temperature and motion will be. The body should look the same regardless of the coordinate system of the observer (i.e., the constitutive relations must be invariant under rigid translation and rotation of the spatial coordinate system). The body must remain invariant under certain rotations and translations of the material coordinate system depending on its material symmetry. The motion of a locality in the body should not be too greatly influenced by motions distant from it in space and time. Finally, the constitutive relations should be admissible under all known thermodynamic and physical laws (Clausius-Duhem relations, conservation of mass, etc.). These observations are equivalent to Eringen's axioms of causality and determinism, objectivity, material invariance, neighborhood and memory, and admissibility, respectively. Once it is decided upon which variables the stress, etc., are to depend, the above axioms may be systematically applied.

In the special case that the material considered is isotropic, the stress may be expressed in terms of the invariants of the independent tensor fields (Cauchy's Theorem). If we further suppose that the constitutive relations are susceptible to polynomial expansion (which is not a strong assumption), we may invoke the Caley-Hamilton Theorem (any matrix satisfies its own characteristic equation) to reduce the polynomials to second degree. The coefficients of the now quadratic constitutive relations will be polynomials in the invariants of the independent tensor fields and the independent scalers. For small perturbations in these independent variables we may make a linear approximation to the

277

constitutive relations by expanding the coefficients and throwing out terms of greater than second order.

These restrictions and techniques are very powerful. For a general discussion see Eringen (1962, 1967). To demonstrate this and to gain insight, we shall *sketch*[1] the derivation of the Navier-Stokes equation from scratch. Here, as in Eringen, majuscules refer to the material coordinate system. This system describes the location of all particles at $t = 0$. Minuscules describe the coordinates of the material particles at some later time in terms of their location in space. The two systems are equivalent to the more standard Lagrangian and Eulerian coordinate systems, respectively.

With complete generality we may write:

$$t(\mathbf{X}, t) = \mathcal{F}[x(\mathbf{X}', t'), \theta(\mathbf{X}', t'), \mathbf{X}, t]$$
$$q(\mathbf{X}, t) = \mathcal{Q}[x(\mathbf{X}', t'), \theta(\mathbf{X}', t'), \mathbf{X}, t]$$
$$\epsilon(\mathbf{X}, t) = \mathcal{E}[x(\mathbf{X}, t'), \theta(\mathbf{X}', t'), \mathbf{X}, t]$$
$$\eta(\mathbf{X}, t) = \mathcal{N}[x(\mathbf{X}', t'), \theta(\mathbf{X}', t'), \mathbf{X}, t]$$

$\mathbf{t}$ = stress tensor
$\mathbf{q}$ = heat flux
ϵ = internal energy/unit mass
η = entropy/unit mass
$\mathcal{F}, \mathcal{Q}, \mathcal{E}, \mathcal{N}$ = functionals
$t' \leq t \quad X' \in B$ (in body)
$\mathbf{x}$ = spatial coordinates
$\mathbf{X}$ = material coordinates
θ = temperature

This follows from the axiom of Determinism and is valid for all thermomechanical materials.

Objectivity requires dependence on a difference vector, independent of time of observation, and on $\tau' = t - t'$ which has a range $0 \leq \tau' \leq \infty$

$$t(\mathbf{X}, t) = \mathcal{F}[x(\mathbf{X}, \tau') - x(\mathbf{X}', \tau'), \theta(\mathbf{X}', \tau'), \mathbf{X}]$$

This comes from making spatial variables invariant under a transformation of the sort:

$$x(\mathbf{X}, \bar{t}) = \mathbf{Q}(t)x(\mathbf{X}, t) + \mathbf{b}(t)$$
$$\mathbf{Q}\mathbf{Q}^T = \mathbf{Q}^T\mathbf{Q} = \mathbf{I}$$
$$\det \mathbf{I} = \pm 1$$
$$\bar{t} = t - a$$

[1]This is a sketch only. We follow Eringen completely.

Neighborhood and Memory can now be applied. We first do *Neigh-borhood*. Provided that the displacements are smooth, we may expand $\mathbf{x}(\mathbf{X}', \tau')$ about $\mathbf{x}(\mathbf{X}, \tau')$:

$$\mathbf{x}(\mathbf{X}', \tau') = \mathbf{x}(\mathbf{X}, \tau') + \frac{X'^K - X^K}{1!}\,\mathbf{x}_{,K}(\mathbf{X}, \tau')$$

$$+ \frac{(X'^K - X^K)(X'^L - X^L)}{2!}\,\mathbf{x}_{,KL}(\mathbf{X}, \tau') + \ldots$$

We can do exactly the same thing for θ.

Thus

$$\mathbf{t}(\mathbf{X}, t) = \mathcal{F}[\mathbf{x}_{,K}, \mathbf{x}_{,KL}, \ldots, \theta(\tau'), \theta_{,K}(\tau'), \ldots, \mathbf{X}, \mathbf{D}_K].$$

$\mathbf{D}_K$ accounts for orientation of anisotropic material. If we restrict ourselves to first order gradients, we may use Cauchy's Theorem (spatially isotropic scaler functions of vectors can be expressed as functions of inner product of vectors and the box product):

$$\delta_{kl} x_{k,K}\, x_{l,L} = C_{KL}(\tau'), \qquad (\det C_{KL})^{1/2} = \rho_0/\rho(\tau')$$

We can apply this to $t_{KL}(\mathbf{X}, t) = t_{kl}(\mathbf{X}, t) x^k_{,K} x^l_{,L}$

Thus:

$$t_{kl}(\mathbf{X}, t) = F_{KL}[\mathbf{C}(\tau'), \rho^{-1}(\tau'), \theta(\tau'), \theta_{,K}(\tau'), \mathbf{X}]\, X^K_{,k}\, X^L_{,l}$$
$$q_k(\mathbf{X}, t) = Q_K[\mathbf{C}(\tau'), \rho^{-1}(\tau'), \theta(\tau'), \theta_{,K}(\tau'), \mathbf{X}]\, X^K_{,k}$$
$$\epsilon(\mathbf{X}, t) = E[\mathbf{C}(\tau'), \rho^{-1}(\tau'), \theta(\tau'), \theta_{,K}(\tau'), \mathbf{X}]$$
$$\eta(\mathbf{X}, t) = H[\mathbf{C}(\tau'), \rho^{-1}(\tau'), \theta(\tau'), \theta_{,K}(\tau'), \mathbf{X}]$$

Now $\tau' = t - t'$, so we can apply the axiom of smooth memory and expand those variables which depend on τ' about t. Note ρ^{-1} could be absorbed in $\mathbf{C}$ as it is proportional to the determinant of $\mathbf{C}$. We absorb all derivatives of ρ^{-1} in the derivatives of $\mathbf{C}$. The Taylor series coefficients τ', $\tau'^2 \ldots$ may be absorbed in the general functional. Thus:

$$t_{kl}(\mathbf{X}, t) = F_{KL}[\mathbf{C}, \dot{\mathbf{C}}, \ddot{\mathbf{C}}, \ldots, \rho^{-1}, \theta, \dot{\theta}, \ddot{\theta}, \ldots, \theta_{,K}, \dot{\theta}_{,K}, \ldots, \mathbf{X}].$$

The internal variables (excepting $\mathbf{X}$) are now functions of $\mathbf{X}$ and t only.

The maximum order of time derivative reflects depth of memory. An elastic material has memory only for its initial or natural state. This it remembers perfectly, however. Thus

$$t_{kl}\,(\mathbf{X},\,t)_{\substack{\text{Simple}\\ \text{Elastic}\\ \text{Body}}} = F_{KL}\,(\mathbf{C},\,\theta,\,\mathbf{X})\,X_{,k}^{K}\,X_{,l}^{L}.$$

A fluid has *no* memory of a natural state—all initial states are equivalent. For this reason a fluid *cannot be anisotropic*.

$$\mathbf{X} \to \mathbf{x}$$
$$C_{KL} = g_{kl}\,x_{,K}^{k}\,x_{,L}^{l} \to g_{kl}$$
$$\dot{C}_{KL} = 2\,d_{kl}\,x_{,K}^{k}\,x_{,L}^{l} \to 2\,d_{kl}$$
$$\theta_{,K} \to \theta_{,k}.$$

Thus

$$t_{kl}(x,\,t) = f_{kl}\,(\mathbf{d},\,\rho^{-1},\,\theta).$$

We have neglected higher order space and time derivatives.

The Stokesian fluid must be *thermodynamically admissible*. In particular, it must not violate the Clausius-Duhem condition:

$$\rho\!\left(\dot{\eta} - \frac{\dot{\epsilon}}{\theta}\right) + \frac{1}{\theta}\,f_{kl}d_{lk} + \frac{1}{\theta^2}\,q_k\theta_{,k} \geq 0.$$

Thus, applying the chain rule,

$$\rho\!\left(\frac{\partial \eta}{\partial d_{kl}} - \frac{1}{\theta}\frac{\partial \epsilon}{\partial d_{kl}}\right)\dot{d}_{kl} - \left(\frac{\partial \eta}{\partial \rho^{-1}} - \frac{1}{\theta}\frac{\partial \epsilon}{\partial \rho^{-1}}\right)\frac{\dot{\rho}}{\rho} + \rho\!\left(\frac{\partial \eta}{\partial \theta} - \frac{1}{\theta}\frac{\partial \epsilon}{\partial \theta}\right)\dot{\theta}$$

$$+ \frac{1}{\theta}\,f_{kl}d_{lk} + \frac{1}{\theta^2}\,g_k\theta_{,k} \geq 0.$$

This inequality is linear in $\mathbf{d}$, $\dot{\theta}$, and $\theta_{,k}$. It cannot be maintained for arbitrary values of these quantities which can be varied independently. Thus:

$$\frac{\partial \eta}{\partial d_{kl}} - \frac{1}{\theta}\frac{\partial \epsilon}{\partial d_{kl}} = 0,$$

$$\frac{\partial \eta}{\partial \theta} - \frac{1}{\theta}\frac{\partial \epsilon}{\partial \theta} = 0,$$

$$g_k = 0.$$

Since $\dot{\rho} = -\rho d_{kk}$

$$\left(\frac{\partial \eta}{\partial \rho^{-1}} - \frac{1}{\theta}\frac{\partial \epsilon}{\partial \rho^{-1}}\right) d_{kk} + \frac{1}{\theta} f_{kl} d_{lk} \geq 0. \quad \text{Clausius-Duhem condition}$$

Let us define a free energy: $\psi(\rho^{-1}, \theta) \equiv \epsilon - \eta\theta$,

$$\eta = -\frac{\partial \psi}{\partial \theta} = -\frac{\partial \epsilon}{\partial \theta} + \eta + \theta\frac{\partial \eta}{\partial \theta},$$

$$\frac{-\partial \psi}{\partial \rho^{-1}} = -\frac{\partial \epsilon}{\partial \rho^{-1}} + \theta\frac{\partial \eta}{\partial \rho^{-1}} \equiv \Pi, \qquad \text{Thermodynamic Pressure}$$

$$_D t_{kl} \equiv t_{kl} + \Pi\delta_{kl}.$$

The Clausius-Duhem condition then reads

$$_D t_{kl} d_{lk} \geq 0.$$

Thus, we may write the constitutive equation of a Stokesian fluid:

$$t_{kl} = -\Pi(\rho^{-1}, \theta)\delta_{kl} + {}_D f_{kl}(\mathbf{d}, \rho^{-1}, \theta),$$

$$q_k = 0,$$

$$\epsilon = \psi(\rho^{-1}, \theta) - \theta\frac{\partial \psi}{\partial \theta},$$

$$\eta = -\frac{\partial \psi}{\partial \theta},$$

where $\Pi(\rho^{-1}, \theta) = -\dfrac{\partial \psi}{\partial \rho^{-1}}$ is the thermodynamic or "retrievable" pressure. Further, $_D\mathbf{f}$ must be spatially invariant.

$$\mathbf{Q}(t)_D\mathbf{f}(\mathbf{d}, \rho^{-1}, \theta)\mathbf{Q}^T(t) = {}_D\mathbf{f}(\mathbf{QdQ}^T, \rho^{-1}, \theta),$$

$$_D f_{kl} d_{lk} \geq 0 \rightarrow {}_d\mathbf{f}(0, \rho^{-1}, \theta) = 0.$$

If we expand $_D\mathbf{f}$ in a polynomial series in $\mathbf{d}$, the Caley-Hamilton Theorem allows us to truncate the series after $\mathbf{d}^2$. The coefficients become functions of ρ^{-1}, θ, and I_d, II_d, III_d, the invariants of $\mathbf{d}$.

Thus

$$_D f(\rho^{-1}, \theta, \mathbf{d}) = \alpha_0(\rho^{-1}, \theta; I_d, II_d, III_d)\mathbf{I} + \alpha_1(\rho^{-1}, \theta, I_d, II_d, III_d)\mathbf{d}$$
$$+ \alpha_2(\rho^{-1}, \theta, I_d, II_d, III_d)\mathbf{d}^2.$$

Further since

$$_Df(\rho^{-1}, \theta, 0) = 0,$$
$$\alpha_0(\rho^{-1}, \theta; 0, 0, 0) = 0.$$

If we linearize this,

$$\alpha_2 = 0,$$
$$\alpha_1 = \alpha_1(\rho^{-1}, \theta),$$
$$\alpha_0 = \alpha_{00} + \alpha_{01}(\rho^{-1}, \theta)I_d; \quad \alpha_{00} = 0.$$

Thus

$$\mathbf{t}(\mathbf{x}, t) = [-\Pi(\rho^{-1}, \theta) + \alpha_{01}(\rho^{-1}, \theta)I_d]\mathbf{I} + \alpha_1(\rho^{-1}, \theta)\mathbf{d}.$$

We may relabel these:

$$t_{kl}(\mathbf{x}, t) = [-\Pi(\rho^{-1}, \theta) + \lambda_v(\rho^{-1}, \theta)I_d]g_{kl} + 2\mu_v(\rho^{-1}, \theta)d_{kl},$$

Navier-Stokes Equation

$$\bar{p} = -\frac{1}{3} t_{kk} = \Pi - \left(\lambda_v + \frac{2}{3}\mu_v\right)I_d.$$

$\bar{p}$ is the mechanical pressure. It is in general *not* equal to the thermo-dynamic pressure, but is rather equal to the sum of the volume deriva-tives of the retrievable energy stored elastically and the viscous energy dissipated in storing that energy.

$$\int \bar{p}dV \qquad = \int \Pi dV - \qquad \int K_v I_d dV$$

Mechanical work done on unit volume	$=$	ψ	$+$	Work done against dissipative viscous forces

We interpret this as follows: As long as $\bar{p} \neq \Pi$ there is dissipation. $\bar{p}$, however, will quickly become equal to Π as the fluid is compressed elastically. Only in the case of elastic oscillations (*P*-waves) will the λ_v-type dissipation be significant, since then $\bar{p}$ will be different from Π essentially all of the time. Under equilibrium conditions or quasi-static conditions, such as is the case in the viscous deformation of the earth by steady surface loads, λ_v-type dissipation will be completely unim-portant. $\bar{p}$ may always be taken equal to Π.

A. An Alternate Derivation of Equations II-22, 23

Let us define τ_{kl}:

$$t_{kl} \equiv -\bar{p}(\mathbf{X})\delta_{kl} + \tau_{kl},$$

where $\bar{p}(\mathbf{X})$ is the zero order hydrostatic pressure in the undeformed coordinate system $\mathbf{X}$. In the quasi-static viscous case, $\bar{p} = \Pi$ and so

$$\tau_{kl} = 2\mu d_{kl}.$$

Also in the viscous case $\mathbf{X} = \mathbf{x}$, as a fluid has no memory of an initial state.

The Cauchy equations of motion read:

$$t_{kl,k} + \rho(f_l - a_l) = 0,$$

and so the viscous equations of motion for quasi-static flow are:

$$\tau_{lk,l} - \bar{p}(\mathbf{x}),_k + \rho(f_k - a_k) = 0. \qquad \text{Viscous Equation of Motion}$$

The constitutive relations are:

$$\tau_{kl} = 2\mu d_{kl} \qquad\qquad \text{Viscous Constitutive Equation}$$

In the elastic case

$$t_{kl} = -\bar{p}(\mathbf{X})\delta_{kl} + \tau_{kl}$$
$$\tau_{kl} = \lambda e_{rr}\delta_{kl} + 2\mu e_{kl}$$

$\mathbf{X} \neq \mathbf{x}$. Rather $\mathbf{X}$ and $\mathbf{x}$ are related by the displacement $\mathbf{u}$:

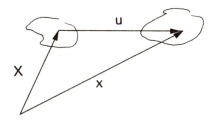

To first order

$$\bar{p}(\mathbf{X}) = \bar{p}(\mathbf{x} - \mathbf{u}) = \bar{p}(\mathbf{x}) - u_k\bar{p},_k$$

The equation of motion thus becomes:

$$\tau_{kl,l} - \bar{p}(\mathbf{x})_{,k} + (u_l \bar{p}_{,l})_{,k} + \rho(f_k - a_k) = 0,$$

Elastic Equation of Motion

subject to the constitutive relation

$$\tau_{kl} = \lambda e_{rr} \delta_{kl} + 2\mu e_{kl}.$$

Elastic Constitutive Relation

The derivation presented in Chapter II is, I believe, entirely equivalent to the above.[2]

[2]The derivation presented here was shown me by Dr. C. Eringen.

Flat Space Runge-Kutta Equations

A. Reduction of the Elastic Equation to Linear Form

The elastic constitutive relations are as usual:

$$\tau_{ij} = \lambda\Theta\delta_{ij} + 2\mu e_{ij}$$

$$e_{ij} = \frac{1}{2}\left(\frac{\partial u_j}{\partial x_i} + \frac{\partial u_i}{\partial x_j}\right)$$

$$\Theta = \frac{\partial u_k}{\partial x_k} \tag{1}$$

$\sigma = \lambda + 2\mu$, τ_{ij} are the components of τ.

Assume u goes to zero at $\pm\infty$. Then Fourier transform (III-1). Let a bar denote the Fourier transform:

$$\bar{u}(k_x, k_y, z) = \int_{-\infty}^{\infty}\int_{-\infty}^{\infty} u(x, y, z)\, e^{-ik_x x} e^{-ik_y y}\, dx dy.$$

Then

$$\partial_x \bar{u} = ik_x \bar{u}.$$

Equation III-1,

$$\nabla \cdot \tau - \rho_0 g_0 \nabla u_z + g_0 \rho_0 \nabla \cdot u\hat{z} = 0$$

becomes:

$$ik_x \bar{\tau}_{xx} + ik_y \bar{\tau}_{yx} + \partial_z \bar{\tau}_{zx} - \rho_0 g_0 ik_x \bar{u}_z = 0$$
$$ik_x \bar{\tau}_{xy} + ik_y \bar{\tau}_{yy} + \partial_z \bar{\tau}_{zy} - \rho_0 g_0 ik_y \bar{u}_z = 0 \tag{2}$$
$$ik_x \bar{\tau}_{xz} + ik_y \bar{\tau}_{yz} + \partial_z \bar{\tau}_{zz} + \rho_0 g_0 (ik_x \bar{u}_x + ik_y \bar{u}_y) = 0.$$

285

We similarly transform the constitutive relations.

$$\bar{\tau}_{xx} = \sigma_i k_x \bar{u}_x + \lambda(ik_y \bar{u}_y + \partial_z \bar{u}_z)$$
$$\bar{\tau}_{yy} = \sigma_i k_y \bar{u}_y + \lambda(ik_x \bar{u}_x + \partial_z \bar{u}_z)$$
$$\bar{\tau}_{zz} = \sigma_i \partial_z \bar{u}_z + \lambda(ik_x \bar{u}_x + ik_y \bar{u}_y).$$
$$\bar{\tau}_{xy} = \mu(ik_y \bar{u}_x + ik_x \bar{u}_y)$$
$$\bar{\tau}_{xz} = \mu(\partial_z \bar{u}_x + ik_x \bar{u}_z)$$
$$\bar{\tau}_{yz} = \mu(\partial_z \bar{u}_y + ik_y \bar{u}_z).$$

(3)

From (3) we see immediately

$$\partial_z \bar{u}_x = -ik_x \bar{u}_z + \mu^{-1}\bar{\tau}_{xz}$$
$$\partial_z \bar{u}_y = -ik_y \bar{u}_z + \mu^{-1}\bar{\tau}_{yz}$$
$$\partial_z \bar{u}_z = \sigma^{-1}\lambda(ik_x \bar{u}_x + ik_y \bar{u}_y) + \sigma^{-1}\bar{\tau}_{zz}.$$

(4)

From (2) we see

$$\partial_z \bar{\tau}_{zz} = -ik_x \bar{\tau}_{xz} - ik_y \bar{\tau}_{yz} - \rho_0 g_0(ik_x \bar{u}_x + ik_y \bar{u}_y).$$ (5)

We also get from (2)

$$\partial_z \bar{\tau}_{xz} = -ik_x \bar{\tau}_{xx} - ik_y \bar{\tau}_{yx} + \rho_0 g_0 ik_x \bar{u}_z,$$
$$\partial_z \bar{\tau}_{yz} = -ik_x \bar{\tau}_{xy} - ik_y \bar{\tau}_{yy} + \rho_0 g_0 ik_y \bar{u}_z.$$

(6)

Since boundary conditions are $\hat{z} \cdot \tau$ continuous, we wish to eliminate $\bar{\tau}_{xy}, \bar{\tau}_{xx} \bar{\tau}_{yy}$. Substituting for these quantitites from the transformed constitutive relation (3), and eliminating $\partial_z \bar{u}_z$ from (4) we get:

$$\partial_z \bar{\tau}_{xz} = ik_x[\sigma ik_x \bar{u}_x + \lambda(ik_y \bar{u}_y - \sigma^{-1}\lambda(ik_x \bar{u}_x + ik_y \bar{u}_y)$$
$$+ \sigma^{-1}\bar{\tau}_{zz})] - ik_y[\mu(ik_y \bar{u}_x + ik_x \bar{u}_y)] + \rho_0 g_0 ik_x \bar{u}_z. \quad (7)$$

$$\partial_z \bar{\tau}_{xz} = (4\sigma^{-1}\mu(\lambda + \mu)k_x^2 + \mu k_y^2)\bar{u}_x + \sigma^{-1}\mu(3\lambda + 2\mu)k_x k_y \bar{u}_y$$
$$- ik_x \sigma^{-1}\lambda \bar{\tau}_{zz} + \rho_0 g_0 ik_x \bar{u}_z.$$

$$\partial_z \bar{\tau}_{yz} = \sigma^{-1}\mu(3\lambda + 2\mu)k_x k_y \bar{u}_x + (4\sigma^{-1}\mu(\lambda + \mu)k_y^2 + \mu k_x^2)\bar{u}_y$$
$$- ik_y \sigma^{-1}\lambda \bar{\tau}_{zz} + \rho_0 g_0 ik_y \bar{u}_z.$$

We can rewrite equations (4), (5), (6) in the form of equation 8 (p. 287).

$$
\partial_z
\begin{bmatrix}
\bar{u}_x \\
\bar{u}_y \\
\bar{u}_z \\
\bar{\tau}_{xz} \\
\bar{\tau}_{yz} \\
\bar{\tau}_{zz}
\end{bmatrix}
=
\begin{bmatrix}
0 & 0 & -ik_x & \mu^{-1} & 0 & 0 \\
0 & 0 & -ik_y & 0 & \mu^{-1} & 0 \\
-\sigma^{-1}\lambda ik_x & -\sigma^{-1}\lambda ik_y & 0 & 0 & 0 & \sigma^{-1} \\
4\mu\sigma^{-1}(\lambda+\mu)k_x^2 + \mu k_y^2 & \sigma^{-1}\mu(3\lambda+2\mu)\cdot k_x k_y & \rho_0 g_0 ik_x & 0 & 0 & -ik_x\sigma^{-1}\lambda \\
\sigma^{-1}\mu(3\lambda+2\mu)\cdot k_x k_y & 4\mu\sigma^{-1}(\lambda+\mu)k_y^2 + \mu k_x^2 & \rho_0 g_0 ik_y & 0 & 0 & -ik_y\sigma^{-1}\lambda \\
-\rho_0 g_0 ik_x & -\rho_0 g_0 ik_y & 0 & -ik_x & -ik_y & 0
\end{bmatrix}
\begin{bmatrix}
\bar{u}_x \\
\bar{u}_y \\
\bar{u}_z \\
\bar{\tau}_{xz} \\
\bar{\tau}_{yz} \\
\bar{\tau}_{zz}
\end{bmatrix}
\tag{8}
$$

Note that in deriving this general result we have not assumed λ, μ to be constant. We have never assumed $\partial_z \lambda$ or $\partial_z \mu = 0$.

1. Separation into Two Decoupled Sets of Equations

Equation (8), a set of six coupled, first order, ordinary differential equations, separates into a set of two and a set of four coupled, first order, ordinary differential equations. To see this let $k_y = 0$.

$$\partial_z \begin{vmatrix} \bar{u}_y \\ \bar{\tau}_{yz} \end{vmatrix} = \begin{vmatrix} 0 & \mu^{-1} \\ \mu k_x^2 & 0 \end{vmatrix} \begin{vmatrix} \bar{u}_y \\ \bar{\tau}_{yz} \end{vmatrix} \tag{9}$$

$$\partial_z \begin{vmatrix} \bar{u}_x \\ \bar{u}_z \\ \bar{\tau}_{xz} \\ \bar{\tau}_{zz} \end{vmatrix} = \begin{vmatrix} 0 & -ik_x & \mu^{-1} & 0 \\ -\sigma^{-1}\lambda ik_x & 0 & 0 & \sigma^{-1} \\ 4\mu\sigma^{-1}(\lambda + \mu)k_x^2 & \partial_0 g_0 ik_x & 0 & -\lambda\sigma^{-1}ik_x \\ -\rho_0 g_0 ik_x & 0 & -ik_x & 0 \end{vmatrix} \begin{vmatrix} \bar{u}_x \\ \bar{u}_z \\ \bar{\tau}_{xz} \\ \bar{\tau}_{zz} \end{vmatrix}$$

Elastic Equation (10)

We would have gotten precisely the same result (with x and y interchanged) had we let $k_x = 0$ instead of k_y. Since the equations are linear, we may superimpose the results. Thus, both (9) and (10) are valid on the full two-dimensional Fourier decomposition on the surface.

Physically we can understand why the motion ought to separate into divergent and non-divergent motions on surfaces perpendicular to $\hat{z}$. Non-divergent motion on such planes can never generate a normal component of displacement. Conversely, a motion which has a normal component of displacement can never induce a non-divergent displacement field on surfaces perpendicular to $\hat{z}$. A normal component of motion can only induce horizontal motions by producing compressions or rarefactions of the media.

Equation (9) further implies that if $\bar{\tau}_{yz}$ and $\bar{u}_y$ are zero on any surface perpendicular to z, they are zero on all such surfaces.

Proof:

Suppose $\bar{\tau}_{yz} = 0$ on S_z

then $\partial_z \begin{vmatrix} \bar{u}_y \\ \bar{\tau}_{yz} \end{vmatrix} = \begin{vmatrix} 0 \\ \mu k_x^2 \bar{u}_y \end{vmatrix}$

i.e. $\partial_z \bar{u}_y = 0$

But if $\bar{u}_y = 0$, from (9) $\bar{\tau}_{yz}$ = constant
$\bar{\tau}_{yz} = 0$ at one point,
therefore $\bar{\tau}_{yz} = 0$ everywhere.

Since in the cases we consider, only normal stress, τ_{zz} ($\tau_{xz} = \tau_{yz} = 0$) will be applied to the free surface, and u_x and u_y are zero at one radius at least, we may restrict our attention solely to (10).

2. The Incompressible Elastic Equations

The incompressible elastic equations follow simply from (10) by letting $\lambda \to \infty$ and remembering $\sigma \equiv \lambda + 2\mu$

$$\partial_z \begin{vmatrix} \bar{u}_x \\ \bar{u}_z \\ \bar{\tau}_{xz} \\ \bar{\tau}_{zz} \end{vmatrix} = \begin{vmatrix} 0 & -ik_x & \mu^{-1} & 0 \\ -ik_x & 0 & 0 & 0 \\ 4\mu k_x^2 & \rho_0 g_0 ik_x & 0 & -ik_x \\ \rho_0 g_0 ik_x & 0 & -ik_x & 0 \end{vmatrix} \begin{vmatrix} \bar{u}_x \\ \bar{u}_z \\ \bar{\tau}_{xz} \\ \bar{\tau}_{zz} \end{vmatrix}$$

B. Reduction of the Viscous Equations

The Runge-Kutta form of the viscous equations may be obtained in an entirely parallel fashion. The only complication is that the constitutive equations relate stress to rate of strain rather than strain. We assume the viscous fluid to be incompressible ($\lambda_v = \infty$, Appendix I shows this is equivalent to assuming quasi-static flow).

$$\nabla \cdot \tau + g_0 u_z \partial_z \rho_0 \hat{z} = 0. \tag{12}$$

The stress-strain relations are the same as (3) and (4) except velocity is substituted for displacement. Let $k_y = 0$.

$$\partial_z \bar{v}_x = ik_x \bar{v}_z + \mu^{-1} \bar{\tau}_{xz},$$
$$\partial_z \bar{v}_z = ik_x \bar{v}_x.$$

From (12) we see:

$$\begin{aligned} \partial_z \bar{\tau}_{zz} &= -ik_x \bar{\tau}_{xz} - g_0 \bar{u}_z \partial_z \rho_0, \\ \partial_z \bar{\tau}_{zx} &= ik_x \bar{\tau}_{xx}, \quad \text{or} \\ \partial_z \bar{\tau}_{xz} &= 4\mu k_x^2 \bar{v}_x - ik_x \bar{\tau}_{zz}. \end{aligned} \tag{13}$$

Since

$$\bar{\tau}_{xx} = \sigma ik_x \bar{v}_x + \lambda \, \partial_z \bar{v}_z$$
$$\partial_z \bar{v}_z = -\sigma^{-1}\lambda \, (ik_x \bar{v}_x) + \sigma^{-1}\bar{\tau}_{zz}$$
$$\bar{\tau}_{xx} = ik_x \bar{v}_x(\sigma - \lambda^2\sigma^{-1}) + \lambda\sigma^{-1}\bar{\tau}_{zz}.$$

Thus, the incompressible viscous equations may be written:

$$\partial_z \begin{vmatrix} \bar{v}_x \\ \bar{v}_z \\ \bar{\tau}_{xz} \\ \bar{\tau}_{zz} \end{vmatrix} = \begin{vmatrix} 0 & -ik_x & \mu^{-1} & 0 \\ -ik_x & 0 & 0 & 0 \\ 4\mu k_x^2 & 0 & 0 & -ik_x \\ 0 & 0 & -ik_x & 0 \end{vmatrix} \begin{vmatrix} \bar{v}_x \\ \bar{v}_z \\ \bar{\tau}_{xz} \\ \bar{\tau}_{zz} \end{vmatrix} + \bar{u}_z \begin{vmatrix} 0 \\ 0 \\ 0 \\ g_0\partial_z\rho_0 \end{vmatrix}$$

$$(14)$$

where $\bar{u}_z = \int_0^t \bar{v}_z \partial t$ is the Fourier transformed total vertical viscous displacement since the motion began.

Reduction of Equation of Motion in Spherical Coordinates to Scaler Form after Backus (1967)

THE REDUCTION of equations to Runge-Kutta form is more difficult in spherical coordinates than in Cartesian coordinates because: (1) Representation of the vector field by scaler fields is more difficult in spherical coordinates, and (2) the spherical metric introduces effects which must be accounted for properly.

Backus (1967) has shown how vectors and tensors may be represented by scalers in spherical coordinates. This method is summed up in his Tangent Tensor and Tangent Vector Representation Theorems. Metric effects are generally handled by distinguishing covariant and contravariant vector (and tensor) components and using covariant or contravariant differentiation. If one is dealing only in spherical coordinates, certain specializations of the usual base vectors and reciprocal base vectors can be made to simplify things (Backus, 1967). Dyadic notation is convenient.

In this appendix we shall briefly review Backus's method. We shall not go into great mathematical detail as this is all covered in Backus's paper. Our system describing elastic deformation is exactly the same as Backus's free oscillation system if $\omega = 0$. The viscous system can be obtained simply by dropping the advection term in the elastic system, being careful to distinguish total viscous displacement (in the body force terms) from viscous rate of displacement (which is related to stress by the constitutive relations). Obtaining the viscoelastic system of coupled Runge-Kutta propagations is straightforward methodologically and is covered in Section III.B.3.a.

Precisely as in the Cartesian case (Appendix II), the motion must separate into divergent and non-divergent parts on surfaces $\perp$ to $\hat{r}$. The gravitational terms split with the divergent system of equations, which is also the only type of motion excited by the boundary conditions pertinent for glacial uplift phenomena. Two more equations are added by the Poisson gravity equation. Thus, a set of six coupled, first order, ordinary differential equations forms the elastic Runge-Kutta system.

$\hat{r}$ does not change as a function of position, so we need only define base vectors and reciprocal base vectors tangent to spheres concentric to the origin. A suitable pair can be defined:

$$\partial_i r = \frac{\partial r}{\partial x^i} \tag{1}$$

$$\nabla_s x^i = \text{surface gradient} = (\nabla - \hat{r}\hat{r} \cdot \nabla)x^i.$$

These vectors transform covariantly and contravariantly, respectively. They are reciprocal:

$$\partial_i \mathbf{r} \cdot \nabla_s x^j = \delta_i^j. \tag{2}$$

Hence, we can define the usual metric tensor

$$g_{ij} = \partial_i \mathbf{r} \cdot \partial_j \mathbf{r} \quad g^{ij} = \nabla_s x^i \cdot \nabla_s x^j. \tag{3}$$

This metric tensor is just like the usual one except that its indices take on two values instead of three. Covariant and contravariant vector or tensor components may be determined:

$$v_i = \mathbf{v} \cdot \partial_i \mathbf{r}$$
$$v^i = \mathbf{v} \cdot \nabla_s x^i$$

so[1]

$$\mathbf{v}_s = v_i \nabla_s x^i = v^i \partial_i \mathbf{r} \tag{4}$$
$$\mathbf{\tau}_s = \tau_{ij} \nabla_s x^i \nabla_s x^j = \tau^{ij} \partial_i \mathbf{r} \partial_j \mathbf{r}.$$

Clearly

$$\mathbf{u} = \hat{r} u_r + \nabla_s x^i u_i. \tag{5}$$

Any tensor may be written on the sum of a scaler, two tangent vectors, and a tangent tensor:

$$\mathbf{\tau} = \hat{r}\hat{r}\tau_{rr} + \hat{r}\tau_{rs} + \tau_{sr}\hat{r} + \tau_{ss}. \tag{6}$$

This can be easily seen just by writing out the dyadic components of an

---

[1]Summing over identical diagonal indices is assumed.

arbitrary tensor:

	scaler	tangent vector	
$\tau =$	$\hat{\mathbf{r}}\tau_{rr}\hat{\mathbf{r}}$	$+ \hat{\mathbf{r}}\tau_{r1}\nabla_s x^1$	$+ \hat{\mathbf{r}}\tau_{r2}\nabla_s x$
tangent	$+ \nabla_s x^1 \tau_{1r}\hat{\mathbf{r}}$	$+ \nabla_s x^1 \tau_{11}\nabla_s x^1$	$+ \nabla_s x^1 \tau_{12}\nabla_s x^2$
vector	$+ \nabla_s x^2 \tau_{2r}\hat{\mathbf{r}}$	$+ \nabla_s x^2 \tau_{21}\nabla_s x^1$	$+ \nabla_s x^2 \tau_{22}\nabla_s x^2$
		tangent tensor	

$$(7)$$

We can also define covariant differentiation on the spherical surfaces. Again, it is entirely analogous to the usual covariant differentiation, the Crystoffel symbols being defined in precisely the same way, except the indices only take on two values ($i, j = 2,3$ in the usual spherical co-ordinates).

$$D_i v_j = \partial_i v_j - \binom{k}{ij} v_k \tag{8}$$

$$D_i v_j = \partial_i v_j + \binom{j}{ik} v^k.$$

Similarly we may define a surface rotator:

$$h_{ij} = \hat{\mathbf{r}} \cdot (\partial_i \mathbf{r} \times \partial_j \mathbf{r}) = g e_{ij} \tag{9}$$
$$h_i^j = \hat{\mathbf{r}} \cdot (\partial_i \mathbf{r} \times \nabla_s x^i)$$
$$h_j^i = \hat{\mathbf{r}} \cdot (\nabla_s x^i \times \partial_j \mathbf{r})$$
$$h^{ij} = \hat{\mathbf{r}} \cdot (\nabla_s x^i \times \nabla_s x^j) = g^{-1} e_{ij}, \quad g = \sqrt{\det. g_{ij}}.$$

Lastly, we may project all vector fields onto the unit sphere. All differentations, etc., may be carried out on the unit sphere and then re-projected back to S_r if desired.

$$\partial_i \mathbf{r} = \mathbf{r}\partial_i \hat{\mathbf{r}}$$
$$\nabla_r = r^{-1}\nabla_1. \tag{10}$$

Let $\tilde{u}_i$ denote the covariant component of u_s on S_1. Then

$$\mathbf{u} = \hat{\mathbf{r}}u_r + \nabla_1 x^i \tilde{u}_i \tag{11}$$
$$= \hat{\mathbf{r}}u_r + \partial_i \hat{\mathbf{r}}\tilde{u}^i.$$
$$\tau = \hat{\mathbf{r}}\hat{\mathbf{r}}\tau_{rr} + \hat{\mathbf{r}}\partial_i \hat{\mathbf{r}}\tilde{\tau}_r^i + \partial_i \hat{\mathbf{r}}\hat{\mathbf{r}}\tilde{\tau}_r^i + \partial_i \hat{\mathbf{r}}\partial_j \hat{\mathbf{r}}\tilde{\tau}^{ij}. \tag{12}$$

Note the tilde may be removed, i.e., the vector or tangent component may be reprojected to S_r, by multiplying by

r^{-1} for contravariant components,
r for covariant components.

Thus, $\tau_{ri} = r\tilde{\tau}_{ri}$ etc.

Now the *Tangent Vector Representation Theorem* states that any tangent vector $\mathbf{u}_s$ can be represented by two unique scaler fields:

$$\mathbf{u}_s = \nabla_1 V - \hat{\mathbf{r}} \times \nabla_1 W \tag{13}$$

or

$$\tilde{u}_i = \tilde{D}_i V + \tilde{h}_i^j \tilde{D}_j W.$$

Similarly, the *Tangent Tensor Representation Theorem* states that any second order tangent tensor can be represented by four unique scaler fields:

$$\tilde{\tau}_{ij} = H\tilde{h}_{ij} + (L - \nabla_1^2 M)\tilde{g}_{ij} + 2\tilde{D}_i\tilde{D}_j M \\ + (\tilde{h}_i^k \tilde{D}_j + \tilde{h}_j^k \tilde{D}_i)\tilde{D}_k N, \tag{14}$$

where $\nabla_1^2 M = g^{ij}D_iD_jM$. If τ_s is symmetric $H = 0$.
Thus, if u = displacement, τ = stress:

$$\mathbf{u} = \hat{\mathbf{r}}U + \nabla_1 x^i[\tilde{D}_i V + \tilde{h}_i^j \tilde{D}_j W] \tag{15}$$

$$\tau = \hat{\mathbf{r}}\hat{\mathbf{r}}P + (\hat{\mathbf{r}}\nabla_1 x^i + \nabla_1 x^i\hat{\mathbf{r}})[\tilde{D}_i Q + \tilde{h}_i^j \tilde{D}_j R] + \nabla_1 x^i \nabla_1 x^j \\ \cdot [(L - \nabla_1^2 M)\tilde{g}_{ij} + 2\tilde{D}_i\tilde{D}_j M + (\tilde{h}_i^k \tilde{D}_j + \tilde{h}_j^k \tilde{D}_i)\tilde{D}_k N]. \tag{16}$$

Any force applied to the surface of the earth may be represented:

$$\mathbf{F} = \hat{\mathbf{r}}P' + \nabla_1 x^i[\tilde{D}_i Q' + \tilde{h}_i^j D_j R'] \tag{17}$$

As in Appendix III,

$$\tau = \lambda_e(\nabla \cdot \mathbf{u})\mathbf{I} + 2\mu\mathbf{e} \tag{18}$$

$$2\mathbf{e} = \nabla\mathbf{u} + (\nabla\mathbf{u})^T \tag{19}$$

The various terms of equations II-22, 23 may be computed; i.e., $\nabla \cdot \tau$, $\nabla(\mathbf{u} \cdot \nabla p_0)$ etc. (See Backus, 1967, Section 5.) Equating similar vector and tensor components, and terms multiplied by similar metric factors

$$
\partial_r
\begin{bmatrix}
\mu^* U \\
\mu^* V \\
P \\
Q \\
\phi_1 \\
g_1
\end{bmatrix}
= r^{-1}
\begin{bmatrix}
-2\tilde{\lambda}\tilde{\beta}^{-1} & -\tilde{\lambda}\tilde{\beta}^{-1}\nabla_1^2 & r\tilde{\beta}^{-1} & 0 & 0 & 0 \\
-1 & 1 & 0 & r\tilde{\mu}^{-1} & 0 & 0 \\
4\left(\tilde{\gamma} r^{-1} - \dfrac{g_0\rho_0}{\mu^*}\right) & \left(2r^{-1}\tilde{\gamma} - \dfrac{\rho_0 g_0}{\mu^*}\right)\nabla_1^2 & -4\tilde{\mu}\tilde{\beta}^{-1} & -\nabla_1^2 & 0 & \rho_0 r \\
\left(\dfrac{\rho_0 g_0}{\mu^*} - 2r^{-1}\tilde{\gamma}\right) & -r^{-1}[(\tilde{\gamma}+\tilde{\mu})\nabla_1^2 + 2\tilde{\mu}] & -\tilde{\lambda}\tilde{\beta}^{-1} & -3 & \rho_0 & 0 \\
\dfrac{-4\pi G\rho_0 r}{\mu^*} & 0 & 0 & 0 & 0 & r \\
0 & \dfrac{-4\pi G\rho_0 \nabla_1^2}{\mu^*} & 0 & 0 & 0 & -r^{-1}\nabla_1^2 - 2
\end{bmatrix}
\begin{bmatrix}
\mu^* U \\
\mu^* V \\
P \\
Q \\
\phi_1 \\
g_1
\end{bmatrix}
\qquad \text{(II-20)}
$$

$$\tilde{\gamma} \equiv \frac{\tilde{\mu}(3\tilde{\lambda} + 2\tilde{\mu})}{\tilde{\lambda} + 2\tilde{\mu}} \qquad \tilde{\mu} \equiv \frac{\mu}{\mu^*}$$

$$\tilde{\beta} \equiv \tilde{\lambda} + 2\tilde{\mu} \qquad \tilde{\lambda} \equiv \frac{\lambda}{\mu^*}$$

$$g_1 \equiv 4\pi G\rho_0 U + \partial_r \phi_1$$

$$
\partial_r
\begin{bmatrix}
\mu^* U \\
\mu^* V \\
P \\
Q \\
\phi_1 \\
\hat{g}_1
\end{bmatrix}
= r^{-1}
\begin{bmatrix}
-2\tilde{\lambda}\tilde{\beta}^{-1} & -\tilde{\lambda}\tilde{\beta}^{-1}\nabla_1^2 & r\tilde{\beta}^{-1} & 0 & 0 & 0 \\[4pt]
-1 & 1 & 0 & r\tilde{\mu}^{-1} & 0 & 0 \\[4pt]
4\left(\tilde{\gamma}r^{-1} - \dfrac{g_0\rho_0}{\mu^*}\right) + \dfrac{4\pi G r\rho_0^2}{\mu^*} & \left(2r^{-1}\tilde{\gamma} - \dfrac{\rho_0 g_0}{\mu^*}\right)\nabla_1^2 & -4\tilde{\mu}\tilde{\beta}^{-1} & -\nabla_1^2 & 0 & \rho_0 r \\[10pt]
\left(\dfrac{\rho_0 g_0}{\mu^*} - 2r^{-1}\tilde{\gamma}\right) & -r^{-1}[(\tilde{\gamma}+\tilde{\mu})\nabla_1^2 + 2\mu] & -\lambda\beta^{-1} & -3 & \rho_0 & 0 \\[8pt]
0 & 0 & 0 & 0 & 0 & r \\[4pt]
\dfrac{-4\pi G}{\mu^*}(\partial_r\rho_0 r + 4\tilde{\mu}\tilde{\beta}^{-1}\rho_0) & \dfrac{-4\pi G\rho_0\nabla_1^2}{\mu^*}(1 - \tilde{\lambda}\beta^{-1}) & \dfrac{-4\pi G r\rho_0\tilde{\beta}^{-1}}{\mu^*} & 0 & -r^{-1}\nabla_1^2 & -r^{-1}\nabla_1^2 - 2
\end{bmatrix}
\begin{bmatrix}
\mu^* U \\
\mu^* V \\
P \\
Q \\
\phi_1 \\
\hat{g}_1
\end{bmatrix}
$$

(II-21)

$$\hat{g}_1 \equiv \partial_r\phi_1$$
$$\hat{g}_1 = g_1 - 4\pi G\rho_0 U$$

(g_{ij}, $h_i^k D_j$ etc.) then leads directly to the Runge-Kutta equations of motion (p. 295, 296).

System (21) is exactly the same as system (20) except that g_1 is defined in a manner which permits more direct physical interpretation.

Note, referring to (15), (16), that W and R have separated out into a 2×2 system that is independent of U, V, P, Q, ϕ_1, g_1.[2] U, V, P, Q are related to **u**, the displacement, and the stress at the surface $\hat{\mathbf{r}} \cdot \tau$:

$$\mathbf{u} = \hat{\mathbf{r}} U + \nabla_1 x^i \tilde{D}_i V, \tag{22}$$
$$\hat{\mathbf{r}} \cdot \tau = \hat{\mathbf{r}} P + \nabla_1 x^i \tilde{D}_i Q, \tag{23}$$

since $\nabla_1 = \nabla_1 x^i \partial_i = \nabla_1 x^i \tilde{D}_i$, this may also be written:

$$\mathbf{u} = \hat{\mathbf{r}} U + \nabla_1 V, \tag{24}$$
$$\hat{\mathbf{r}} \cdot \tau = \hat{\mathbf{r}} P + \nabla_1 Q \tag{25}$$

Of course, the scaler fields may be expressed as a sum of spherical harmonics,

$$U = \Sigma U_l^m(r) Y_l^m(x^1, x^2),$$
$$P = \Sigma P_l^m(r) Y_l^m(x^1, x^2), \tag{26}$$

etc. So for each harmonic (24, 25) become:

$$\mathbf{u}_l^m = \hat{\mathbf{r}} U_l^m Y_l^m + V_l^m \nabla_1 Y_l^m \tag{27}$$

$$(\hat{\mathbf{r}} \cdot \tau)_l^m = \hat{\mathbf{r}} P_l^m Y_l^m + Q_l^m \nabla_1 Y_l^m. \tag{28}$$

The first term in (27, 28) is the scaler expansion coefficient (i.e., U_l^m) times a unit vector. The second term involves the product of V_l^m with a vector, $\nabla_1 Y_l^m$, but $\nabla_1 Y_l^m$ is not of unit magnitude. To interpret (27, 28) correctly we should "normalize" $\nabla_1 Y_l^m$. We consider this problem in Appendix V. Note that $\nabla_1^2 Y_l^m = -l(l + 1) Y_l^m$ so ∇_1^2 in all equations may be replaced by $-l(l + 1)$ providing that we consider surface harmonics of order l. This we do implicitly in the text, so P, Q, U, V should be considered harmonic coefficients of order l: P_l^m, Q_l^m, U_l^m, V_l^m.

---

[2]L, M, N have been eliminated by substitution from the constitutive equation as in the Cartesian case.

Boundary Conditions at the Fluid Core

IN THE CORE (fluid core), Backus (1967) shows the equations of motion reduce:

$$\partial_r U = -2Ur^{-1} - r^{-1}\nabla_1^2 V + \lambda^{-1}P \tag{1}$$

$$\partial_r P = -4r^{-1}\rho_0 g_0 U + \rho_0 g_1 - r^{-1}\rho_0 g_0 \nabla_1^2 V \tag{2}$$

$$\partial_r \phi_1 = -4\pi G\rho_0 U + g_1 \tag{3}$$

$$\partial_r g_1 = -4\pi Gr^{-1}\rho_0 \nabla_1^2 V - r^{-2}\nabla_1^2 \phi_1 - 2r^{-1}g_1 \tag{4}$$

$$P - \rho_0\phi_1 - U\rho_0 g_0 = 0 \tag{5}$$

Following Longman (1962, 1963) (but not too closely), we may reduce this system to a simple Runge-Kutta form, gaining considerable insight in so doing. First, we should note that the above equations require the satisfaction of the Adams-Williamson equation. This is not only not surprising but inescapable in a self-gravitating, compressible system initially in hydrostatic equilibrium.

Differentiate (5):

$$\partial_r P - \rho_0 g_0 \partial_r U = U\partial_r(\rho_0 g_0) + \phi_1\partial_r\rho_0 + \rho_0\partial_r\phi_1. \tag{6}$$

Form the similar expression from (1), (2):

$$\partial_r P - \rho_0 g_0 \partial_r U = -2r^{-1}\rho_0 g_0 U + \rho_0 g_1 - \rho_0 g_0 \lambda^{-1}P. \tag{7}$$

Equating (6), (7):

$$\partial_r\phi_1 = -\left(2r^{-1}g_0 + \dot{g}_0 + g_0\frac{\dot{\rho}_0}{\rho_0}\right)U - g_0\lambda^{-1}P - \frac{\dot{\rho}_0}{\rho_0}\phi_1 + g_1 \tag{8}$$

$$\overset{\text{by (3)}}{\partial_r\phi_1} = -4\pi G\rho_0 U + g_1. \tag{9}$$

The dot ($\cdot$) indicates ∂_r.

Using (5) we may substitute for ϕ_1:

$$\phi_1 = \frac{P}{\rho_0} - Ug_0. \tag{10}$$

(8), (9), (10) imply

$$(\dot{g}_0 + 2r^{-1}g_0 - 4\pi G\rho_0)U + \left(\frac{\dot{\rho}_0}{\rho_0^2} + g_0\lambda^{-1}\right)P = 0. \tag{11}$$

Since

$$\nabla^2\phi_0 = 4\pi G\rho_0 = \nabla \cdot \nabla \phi_0 = \nabla \cdot g_0\hat{r} = \dot{g}_0 + 2r^{-1}g_0$$

the first parenthesis is zero. Thus:

$$\frac{\dot{\rho}_0}{\rho_0} + \frac{g_0\rho_0}{\lambda} = 0, \tag{12}$$

which is the Adams-Williamson equation.

Equation (2) states

$$\lambda^{-1}P = \Theta \tag{13}$$

where

$$\Theta = \nabla \cdot \mathbf{u} = \partial_r U + 2r^{-1}U + r^{-1}\nabla_1^2 V.$$

Substitute (3) into (4)

$$\ddot{\phi}_1 + 2r^{-1}\dot{\phi}_1 + r^{-2}\nabla_1^2\phi_1 =$$
$$-4\pi G\rho_0(\dot{U} + r^{-1}\nabla_1^2 V + 2r^{-1}U) - 4\pi GU\dot{\rho}_0$$

$$\ddot{\phi}_1 + 2r^{-1}\dot{\phi}_1 + r^{-2}\nabla_1^2\phi_1 = -4\pi G(\rho_0\Theta + \dot{\rho}_0 U)$$
$$\tag{14}$$

$$\overset{(13)}{\underset{(12)}{\rho_0\Theta}} + \rho_0 U = \lambda^{-1}P\rho_0 - Ug_0\rho_0^2\lambda^{-1}$$

$$\overset{(5)}{=} \lambda^{-1}\rho_0^2\phi_1.$$

Thus

$$\partial_r \dot\phi_1 = -\left(r^{-2}\nabla_1^2 + \frac{4\pi\rho_0^2 G}{\lambda}\right)\phi_1 - 2r^{-1}\dot\phi_1$$

$$\partial_r \phi_1 = \dot\phi_1 \tag{15}$$

$$P = \rho_0\phi_1 + U\rho_0 g_0$$

$$\partial_r \begin{vmatrix} \phi_1 \\ \dot\phi_1 \end{vmatrix} = \begin{vmatrix} 0 & 1 \\ -\left(r^{-2}\nabla_1^2 + \frac{4\pi\rho_0^2 G}{\lambda}\right) & -2r^{-1} \end{vmatrix} \begin{vmatrix} \phi \\ \dot\phi_1 \end{vmatrix} \tag{16}$$

Two boundary conditions are sufficient to make (16) determinant. We know: (1) ϕ_1 must be finite at $r = 0$, and (2) $\dot\phi_1 + 4\pi G\rho_0 U$ must be continuous across the core-mantle boundary (II-33). The system (16) is thus determined. (15) interprets ϕ_1 and U in terms of P.

To apply these results we normalize by letting $r = ax$.

$$\partial_x \begin{vmatrix} \phi_1 \\ \dot\phi_1 \end{vmatrix} = \begin{vmatrix} 0 & 1 \\ -\left(\dfrac{\nabla_1^2}{x^2} + \dfrac{4\pi G\rho_0^2 a^2}{\lambda}\right) & \dfrac{-2}{x} \end{vmatrix} \begin{vmatrix} \phi_1 \\ \dot\phi_1 \end{vmatrix} \tag{17}$$

where ($\cdot$) now indicates ∂_x.

Similarly normalized, (14) becomes

$$\frac{1}{a^2}\ddot\phi_1 + \frac{2}{a^2 x}\dot\phi_1 + \frac{1}{a^2 x^2}\nabla_1^2\phi_1 = \frac{4\pi G\rho_0^2\phi_1}{\lambda}. \tag{18}$$

We now solve (18) for a homogeneous sphere of radius ϵ. Let

$$\phi_1 = \sum_{i=0} a_i \epsilon^{i+k}.$$

Substituting into (18) gives:

$$\sum_{i=0}^{\infty} [(i+k)(i+k-1) + 2(i+k) - n(n+1)]a_i \epsilon^{i+k-2}$$

$$+ \frac{4\pi G\rho_0^2 a^2}{\lambda}\sum_{i=0}^{\infty} a_i \epsilon^{i+k}.$$

The indical equation is

$$(k)(k - 1) + 2k - n(n + 1) = 0,$$

or

$$k(k + 1) - n(n + 1) = 0,$$
$$k = n,$$
$$k = -n - 1.$$

We rule out the second possibility since ϕ_1 must be finite at $r = 0$. Each coefficient of each individual power of ϵ must vanish, so:

$$\frac{a_{i+2}}{a_i} = -\frac{4\pi G\rho_0^2 a^2}{\lambda} \Big/$$

$$[(i + n + 2)(i + n + 1) + 2(i + n + 2) - n(n + 1)], \quad (19)$$

$$\frac{a_{i+2}}{a_i} = -\frac{4\pi G\rho_0^2 a^2}{\lambda} \Big/ [i^2 + (2n + 5)i + 4n + 6].$$

Thus:

$$\phi_1 = a_0\epsilon^n + a_2\epsilon^{n+2} + a_4\epsilon^{n+4} + \cdots \qquad (20)$$

$$\dot{\phi}_1 = na_0\epsilon^{n-1} + (n + 2)a_2\epsilon^{n+1} + (n + 4)a_4\epsilon^{n+3} + \cdots \qquad (21)$$

ϵ is the radius of a homogeneous fluid center core. Two possibilities now present themselves:

(1) We could take ϵ small and use (17) to propagate starting vectors (20), (21) to the core-mantle boundary. Longman advocates this.
a) a_0 undetermined.
b) At core-mantle boundary use (15) to decode ϕ_1, $\dot{\phi}_1$:

$$\phi_1 = \phi_1$$
$$P = \rho\phi_1 + U\rho_c g_0$$

$$g_1^+ = \frac{\dot{\phi}_1}{r^*} + 4\pi G\rho_c U \qquad (22)$$

$$Q = 0.$$

(2) Alternatively, we could take ϵ = full core radius, evaluate the power series solution (20), (21) and use (22) to interpret (20) and (21). Kaula (1963) and Takeuchi (1962) take this approach.

These two possibilities lead to the following starting vectors:

$$
\begin{vmatrix} \mu^*U \\ \mu^*V \\ P \\ Q \\ \phi_1 \\ g_1 \end{vmatrix}
=
\begin{vmatrix} \mu^* \\ 0 \\ \rho_c g_0 \\ 0 \\ 0 \\ 4\pi G\rho_c \end{vmatrix} A
+
\begin{vmatrix} 0 \\ \mu^* \\ 0 \\ 0 \\ 0 \\ 0 \end{vmatrix} B
+
\begin{vmatrix} 0 \\ 0 \\ \rho[\phi_1] \\ 0 \\ [\phi_1] \\ \dfrac{[\dot{\phi}_1]}{r^*} \end{vmatrix} a_0
\tag{23}
$$

where ϕ_1, $\dot{\phi}_1$ come from (20), (21) with a_0 factored out. (23) can be propagated to the free surface where the three surface boundary conditions may be used to solve for A, B, a_0.

From (II-39) and (II-42) our physically determined starting propagator vectors look:

$$
\begin{vmatrix} \mu^*U \\ \mu^*V \\ P \\ Q \\ \phi_1 \\ g_1 \end{vmatrix}
=
\begin{vmatrix} \mu^* \\ 0 \\ \rho_c g_0 \\ 0 \\ 0 \\ 4\pi G\rho_c \end{vmatrix} A
+
\begin{vmatrix} 0 \\ \mu^* \\ 0 \\ 0 \\ 0 \\ 0 \end{vmatrix} B
+
\begin{vmatrix} 0 \\ 0 \\ \rho_0 \\ 0 \\ 1 \\ \dfrac{n}{a} \end{vmatrix} C
\tag{24}
$$

Starting vectors (24a) are defined the same as above, except ρ_0 in the last vector is zero, so the last vector reads $0 \quad 0 \quad 0 \quad 0 \quad 1 \quad \dfrac{n}{a}$.

We can calculate the deformation of a model earth using different types of starting vectors. In the table which follows, IGCORE is used as a code: If IGCORE = 0, starting vectors (24a) are used. If IG-CORE = 1, starting vectors (24) are used. Finally, if IGCORE = 2, starting vectors (23) involving a power-series approximation are used. In all cases the "Homogeneous E" model of Table III-1 is used.

As can be seen from Table App. IV-1, the power-series starting vectors (IGCORE = 2) give virtually identical results to our physical boundary conditions (IGCORE = 1). The importance of the gravita-

"Homogeneous E" Model

IGCORE	x	$U \times 10^{-4}$	P	$\phi_1 \times 10^{-1}$	$g_1 \times 10^{-9}$
0	.5	−.778	−1.04	−.118	−.948
	1.0	−1.34	−1.00	−.549	−.580
1	.5	−.705	−1.10	−.145	−.883
	1.0	−1.30	−1.00	−.577	−.567
2	.5	−.704	−1.10	−.147	−.881
	1.0	−1.30	−1.00	−.577	−.566

App., Table IV-1 Core and surface displacement, U, stress, P, and gravitational parameters, showing that power series and physical core-mantle boundary conditions agree well.

n	$-h'_C$	$-h'_L$	$-h'_K$	$-h'_T$
2	1.0022	1.007	.981	1.034
3	1.0491	1.059	1.050	1.078
4	1.0466	1.059	1.058	1.083
5	1.0786	1.093	1.093	1.121
6	1.1338	1.152	1.153	1.185
7	1.2026	1.223	1.223	1.260
10	1.4069	1.439	1.439	1.486

n	$-k'_C$	$-k'_L$	$-k'_K$	$-k'_T$
2	.3105	.310	.303	.312
3	.1968	.197	.197	.191
4	.1345	.133	.134	.126
5	.1060	.104	.105	.096
6	.0918	.090	.090	.071
7	.0843	.082	.082	.072

h'_C our calculations
h'_L Longman's calculations
h'_K Kaula's calculations
h'_T Takeuchi's calculations

Same notation for k's. h, k are load deformation coefficients as defined in Munk and MacDonald (1960b).

App. Table IV-2 Load deformation coefficients calculated by others compared to those calculated here. Data are from Longman (1966). A Gutenberg earth model is assumed.

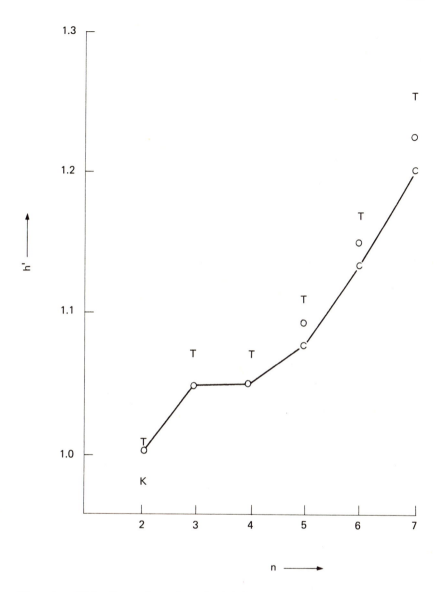

Figure App. IV-1. Comparison of our load deformation coefficients (*c*) to those of Takeuchi (*T*) and Kaula and Longman (⊙). Data are listed in Table App. IV-2.

tional interaction between the mantle and the core is shown by comparing IGCORE = 1 or 2 to IGCORE = 0.

It is also instructive to compare our determinations of load deformation coefficients with values published in the literature. Longman (1966) has made a useful compilation of calculations based on the Gutenberg Earth Model. Table App. IV-2 compares calculations using our boundary conditions with the calculations of Longman, Kaula, and Takeuchi. Figure App. IV-1 is a plot of the first part of Table App. IV-2. We use a Gutenberg Earth Model also; the effect of the earth model chosen can be seen by comparing the n = 2 entry in Table App. IV-2 to h_2' and k_2' calculated for an HB_1 model earth in Section III.B.4.

Figure App. IV-1 and Table App. IV-2 show our calculations to be in excellent agreement with Longman's. The $\sim 3\%$ discrepancies at the higher order numbers are probably due to the relative coarseness of our $\lambda(r)$, $\mu(r)$, $\rho_0(r)$, $g_0(r)$ definition.

Longman suggests the difference between Takeuchi's and his results may be attributed to Takeuchi's assumption of a homogeneous core. He correctly comments that a constant density core is impossible for a self-gravitating compressible fluid. Our approach, however, indicates that density stratification in the core will have no first order effect on the boundary conditions at the core-mantle boundary. The only way such stratification could affect the boundary conditions is through a gravitational interaction with the mantle, but this interaction would be a *small* second order contribution. It seems likely that the differences between Kaula and Takeuchi are attributable to differences in the models assumed and not due to computational techniques.

The power-series boundary conditions and physical boundary conditions agree to within 1.4% of the value of ϕ_1 at the core-mantle boundary (Cathles, 1971, App. IV). This discrepancy has a .13% effect on the vertical displacement at the core-mantle boundary. We take these discrepancies to be insignificant and adopt the physical boundary conditions for two reasons: (1) They are easier to calculate; (2) their meaning is more obvious.

Spherical Harmonics

IN THIS APPENDIX we shall discuss two different subjects. First, we shall show how vector spherical harmonics are suitably normalized. For this purpose we shall use fully normalized spherical harmonics. Second, we shall discuss the translation of data into a spherical harmonic description assuming digital techniques are used. There we shall normalize our spherical harmonics slightly differently to follow Ellsaesser (1966) and agree with our computer programs. The filtering of data is discussed.

A. The Normalization of Vector Spherical Harmonics

Suppose $\nabla^2 \zeta = 0$ $\qquad$ (1)

Let $\zeta = R(r)\Theta(\theta)\Phi(\phi)$ and $\mu = \cos\theta$. $\qquad$ (2)

(1) separates into three parts and can be satisfied for any unitary values of $n, m, |m| \leq n, n \geq 0$. If

$\Phi = e^{im\phi}$, then $\qquad$ (3)

Θ = solution to the generalized Legendre equation = $P_l^m(\mu)$, $\qquad$ (4)

$$R = Ar^n + \frac{B}{r}^{\,n+1}.$$ $\qquad$ (5)

Surface spherical harmonics can be constructed from Φ, Θ. Since

$$\int_{-1}^{1} P_l^m(\mu)P_{l'}^m(\mu)d\mu = \frac{2}{2l+1}\frac{(l+m)!}{(l-m)!}\delta_{l'l}$$ $\qquad$ (6)

$$\int_0^{2\pi} d\phi = 2\pi,$$ $\qquad$ (7)

$$Y_l^m(\theta, \phi) = \sqrt{\frac{2l + 1}{4\pi} \frac{(l - m)!}{(l + m)!}} P_l^m(\mu) e^{im\phi}, \tag{8}$$

$$\int_0^{2\pi} d\phi \int_0^\pi \sin \theta \, d\theta \, Y_l^{*m'} Y_l^m = \delta_{l'l} \delta_{m'm}, \tag{9}$$

also

$$Y_l^{-m} = (-1)^m Y_l^{*m} \tag{10}$$

The * indicates complex conjugate. Y_l^m are *normalized spherical har-monics.* Any scaler field over the unit sphere may be expressed as a sum of normalized spherical harmonics:

$$g(\theta, \phi) = \sum_{l=0}^\infty \sum_{m=-l}^l A_l^m Y_l^m(\theta, \phi), \tag{11}$$

$$A_l^m = \int Y_l^{*m}(\theta, \phi) g(\theta, \phi) \sin \theta d\theta d\phi. \tag{12}$$

Asymptotically, for large $n, m = 0, \theta \simeq 90°$

$$Y_n^0 = \frac{\sqrt{2}}{\pi} \sin\left(\left(n + \frac{1}{2}\right)\theta + \frac{\pi}{4}\right). \tag{13}$$

Thus, at large n, the spherical description approaches the flat space one, with $k_x = \dfrac{n + \frac{1}{2}}{r}$.

We can easily get a feel for what the normalization of the term should be, if we assume for a moment $m = 0$.

$$\nabla_1 Y_l^0 = \frac{\partial}{\partial \theta} Y_l^0 \tag{14}$$

$$\frac{\partial}{\partial \theta} = \frac{\partial \mu}{\partial \theta} \frac{\partial}{\partial \mu} = -\sin \theta \frac{\partial}{\partial \mu}, \tag{15}$$

$$P_l^m(\mu) \equiv (-1)^m (1 - \mu^2)^{m/2} \frac{d^m}{d\mu^m} P_l(\mu), \tag{16}$$

$$P_l(\mu) \equiv \frac{1}{2^l l!} \frac{d^l}{d\mu^l} (\mu^2 - 1)^l. \tag{17}$$

Thus $\dfrac{d}{d\theta} P_l(\mu) = P_l^1(\mu),$

$$\sqrt{\frac{4\pi}{2l+1}} \frac{\partial}{\partial \theta} Y_l^0 = \sqrt{\frac{4\pi}{2l+1} \frac{(l+1)!}{(l-1)!}} Y_l^1, \tag{18}$$

$$\frac{\partial}{\partial \theta} Y_l^0 = \sqrt{l(l+1)} Y_l^1. \tag{19}$$

Since, for large l, $|Y_1^0| \cong |Y_l^1|$, we might expect a suitable *normalized vector harmonic* to be defined:

$$\tilde{\mathbf{B}}_l^m \equiv \frac{1}{\sqrt{l(l+1)}} \nabla_1 Y_l^m(\theta, \phi). \tag{20}$$

Similarly, we could define

$$\tilde{\mathbf{A}}_l^m \equiv Y_l^m(\theta, \phi)\hat{r}, \tag{21}$$

so

$$\mathbf{u} = U_l^m \tilde{\mathbf{A}}_l^m + \sqrt{l(l+1)} V_l^m \tilde{\mathbf{B}}_l^m \tag{22}$$

$$(\hat{\mathbf{r}} \cdot \tau)_l^m = P_l^m \tilde{\mathbf{A}}_l^m + \sqrt{l(l+1)} Q_l^m \tilde{\mathbf{B}}_l^{\bar{m}} \tag{23}$$

The coefficients of $\tilde{\mathbf{A}}_l^m$ and $\tilde{\mathbf{B}}_l^m$ may now be meaningfully compared, since $\tilde{\mathbf{B}}_l^m$ and $\tilde{\mathbf{A}}_l^m$ are of comparable magnitudes.

Working through several identities, we may derive an expression for $\tilde{\mathbf{B}}_l^m$ in the general case when $m \neq 0$.[1]

$$\tilde{\mathbf{B}}_l^m = \frac{\sqrt{l(l+1)}}{(2l+1)\sin\theta} \left\{ \hat{\mathbf{a}}_\theta \left[\frac{l-m}{l} \sqrt{\frac{2l+1}{2l-1}} Y_{l-1}^m \right. \right.$$

$$\left. \left. - \frac{l+m+1}{l+1} \sqrt{\frac{2l+1}{2l+3}} Y_{l+1}^m \right] + \hat{\mathbf{a}}_\phi \frac{im(2l+1)}{l(l+1)} Y_l^m \right\}, \tag{24}$$

$$\tilde{\mathbf{B}}_l^0 = \hat{\mathbf{a}}_\theta Y_l^1(\theta, \phi). \tag{25}$$

---

[1] Note the generalized Rodrigues formula cited in Equation (17) is variously defined. Jackson (1967, pp. 54–69) and Abramowitz and Stegun (1965, p. 334) define it as we do here. Jahnke and Emde (1945) and Morse and Feshbach (1953, Vol. II, pp. 1899, 1325) leave out the $(-1)^m$. For this reason, our Equation (25) differs from that cited in Morse and Feshbach.

B. Description of Data by Spherical Harmonics

Let the unnormalized spherical harmonics be written P_{lm}. From (6)
we see that

$$\int_{-1}^{1} P_n^m P_{n'}^m \, d\mu \; = \; 2\delta_{nn'}, \tag{26}$$

provided

$$P_n^m \; = \; P_{nm}/\sqrt{N}, \tag{27}$$

$$N \; = \; \frac{1}{2n + 1} \frac{(n + m)!}{(n - m)!}. \tag{28}$$

(11) may be written:

$$S_r(\theta, \phi) \; = \; \sum_{m=0}^{r} \sum_{n=m}^{r} (A_n^m \cos m\phi \, + \, B_n^m \sin m\phi) \, P_n^m(\theta). \tag{29}$$

The usual least squares technique may be used to estimate A_n^m, B_n^m:
Minimize M_r:

$$M_r \; = \; \frac{1}{4\pi} \int_{-1}^{1} \int_{0}^{2\pi} [S_r - f(\theta, \phi)]^2 \, d\mu d\phi. \tag{30}$$

For M_r to be a minimum

$$dM_r \; = \; 0 \; = \; \frac{\partial M_r}{\partial A_0} dA_0 + \sum_{m=1}^{r} \sum_{n=m}^{r} \frac{\partial M_r}{\partial A_n^m} dA_n^m + \frac{\partial M_r}{\partial B_n^m} dB_n^m.$$

Because of the independence of the coefficients

$$\frac{\partial M_r}{\partial A_0} \; = \; \frac{\partial M_r}{\partial A_n^m} \; = \; \frac{\partial M_r}{\partial B_n^m} \; = \; 0. \tag{31}$$

Using the orthogonality relations for sine and cosine and relation (26)
for the Legendre functions (31) leads directly to:

$$A_0 \; = \; \frac{1}{4\pi} \iint f(\theta, \phi) \, d\mu d\phi, \tag{32}$$

$$
\begin{matrix} A_n^m \\ B_n^m \end{matrix} = \frac{1}{2\pi} \iint f(\theta, \phi) \begin{matrix} \cos m\phi \\ \sin m\phi \end{matrix} P_n^m(\theta)\, d\mu\, d\phi. \tag{33}
$$

(33) has multiplier $\frac{1}{2}$ instead of $\frac{1}{4}$ because of the way we have normalized P_n^m in (26). Equations (32) and (33) are generally valid and provided the integrals can be evaluated will yield undistorted harmonic coefficients from the data $f(\theta, \phi)$.

We may simplify by noting (32), (33) may be written

$$
\begin{matrix} A_n^m \\ B_n^m \end{matrix} = \frac{1}{2} \int_{-1}^{1} \left[\frac{1}{\pi\delta_m} \int_0^{2\pi} f(\theta, \phi) \begin{matrix} \cos m\phi \\ \sin m\phi \end{matrix} d\phi \right] P_n^m(\mu)\, d\mu.
$$

$$
\delta_m = 2 \text{ if } m = 0 \quad \delta_m = 1 \text{ if } m \neq 0.
$$

The brackets are just a simple Fourier transform of $f(\theta, \phi)$ with respect to θ. If we define

$$
\begin{matrix} A_m(\theta) \\ B_m(\theta) \end{matrix} \equiv \frac{1}{\pi\delta_m} \int_0^{2\pi} f(\theta, \phi) \begin{matrix} \cos m\phi \\ \sin m\phi \end{matrix} d\phi. \tag{34}
$$

Then

$$
\begin{matrix} A_n^m \\ B_n^m \end{matrix} = \frac{1}{2} \int_{-1}^{1} \begin{matrix} A_m \\ B_m \end{matrix} P_n^m(\mu)\, d\mu. \tag{35}
$$

The harmonic coefficient may thus be calculated by first Fourier transforming latitude circles and then Legendre transforming the Fourier-transformed latitude circles around selected meridian circles.

The only problem is that the integrals in (34), (35) generally cannot be evaluated. One way or another we usually wind up approximating the integrals digitally. Digitization breaks down the orthogonality relation upon which (35) is based. Interestingly, with slight modification the Fourier orthogonality relation is still valid.

$$
\int_{-\pi}^{\pi} \cos mx \cos nx\, dx = \pi\delta_{mn}\delta_m. \tag{36}
$$

Let $dx = 2\pi/p$; i.e., 2π divided into p intervals

$$
x_j = j\Delta x = j\frac{2\pi}{p} \tag{37}
$$

$$\sum_{j=1}^{p} \cos \frac{2\pi mj}{p} \cos \frac{2\pi nj}{p} \stackrel{?}{=} \frac{p}{2} \delta_{mn} \delta_m . \tag{38}$$

This may be re-expressed:

$$\frac{1}{2} \sum_{j=1}^{p} \left[\cos\left(\frac{2\pi j}{p}(m+n) \right) + \cos \frac{2\pi j}{p}(m-n) \right] \stackrel{?}{=} \frac{p}{2} \delta_{mn} \delta_m . \tag{39}$$

In general

$$\sum_{j=1}^{p} \exp(i2\pi jm/p) = \begin{cases} p & \text{if } \dfrac{n}{p} = \text{Integer} \\[2mm] 0 & \text{if } \dfrac{n}{p} \neq \text{Integer}. \end{cases} \tag{40}$$

(40) can be seen most easily graphically. It just adds phasers with incremental angles $\Delta\theta$. The total $\sum_{j=1}^{p} \Delta\theta = 2\pi m$, which brings us full circle and back to zero. The only way out of this dilemma is if $\Delta\theta = 2\pi$. Then the sum is $\sum_{j=1}^{p} (j) = p$.

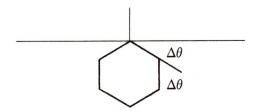

Returning to (39) we see that if $m = n$ the second factor is 1, $\frac{1}{2} \sum_{1}^{p} (1) = \frac{1}{2} p$, and the relation is indeed true. However, if $m = n = \frac{p}{2}$ or p, then (40) indicates the relation on the left of (39) is twice the value on the right. Thus

$$\sum_{j=1}^{p} \cos \frac{2\pi mj}{p} \cos \frac{2\pi nj}{p} = \frac{p}{2} \Delta m \delta_{mn} , \tag{41}$$

where

$$\Delta m = 2 \quad m = 0, \frac{p}{2}, p,$$

$$\Delta m = 1 \quad m \neq \text{above.}$$

(34) becomes

$$\frac{A_m(\theta)}{B_m(\theta)} = \frac{2}{p\Delta m} \sum_{j=1}^{p} f(\theta_j, \phi) \frac{\cos \frac{2\pi mj}{p}}{\sin \frac{2\pi mj}{p}}. \tag{42}$$

The Legendre orthogonality relation (26) needs much more serious alteration. Let

$$\Delta\theta = \frac{\pi}{q} \quad \theta_j = \frac{j\pi}{q},$$

$$d\mu = -\sin\theta d\theta = \frac{-\pi}{q} \sin\theta_j.$$

so (26) becomes

$$\sum_{j=1}^{q} P_n^m(\theta_j) P_{n'}^m(\theta_j) \frac{\pi}{q} \sin\theta_j \overset{?}{=} 2\delta_{nn'}. \tag{43}$$

Relation (43) is almost but not quite valid.

Ellsaesser's approach, and the approach we shall employ, is to restore the exact validity of (43) by slightly altering

$$\frac{\pi}{q} \sin\theta_j.$$

Call this slightly altered factor W_j. Then (43) reads

$$\sum_{j=1}^{q} P_n^m(\theta_j) P_{n'}^m(\theta_j) W_j = 2\delta_{nn'}.$$

(35) then becomes

$$\frac{A_n^m}{B_n^m} = \frac{1}{2} \sum_{k=1}^{q} \frac{A_m(\theta_k)}{B_m(\theta_k)} W_k P_n^m(\mu_k). \tag{44}$$

Ellsaesser (1966) describes how the W_k are determined in his article. He gives the values of W_k for various grids. The main advantage of the Ellsaesser-Neumann method is that, by retrieving the orthogonality of the Legendre functions, the spherical coefficients A_n^m, B_n^m are mutually independent of one another and independent of the point of truncation of the Legendre series (i.e., r in equation (29)).

1. Evaluation of the Significance of Ellsaesser's Coefficients

To determine W_k, n_{max} must be equal to or less than $\frac{q}{4}$. Only if this is so can the equations Ellsaesser uses to determine W_k be solved to yield a unique set W_k. However, such a restriction costs considerable resolution. For example if we have a grid of 72 × 72 points covering a sphere, 72 points around each of 72 meridian circles symmetrically distributed about the equator, ordinarily we could determine spherical harmonic coefficients out to order 36. It thus becomes of great interest to determine exactly how important it is to use Ellsaesser's W_k.

Table App. V-1 compares W_k and $\frac{\pi}{q} \sin \theta_k$ appropriate to the distribution of points mentioned above. As can be seen, W_k differ only very slightly from $\frac{\pi}{q} \sin \theta_k$; and significantly only for the two meridian cir-

k	θ_k	$\frac{\pi}{72} \sin \theta_k$	W_k
1	1.25	.000952	.000831
2	3.75	.002854	.00292
3	6.25	.004747	.00470
4	8.75	.006637	.00667
5	11.25	.008512	.00848
8	18.75	.01402	.01404
12	28.75	.02098	.02099
15	36.25	.02580	.02579
20	48.75	.03280	.03280
25	61.25	.03825	.03825
30	73.75	.04188	.04189

App. Table V-1 Comparison of Ellsaesser's weights, W_k, and $\frac{\pi}{q} \sin \theta_k$.

cles nearest each pole. Assuredly W_k are preferable to $\frac{\pi}{q}\sin\theta_k$. However, the error involved in evaluating coefficients out to order 36 using W_k determined from $q = 72$, as compared to evaluating the same coefficients using W_k determined from $q = 144$, is probably small. In any case it is an error we shall accept in the interest of resolution. In this book we shall evaluate Legender coefficients out to order 36 from data on a 72 × 72 grid using W_k determined for a 144 × 144 grid.

2. Filtering of Data to Eliminate the Gibbs Phenomenon

It is well known that the retransform of the Fourier transform of a function with sharp edges tends to overshoot the original function at sharp edges. For example the retransform of the transform of a boxcar function

$$f(x) = 1 \quad -a \leq x \leq a$$
$$f(x) = 0 \quad \text{otherwise}$$

will look

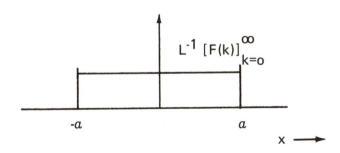

The overshoot is about .2 the height of the original step. This overshoot is known as the Gibbs phenomenon.

If we were to retransform using Fourier coefficients only out to a certain frequency k_c, the Gibbs phenomenon would still be present, only more smeared out. Clearly it is important to eliminate the Gibbs effect. We must eliminate these ridges and troughs which are mathematical artifacts completely unrelated to the physical analysis of our problem or to what would actually happen in the real world. The Gibbs effect is produced by the high frequencies in the sharp edges of $f(x)$. Thus to eliminate it we wish to smooth the edges of $f(x)$.

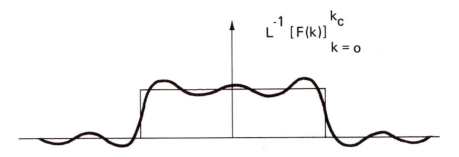

It is simplest to look at the problem from a slightly different perspective, however. What we really want is to obtain a function $F(k)$ whose transform $L^{-1} (F(k))_{k=0}^{k}$ will have suitably low sidelobes. If $F(k)$ is Gaussian, the transform of $F(k)$ will have no sidelobes at all! The problem with a Gaussian distribution is that it extends to infinite frequencies. We must truncate our series at some finite frequency. Consequently, we will actually always take the retransform of a Gaussian and a boxcar of some height that depends on where the Gaussian is truncated. In polar coordinates this would look:

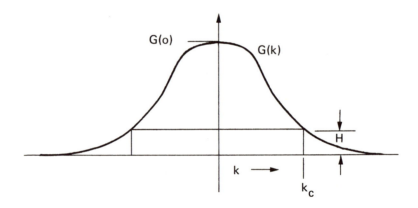

The sidelobes of the transform of a boxcar are about $.2H$ where $H =$ height of the boxcar. Thus we may make the Gibbs warbles as small as we like, by choosing H (at k_c) such that $.2H$ is less than the maximum warble amplitude which can be tolerated. Table App. V-2 gives $G(n) = e^{-\alpha n^2}$ arranged so that $.2e^{-\alpha n^2} \leq (2\%, 1\%,$ and $.5\%)G(0)$. The values of $G(n)$ in Table App. V-2 may be used directly as filter weights to smooth the transform of a given function so as to eliminate the Gibbs warble.

n	2%	1%	.5%	n	.5%	n	.5%
1	.995	.990	.990	1	1.0	19	.43
2	.970	.961	.961	2	1.0	20	.39
3	.972	.923	.905	3	.98	21	.36
4	.896	.861	.835	4	.96	22	.33
5	.835	.795	.748	5	.94	23	.29
6	.779	.719	.664	6	.92	24	.26
7	.712	.638	.571	7	.90	25	.24
8	.638	.554	.482	8	.86	26	.21
9	.560	.472	.395	9	.82	27	.18
10	.491	.395	.320	10	.79	28	.16
11	.423	.301	.251	11	.75	29	.14
12	.361	.264	.194	12	.72	30	.12
13	.301	.210	.145	13	.68	31	.11
14	.249	.163	.105	14	.64	32	.09
15	.201	.124	.077	15	.59	33	.08
16	.162	.093	.054	16	.55	34	.07
17	.128	.069	.037	17	.50	35	.06
18	.100	.050	.025	18	.47	36	.05

App. Table V-2 2%, 1% and .5% Gaussian filters. The filtering of a spike is shown in Figure App. V-1.

In Figure App. V-1 the weights of Table App. V-2 are applied to a delta function. It can be seen that the Gibbs warble is eliminated to the percentage of the amplitude of the function predicted. No claim is made that such truncated Gaussian filters are optimum in any sense. However even small efforts to improve the filter by rounding the sharp edge at the truncation point (n_c) appears always to result in larger, not smaller, warbles in the transformed function.

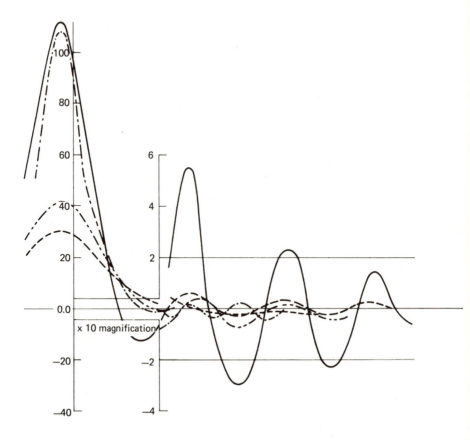

Figure App. V-1. Effects of filtering a spike. (—) Unfiltered, (-·-·-) .5% Gaussian filter $n_{max} = 36$, (·····) 2% Gaussian filter $n_{max} = 18$, (----) .5% Gaussian filter $n_{max} = 18$. The longer dashed curve is the unmagnified extension of the unfiltered spike.

The Isostatic Adjustment of a Layered Viscous Half-Space

WE FIRST CALCULATE some simple cases involving flow in a thin channel or in a uniform viscous half-space. The half-space solutions are used to generate some useful approximations that we have used extensively in the text. We next calculate some simple examples involving viscous layering and channel flow taking into account the effects of an elastic (non-fluid) surface layer.

A. Thin Channel or Half-space Flow with No Lithosphere

Suppose the surface of the fluid has an initial form, f, that is either symmetric about the y axis ($\hat{z}$ vertical) or cylindrically symmetric. The Fourier or Hankel transforms of that surface form are:

$$F(k) = \frac{1}{\sqrt{2\pi}} \int_{-\infty}^{\infty} f(x) e^{-ikx} dx, \tag{1}$$

$$F(k) = \int_{r=0}^{\infty} f(r) J_0(kr) r \, dr. \tag{2}$$

J_0 is a Bessel Function of zero order, k is the wave number, $\frac{2\pi}{\lambda}$, and in the cylindrically symmetric case equals $\sqrt{k_x^2 + k_y^2}$. In the case of (1) $k = k_x$, $k_y = 0$. For the case of uniform half-space viscosity, each component wavelength decays exponentially:

$$e^{-\rho g t / 2\eta k}, \tag{3}$$

(see equation III-17). For thin channel flow, equation III-25 shows each harmonic decays exponentially according to

$$e^{-\rho g k^2 H^3 / 3\eta} \tag{4}$$

In the above η is the viscosity of the half-space or channel, ρ is the density of the fluid, g is the gravitational constant and H is the thickness of the channel.

If we now make the problem dimensionless by letting

$$\tilde{r} = \frac{r}{\sigma}, \tag{5a}$$

$$\tilde{k} = k\sigma, \tag{5b}$$

$$t'_{HS} = \frac{\rho g \sigma}{2\eta} t, \tag{5c}$$

$$t'_C = \frac{\rho g H^3}{3\eta\sigma^2} t, \tag{5d}$$

we see the time dependent uplift of any symmetric surface depression of a half-space may be expressed:

$$f(\tilde{x}, t'_{HS}) = \frac{-1}{\sqrt{2\pi}} \int_{-\infty}^{\infty} F(\tilde{k}) e^{i k \tilde{x}} e^{-t'_{HS}/\tilde{k}} d\tilde{k},$$

$$f(\tilde{r}, t'_{HS}) = \int_{\tilde{k}=0}^{\infty} F(\tilde{k}) e^{-t'_{HS}/\tilde{k}} J_0(\tilde{k}\tilde{r}) \tilde{k} \, d\tilde{k}.$$

Half-Space Uplift (6)

Similarly, the time-dependent uplift of any symmetric surface depression assuming channel flow can be expressed

$$f(\tilde{x}, t'_C) = \frac{-1}{\sqrt{2\pi}} \int_{-\infty}^{\infty} F(\tilde{k}) e^{i k x} e^{-t'_C \tilde{k}^2} d\tilde{k},$$

$$f(\tilde{r}, t'_C) = \int_{\tilde{k}=0}^{\infty} F(\tilde{k}) e^{-t'_C \tilde{k}^2} J_0(\tilde{k}\tilde{r}) \tilde{k} \, d\tilde{k}.$$

Thin Channel Uplift (7)

For simple initial surface depression, some of the integrals indicated in (6) and (7) can be solved analytically, some must be solved nu-

Initial Surface Form	Fourier or Hankel Transform of Surface	Channel Response	Half-Space Response
Gaussian Trough $h(\tilde{x}) = -e^{-\tilde{x}^2}$	$-\dfrac{1}{\sqrt{2}} e^{-\tilde{k}^2/4}$	$h(\tilde{x}, t'_c) = \dfrac{-1}{\sqrt{1+4t'_c}} e^{-\tilde{x}^2/(1+4t'_c)^{1/2}}$	Solved numerically by Burgers and Collette (1958) and here.
Square-edged Trough $h(\tilde{x}) = -H(|\tilde{x}| - 1)$	$-\sqrt{\dfrac{2}{\pi}}\, \dfrac{\sin \tilde{k}}{\tilde{k}}$	$h(\tilde{x}, t'_c) = -\dfrac{1}{2}\left[\text{erf}\left(\dfrac{1+\tilde{x}}{2\sqrt{t'_c}}\right) - \text{erf}\left(\dfrac{\tilde{x}-1}{2\sqrt{t'_c}}\right)\right]$ same as Jefferies, 1970, p 423, except for $-\dfrac{1}{2}$	Can be expressed analytically. Given by Haskell (1936)
Cylindrically Symmetric Gaussian $h(\tilde{r}) = -e^{-\tilde{r}^2}$	$\dfrac{J_1(\tilde{k})}{\tilde{k}}$	$h(\tilde{r}, t'_c) = \dfrac{-1}{1+4t'_c} e^{-\tilde{r}^2/(1+4t'_c)}$	Solved numerically and by series expansion by Burgers and Collette (1958)
Square-edged Cylindrical Depression $h(\tilde{r}) = -H(\tilde{r} - 1)$	$\dfrac{1}{2} e^{-\tilde{k}^2/4}$	Solved numerically here	Solved numerically by Cathles (1965) and given here in Tables I, II.

App. Table VI-1 The isostatic adjustment of various initial surface forms for channel flow (heat conduction analogy) and deep flow models.

merically. Table App. VI-1 lists some simple initial surface depressions that have been considered in the past, and either gives an expression for the proper solution or gives references to where it may be found.

The main difficulty in evaluating the integrals shown in (6) and (7) for the simple forms of Table App. VI-1 is convergence. Particularly in the cylindrically symmetric square-edged cases this is a problem. A natural solution is simply to filter the deformation. This is done naturally by the lithosphere, as shown in the next section. Another easy solution for the half-space flow case is simply to calculate the integral out to short enough wavelengths (high enough k) that very little adjustment has taken place. Since the integral out to this value of k can be calculated for $t'_{HS} = 0$, and since the integral to infinity is known analytically, the residual that must be added to make the integral correct can be calculated. This method was used by Cathles (1965) for the case of a cylindrically square-edged depression. The results are tabulated in Tables App. VI-2 and App. VI-3 and plotted in Figures App. VI-1 and App. VI-2. The residuals indicate the maximum possible error in the calculation, so these tables are good to at least $\pm 1\%$ of the initial surface depression (i.e. $\pm .01$).

Figure App. VI-3 shows the natural log of the central uplift remaining plotted against t'_{HS} for the case of a square-edged trough and a cylindrically square-edged depression. It can be seen that in both cases the uplift is convincingly exponential. The decay time indicated suggests the center of a square edged depression uplift as if it had a wave number $\bar{\bar{k}}$:

$$\bar{\bar{k}} = 1.7\left(\frac{1}{\tilde{L}^2} + \frac{1}{\tilde{M}^2}\right)^{1/2}. \tag{8}$$

Here $\tilde{L}$ and $\tilde{M}$ are the characteristic dimensions of the surface depression in normalized units (5). For a cylindrical depression $L = M = 2R_0$. For an infinite trough, $L =$ width of the trough, $M = \infty$. Heiskanen and Vening Meinesz (1958, p. 369) proposed that a relation similar to (8) might be useful but did not definite it precisely.

For the case of a cylindrical square-edged depression (8) implies:

$$\boxed{\bar{\lambda} = 5.2\, R_0} \tag{9}$$

$$\boxed{\bar{k} = 1.7\left(\frac{1}{L^2} + \frac{1}{M^2}\right)^{1/2}} \tag{10}$$

Residuals

	R = 0.0	R = 0.1	R = 0.3	R = 0.5	R = 0.7	R = 0.9	R = 1.1	R = 1.3	R = 1.5	R = 1.7	R = 1.9	
	-0.056	+0.001	+0.009	-0.013	+0.024	-0.003	-0.019	+0.013	-0.011	+0.003	-0.001	

Tabulated Vertical Displacements

R = 0.0	R = 0.1	R = 0.3	R = 0.5	R = 0.7	R = 0.9	R = 1.1	R = 1.3	R = 1.5	R = 1.7	R = 1.9	Time
-1.000	-1.000	-1.000	-1.000	-1.000	-1.000	+0.	0.	+0.	0.	+0.	0.0
-0.832	-0.838	-0.836	-0.843	-0.856	-0.878	+0.081	+0.061	+0.048	+0.040	+0.034	0.2
-0.702	-0.709	-0.708	-0.723	-0.746	-0.787	+0.133	+0.096	+0.074	+0.059	+0.049	0.4
-0.596	-0.603	-0.605	-0.624	-0.657	-0.715	+0.170	+0.118	+0.088	+0.069	+0.055	0.6
-0.508	-0.515	-0.519	-0.542	-0.583	-0.655	+0.197	+0.132	+0.095	+0.072	+0.056	0.8
-0.434	-0.440	-0.446	-0.473	-0.520	-0.605	+0.216	+0.140	+0.098	+0.073	+0.055	1.0
-0.371	-0.377	-0.384	-0.414	-0.466	-0.562	+0.230	+0.144	+0.098	+0.071	+0.052	1.2
-0.317	-0.323	-0.331	-0.363	-0.419	-0.526	+0.240	+0.146	+0.096	+0.068	+0.048	1.4
-0.270	-0.277	-0.285	-0.319	-0.378	-0.494	+0.246	+0.145	+0.093	+0.064	+0.044	1.6
-0.231	-0.237	-0.245	-0.280	-0.343	-0.466	+0.251	+0.143	+0.089	+0.059	+0.040	1.8
-0.196	-0.202	-0.211	-0.247	-0.311	-0.441	+0.253	+0.140	+0.084	+0.055	+0.036	2.0
-0.166	-0.172	-0.181	-0.217	-0.283	-0.419	+0.253	+0.136	+0.080	+0.050	+0.032	2.2
-0.140	-0.146	-0.154	-0.191	-0.258	-0.399	+0.253	+0.131	+0.075	+0.046	+0.028	2.4
-0.118	-0.123	-0.131	-0.168	-0.236	-0.381	+0.252	+0.127	+0.070	+0.042	+0.025	2.6
-0.098	-0.103	-0.111	-0.148	-0.216	-0.365	+0.250	+0.122	+0.065	+0.038	+0.022	2.8
-0.081	-0.086	-0.094	-0.130	-0.198	-0.350	+0.247	+0.117	+0.060	+0.034	+0.019	3.0
-0.067	-0.071	-0.078	-0.114	-0.181	-0.336	+0.244	+0.112	+0.056	+0.031	+0.016	3.2
-0.054	-0.058	-0.065	-0.100	-0.166	-0.323	+0.240	+0.107	+0.051	+0.027	+0.014	3.4
-0.043	-0.046	-0.053	-0.087	-0.153	-0.312	+0.237	+0.102	+0.047	+0.024	+0.011	3.6
-0.033	-0.036	-0.042	-0.076	-0.141	-0.301	+0.233	+0.097	+0.043	+0.021	+0.009	3.8
-0.025	-0.028	-0.033	-0.066	-0.129	-0.290	+0.229	+0.093	+0.040	+0.019	+0.008	4.0

App. Table VI-2 Theoretical uplift at various radii and times (t_{HS}) for a cylindrical square-edged depression in a fluid of uniform viscosity. Calculated from equation 6b. "Residuals" are discussed in the text and are added to the integral to correct for lack of convergence.

Rate of Uplift at Time T and Radius R

$R = 0.0$	$R = 0.1$	$R = 0.3$	$R = 0.5$	$R = 0.7$	$R = 0.9$	$R = 1.1$	$R = 1.3$	$R = 1.5$	$R = 1.7$	$R = 1.9$	Time
0.840	0.810	0.821	0.783	0.719	0.610	0.403	0.303	0.242	0.201	0.170	0.1
0.651	0.647	0.636	0.604	0.550	0.455	0.264	0.178	0.128	0.096	0.074	0.3
0.530	0.529	0.518	0.492	0.446	0.362	0.185	0.111	0.071	0.047	0.031	0.5
0.441	0.441	0.431	0.410	0.372	0.298	0.133	0.069	0.036	0.019	0.007	0.7
0.371	0.371	0.364	0.346	0.315	0.250	0.096	0.041	0.014	0.001	−0.006	0.9
0.315	0.316	0.310	0.296	0.270	0.213	0.069	0.021	−0.000	−0.009	−0.014	1.1
0.269	0.270	0.266	0.255	0.233	0.183	0.049	0.007	−0.010	−0.016	−0.018	1.3
0.231	0.232	0.229	0.221	0.204	0.160	0.033	−0.003	−0.016	−0.020	−0.020	1.5
0.199	0.200	0.198	0.192	0.179	0.140	0.021	−0.011	−0.020	−0.022	−0.021	1.7
0.172	0.174	0.172	0.168	0.158	0.124	0.011	−0.016	−0.023	−0.022	−0.021	1.9
0.149	0.151	0.150	0.148	0.140	0.111	0.004	−0.020	−0.024	−0.022	−0.020	2.1
0.130	0.131	0.131	0.130	0.125	0.100	−0.002	−0.022	−0.025	−0.022	−0.019	2.3
0.113	0.114	0.115	0.115	0.112	0.090	−0.007	−0.024	−0.025	−0.021	−0.017	2.5
0.098	0.099	0.100	0.102	0.100	0.082	−0.011	−0.025	−0.024	−0.020	−0.016	2.7
0.085	0.086	0.088	0.090	0.090	0.075	−0.013	−0.025	−0.024	−0.019	−0.015	2.9
0.074	0.075	0.077	0.080	0.082	0.069	−0.016	−0.025	−0.023	−0.018	−0.013	3.1
0.064	0.066	0.068	0.071	0.074	0.063	−0.017	−0.025	−0.022	−0.016	−0.012	3.3
0.056	0.057	0.059	0.063	0.067	0.059	−0.019	−0.024	−0.021	−0.015	−0.011	3.5
0.048	0.050	0.052	0.056	0.061	0.054	−0.020	−0.024	−0.020	−0.014	−0.010	3.7
0.042	0.043	0.046	0.050	0.056	0.051	−0.020	−0.023	−0.018	−0.013	−0.009	3.9

App. Table VI-3   Theoretical rate of uplift at various radii and times (t/t_{HS}) for a cylindrical square-edged depression in a fluid of uniform viscosity. Calculated from Table App. VI-2.

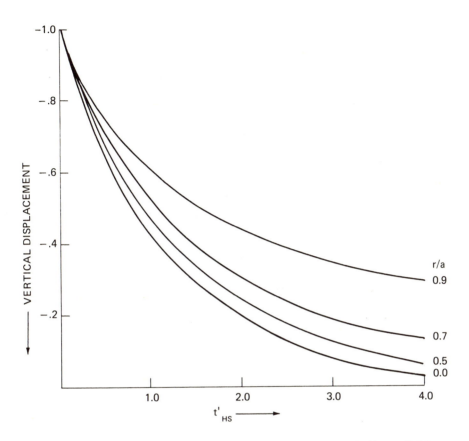

Figure App. VI-1. Uplift as a function of time (t'_{HS}) at various radii inside a cylindrical square-edged depression extending to $\frac{r}{a} = 1.0$ in a fluid of uniform viscosity.

The uplift of the center of the deformation may be described

$$\frac{h}{h_0} = e^{-\rho g t / 2\eta \bar{k}} \qquad (11)$$

Note that we have switched back to dimensional units by using (5). L is the length, and M is the width of the initial surface depression.

We can test (11) further using Haskell's (1936) calculations for an infinite square-edged load. If $\alpha \equiv \frac{x}{l}$, $\beta \equiv \frac{z}{l}$, Haskell showed the total

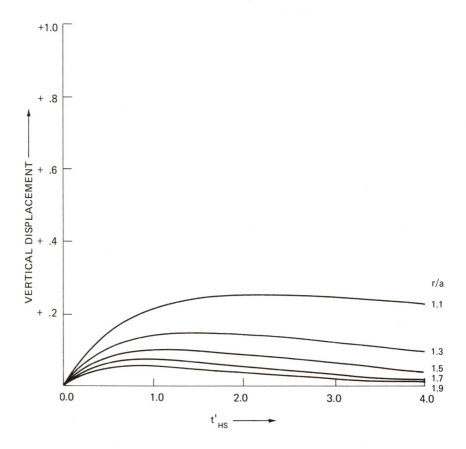

Figure App. VI-2. Uplift as a function of time (t'_{HS}) at various radii external to a cylindrical square-edged depression in a fluid of uniform viscosity.

ultimate vertical displacement at a depth of β = 2.25 was half the total ultimate vertical displacement at z = 0. Further, the depth beneath which half the ultimate fluid transport across a vertical plane occurred is β = 2.25 at the center of the load. These two depths are identical as they logically must be since the source of material for horizontal flow is the vertical displacement caused by the load.

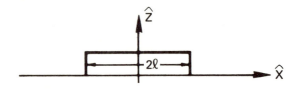

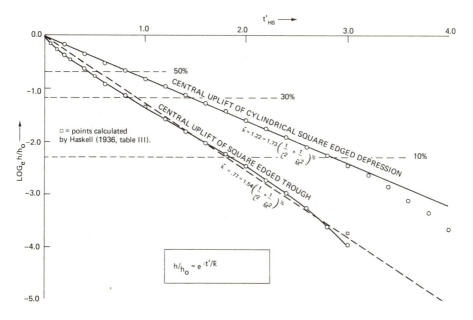

Figure App. VI-3. The logarithm of uplift remaining for the central regions of a square-edged trough and a cylindrical square-edged depression plotted versus t'_{HS}. It can be seen that the central uplift is exponential and similar to that of a harmonic surface deformation with wave number $\tilde{k} = 1.7\left(\dfrac{1}{\tilde{L}^2} + \dfrac{1}{\tilde{M}^2}\right)$. $\tilde{L}$ and $\tilde{M}$ are characteristic dimensions of the initial deformation in normalized units (see equation 5). The values calculated numerically for the square-edged trough are compared to the values calculated analytically and given by Haskell. A uniform viscosity half-space is assumed.

Using equation III-13, it is easy to calculate the depth at which the vertical displacement is half its surface value or at which the horizontal transport across a vertical plane is equal above and below a certain horizontal plane. Call such a depth $z_{.5}$. From (III-13) or Figure III-2.

$$z_{.5}\bar{k} = 1.66$$

or

$$z_{.5} = \frac{1.66}{\bar{k}}.$$

Using our approximation for $\bar{k} = \dfrac{1.7}{2l}$,

$$z_{.5}_{\text{Trough}} = 2.0\,l, \tag{12}$$

or in Haskell's notation

$$\beta_0 = 2.0.$$

This compares well with 2.25 obtained by Haskell's exact calculation. Since for a cylindrical load

$$\bar{k} = \frac{1.70}{\sqrt{2}}\left(\frac{1}{R_0}\right) = \frac{1.2}{R_0},$$

$$z_{.5}_{\text{cyl}} = \frac{1.66}{k} = 1.4\,R_0. \tag{13}$$

From (III-13) we may also calculate the depth at which the vertical displacement or below which the total horizontal flow is .2 and .1, the total vertical displacement at $z = 0$ or the total horizontal flow.

$$z_{.2}_{\text{cyl}} = 2.5\,R_0 \tag{14}$$

$$z_{.1}_{\text{cyl}} = 3.3\,R_0 \tag{15}$$

Thus, if we approximate the Fennoscandian glacial load by a cylindrical load 550 km in radius, we see

$$z_{.5}_{\substack{\text{cyl} \\ \text{Fennoscandia}}} = 825\,\text{km}$$

$$z_{.2}_{\substack{\text{cyl} \\ \text{Fennoscandia}}} = 1475\,\text{km}.$$

Substantial flow thus persists to great depths into the mantle, even in the Fennoscandian case. This was pointed out by Haskell 35 years ago.

B. More Complicated Layered Viscosity Cases that Include the Effects of the Lithosphere

The cases: 1) channel flow where the channel is not necessarily thin with respect to the size of the load, 2) a less (or more) viscous channel overlying a viscous half space, or 3) an elastic layer overlying the viscous half-space, can be easily computed given the foundations laid in Section III.A.2.c,d, and e of the text.

It was shown in Section III.A.2.e that the effects of a lithosphere would be to filter the load.

$$F(k) \rightarrow \frac{F(k)}{\alpha(k)}, \tag{16}$$

and to increase the rate of decay of short wavelength load harmonic:

$$e^{-\rho g t / z \eta k} \rightarrow e^{-\frac{\rho g t \alpha(k)}{2 \eta k}}, \tag{17}$$

where

$$\alpha(k) \equiv 1 + \frac{k^4 D}{\rho g}. \tag{18}$$

D is the flexural rigidity of the lithosphere. For a channel of any thickness the decay time

$$\tau = \frac{2 \eta k}{\rho g} \tag{19}$$

is altered by a factor $\mathcal{L}(k, H)$:

$$\tau = \frac{2 \eta k}{\rho g} \mathcal{L}(k, H), \tag{20}$$

where $\mathcal{L}(k, H)$ is defined by equation III-27 and depends on k and the channel thickness H.

For the case of a layer of different viscosity overlying a viscous half-space (19) is altered:

$$\tau = \frac{2 \eta k}{\rho g} \mathcal{R}(k, H, \tilde{\eta}), \tag{21}$$

where $\mathcal{R}(k, H, \tilde{\eta})$ is defined in Equation III-21. $\tilde{\eta}$ is the ratio of the channel viscosity to the viscosity of the half-space beneath. When kH becomes small, (20) becomes $\dfrac{3 \eta}{\rho g k^2}$ as shown in (III-23). When kH becomes large, (21) becomes $\dfrac{2 \tilde{\eta} \eta k}{\rho g}$ which is the solution appropriate for short enough wavelengths that the channel appears infinitely thick. $\tilde{\eta}\eta$ is the viscosity of the channel and H is the thickness of the channel.

If the thickness of the channel is normalized:

$$\tilde{H} = \frac{H}{\sigma} \tag{22}$$

$\alpha(k)$ becomes

$$\alpha(\tilde{k}) = 1 + \frac{\tilde{k}^4 D}{\rho g \sigma^4} \tag{23}$$

and the inverse transform (6) can be written for the channel flow case taking into account (16), (17), (20), (22), (23):

$$f(\tilde{x}, t'_{HS}) = \frac{-1}{\sqrt{2\pi}} \int_{-\infty}^{\infty} \frac{F(\tilde{k})}{\alpha(\tilde{k})} e^{ik\tilde{x}} e^{-t'_{HS}\alpha(\tilde{k})/\mathcal{L}\tilde{k}} \, d\tilde{k}$$

$$f(\tilde{r}, t'_{HS}) = \int_{\tilde{k}=0}^{\infty} \frac{F(\tilde{k})}{\alpha(\tilde{k})} e^{-t'_{HS}\alpha(\tilde{k})/\tilde{k}} J_0(\tilde{k}\tilde{r}) \tilde{k} \, d\tilde{k}$$

Channel of any Thickness (24)

(24) describes the response to unloading of a channel that is not necessarily thin with respect to the load size. The lithosphere is taken into account. When $\tilde{k}\tilde{H}$ becomes small, (24) is equivalent to (7) with the lithosphere included.

The case of a viscous channel overlying a viscous half-space and being overlain by an elastic lithosphere can be written, with the aid of (16), (17), (21), (22), (23):

$$f(\tilde{x}, t'_{HS}) = \frac{-1}{\sqrt{2\pi}} \int_{-\infty}^{\infty} \frac{F(\tilde{k})}{\alpha(\tilde{k})} e^{ik\tilde{x}} e^{-t'_{HS}\alpha(\tilde{k})/\tilde{k}\tilde{k}} \, d\tilde{k}$$

$$f(\tilde{r}, t'_{HS}) = \int_{\tilde{k}=0}^{\infty} \frac{F(\tilde{k})}{\alpha(\tilde{k})} e^{-t'_{HS}\alpha(\tilde{k})/\tilde{k}\tilde{k}} J_0(\tilde{k}\tilde{r}) \tilde{k} \, d\tilde{k}$$

Viscous Channel Overlying a Viscous Half-Space (25)

(25) becomes equivalent to (6) with the lithosphere included and

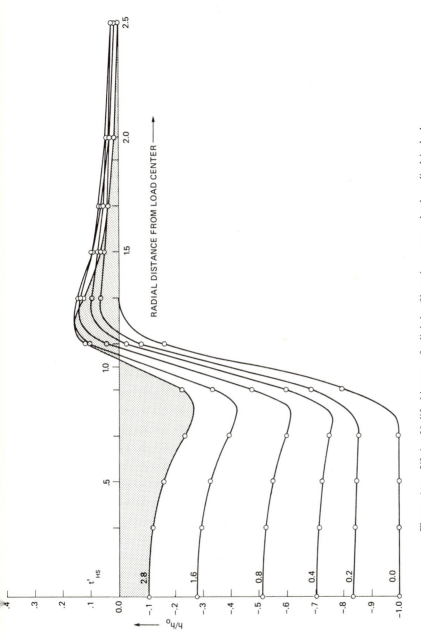

Figure App. VI-4. Uplift history of slightly filtered, square-edged, cylindrical depression:

- In isostatic equilibrium under load at $t'_{HS} = 0$
- Uniform viscous half-space
- No lithosphere

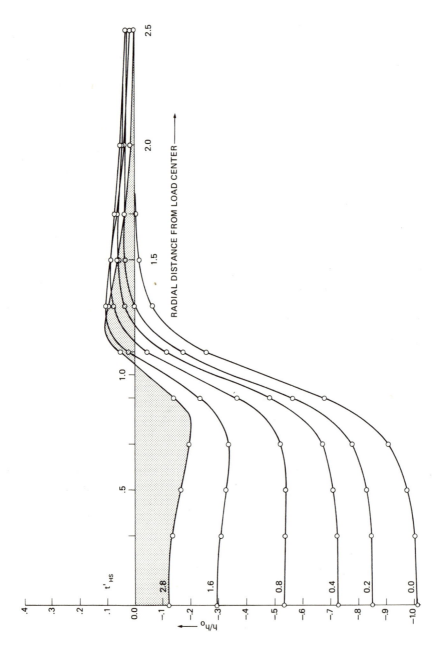

Figure App. VI-5. Uplift history of slightly filtered square-edged cylindrical depression:

- Load removed after having been applied for a time $t_{HS} = T = 5.6$ (i.e. ~20,000 years in Fennoscandian case)
- Uniform viscous half-space

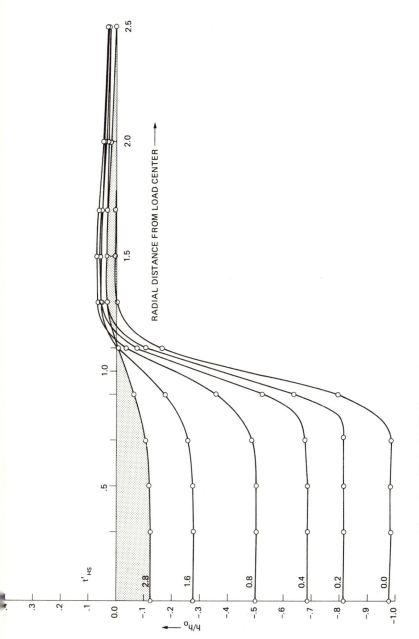

Figure App. VI-6.   Uplift history of slightly filtered square-edged cylindrical depression:

- Load removed at $t_{HS} = 0$ after having been applied for a time $t_{HS} = 5.6$.
- Viscous channel $.136\,\sigma$ thick (i.e., 75 km in Fennoscandia case) and $.04$ times the viscosity of the viscous half-space beneath.
- No lithosphere

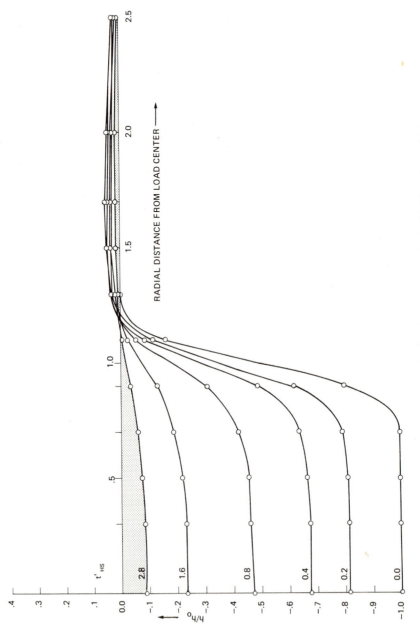

Figure App. VI-7. Uplift history of slightly filtered square-edged cylindrical depression:

- Load removed at t'_{HS} = 0 after application for t'_{HS} = 5.6.
- Viscous channel .182 σ thick (i.e., 100 km in Fennoscandia case) and .04 times the viscosity of the half-space beneath.
- No lithosphere

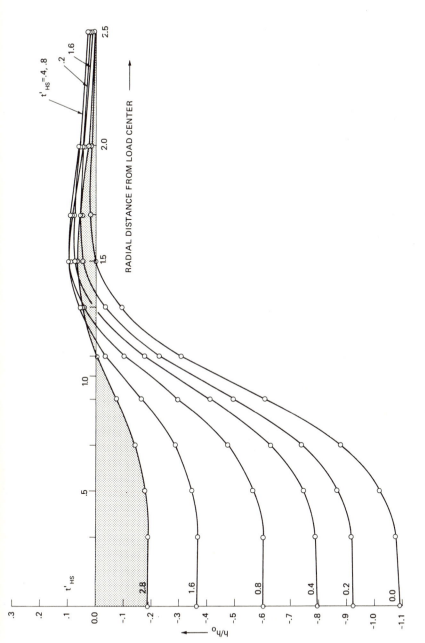

Figure App. VI-8. Uplift history of slightly filtered, square-edged, cylindrical depression:

- Isostatic equilibrium under load at $t_{HS} = 0$.
- Uniform viscous half-space.
- Lithosphere with flexural rigidity 50×10^{23} N-m (For Fennoscandian-sized load).

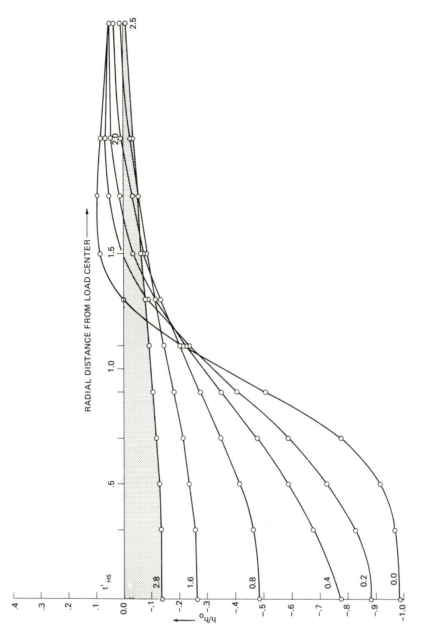

Figure App. VI-9. Uplift history of slightly filter, square-edged, cylindrical depression:

- Load removed after having been applied for t'_{HS} = 5.6.
- Flow restricted to channel .182 σ thick (i.e., 100 km for Fennoscandian case). Channel viscosity is 10^{20} poise (i.e. $\tilde{\eta}$ = 10^{-2}).
- Lithosphere with flexural rigidity 50 × 10^{23} N-m (for Fennoscandian-sized load).

$\eta = \tilde{\eta}\eta$ when $\tilde{k}\tilde{H}$ becomes large, and is equivalent to (6) with the lithosphere included and $\eta = \eta$ when $\tilde{k}\tilde{H}$ is small. Note that $\mathcal{L}$ and $\mathcal{R}$ contain only factors kH, so they are the same normalized or un-normalized.

A load cycle may be taken into if the time decay expressions in (24) and (25) are altered:

$$e^{-t'_{HS}\alpha(\tilde{k})/\mathcal{R}\tilde{k}} \rightarrow e^{-t'_{HS}\alpha(\tilde{k})/\mathcal{R}\tilde{k}} - e^{-(t'_{HS}+T)\alpha(\tilde{k})/\mathcal{R}\tilde{k}} \qquad (26)$$

A similar expression holds for (24), only $\mathcal{R}$ is replaced by $\mathcal{L}$. T is the time the load is applied prior to sudden removal at $t'_{HS} = 0$.

Figures App. VI-4 to 9 illustrate the use of (24)–(26) for cases of interest. The figures are self explanatory. Since $t'_{HS} = 2.8$ implies (by 5c) $t = 10,000$ years if $\rho = 3.313$, $\sigma = 550$ km and $\eta = 10^{22}$ poise, $t'_{HS} = 2.8$ corresponds to present time for the case of the Fennoscandian uplift. The uplift remaining is shaded in the figures. The depressions were slightly filtered to eliminate convergence problems. (No residuals were added to the inverse integral).

Glacial Uplift in Canada

A. Andrews' Analysis[1]

AFTER INVESTIGATING 21 sites in Canada for which the uplift was known, Andrews found the uplift could be well characterized by a formula of the form

$$u(\bar{t}) = u_\infty e^{-k\bar{t}}, \tag{1}$$

where $u(\bar{t})$ is the amount of uplift remaining at any time, $\bar{t} = T - t$. $\bar{t}$ is zero T thousands of years before present. Andrews found that k was quite similar for the 21 sites, being equal to $.45 \pm .05$, if $\bar{t}$ was measured in thousands of years.

If the normalized uplift of the 21 sites was averaged, the k that best characterized the entire uplift was:

$$k = .390.$$

Andrews then used $k = .390$ to characterize the uplift in all of Canada. It is not immediately clear why .390 is preferable to .450.

The prime advantage of assuming a constant form for the exponential decay, k, for the glacial uplift is that then only *one* variable, u_∞, in equation (1) remains undetermined. The uplift over any interval of time determines this constant. The amount of uplift from $\bar{t} = 0$ to present, the rate of uplift at any time, or the uplift over any period can be calculated:

$$\begin{matrix} \text{Uplift over last} \\ t \text{ thousand years} \end{matrix} = u_\infty (e^{-(T-t)k} - e^{-Tk}), \tag{2}$$

$$\begin{matrix} \text{Uplift remaining} \\ \text{at present } (\bar{t} = T) \end{matrix} = u_\infty e^{-kT}, \tag{3}$$

[1] The best reference for this discussion is Andrews (1970).

$$\frac{\text{Rate of uplift}}{\text{at present}} = \left. \frac{\partial u(\bar{t})}{\partial t} \right|_{\bar{t}=T} = u_\infty k e^{-kT}. \tag{4}$$

Where t = thousands of years before present,

      T = constant number of years before present which Andrews associates with the time of deglaciation,

      $\bar{t}$ = $(T - t)$ in equation (1).

Equation (1) can be expressed in an alternate form:

$$u(\bar{t}) = u_\infty i^{\bar{t}} \tag{5}$$

Where $k = -\ln i$. $i = .677$ if $k = .390$.

In fact equation (5) may be related to an infinite series: Suppose the uplift at any time, $\bar{t}$, is given by

Uplift at $\bar{t} = A + Ai + Ai^2 + \ldots + Ai^{\bar{t}} =$

$$\frac{A(1 - i^{\bar{t}})}{1 - i}.$$

The sum of the above series with $\bar{t} = \infty$ is:

Total uplift $(\bar{t} = \infty) = \dfrac{A}{1 - i}.$

Thus the uplift remaining at any time, $u(t)$ is just

$$u(\bar{t}) = \frac{A}{1 - i} - \frac{A(1 - i^{\bar{t}})}{1 - i} = \frac{A}{1 - i} i^{\bar{t}}$$

and we see that

$$u_\infty = \frac{A}{1 - i}. \tag{6}$$

Andrews determined A from the uplift curves since the date of deglaciation for 58 sites in Arctic Canada. We repeat his data in Table App. VII-1. T is the date of deglaciation in thousands of years BP given by Andrews. This he picked as the start of his uplift interval. The end of the interval is present. U_p is the amount of post-glacial uplift from T to present. Andrews determined A by fitting the standard

Appendix Table VII-1

	Data from Andrews (1970a)					Calculated from Andrews' Data					
Site	Length of post-glacial recovery T [10^3 yrs]	A	Uplift from T to present U_p [m]	Ultimate total uplift from T U_∞ m	Uplift remaining U_r m	Present rate of uplift U_T m/1000 yrs	Uplift in last 2000 years U_{2k} m	Uplift in last 4000 years U_{4k} m	Uplift in last 6000 years U_{6k} m	Uplift in last 7000 years U_{7k} m	Uplift in last 8000 years U_{8k} m
1	7.3	55.3	158	171	10	3.87	12	37	93	142	215
2	8.4	84.3	245	261	10	3.84	12	37	92	141	213
3	10.2	80.9	240	250	5	1.82	6	18	44	67	102
4	9.2	68.0	200	211	6	2.27	7	22	55	83	126
5	8.8	71.5	209	221	7	2.78	8	27	67	103	155
6	11.3	41.9	125	130	2	.62	2	6	15	23	34
7	9.2	47.6	140	147	4	1.58	5	15	38	58	88
8	8.8	46.2	135	143	5	1.80	5	17	43	66	100
9	6.8	56.1	159	174	12	4.78	14	46	115	176	265
10	7.5	67.2	193	208	11	4.35	13	42	105	160	242
11	6.9	52.8	150	163	11	4.31	13	42	104	159	240
12	8.0	30.3	87	94	4	1.61	5	16	39	59	90
13	8.4	36.1	105	112	4	1.65	5	16	40	61	91
14	9.0	25.9	76	80	2	.93	3	9	22	34	52
15	7.9	28.7	83	89	4	1.59	5	15	38	58	88
16	6.5	25.4	71	79	6	2.44	7	23	58	89	135
17	8.0	28.1	81	87	4	1.50	5	14	36	55	83
18	8.0	57.1	164	177	8	3.05	9	29	73	112	169
19	7.9	104.4	300	323	15	5.78	18	56	139	213	321
20	6.8	37.7	105	115	8	3.16	10	31	77	118	178
21	6.2	33.8	94	105	9	3.65	11	35	87	134	202

22	7.4	59.3	170	184	10	4.00	12	39	96	147	222
23	9.2	21.8	64	67	2	.72	2	7	18	27	40
24	6.0	42.1	116	130	13	4.88	15	47	118	180	272
25	7.9	57.3	165	177	8	3.17	10	31	76	117	176
26	8.8	27.4	80	85	3	1.07	3	10	26	39	59
27	8.2	35.9	104	111	5	1.77	5	17	43	65	98
28	12.5	41.3	125	129	1	.38	1	4	9	14	21
29	11.5	39.2	108	121	1	.36	2	5	13	20	30
30	13.0	44.8	134	139	1	.34	1	3	8	12	19
31	12.7	27.1	81	84	1	.23	1	2	6	8	13
32	12.7	72.8	218	225	2	.62	2	6	15	23	34
33	10.0	66.9	198	207	4	1.63	5	16	39	60	91
34	11.5	77.1	230	239	3	1.05	3	10	25	39	59
35	11.5	58.6	175	181	2	.79	2	8	19	29	44
36	10.9	61.5	183	190	3	1.06	3	10	25	39	58
37	10.9	49.7	148	154	2	.86	3	8	21	31	47
38	7.5	65.7	196	203	11	4.25	13	41	102	156	236
39	8.0	44.5	128	138	6	2.38	7	23	57	87	132
40	6.9	40.8	116	126	9	3.33	10	32	80	123	185
41	8.2	26.2	76	81	3	1.29	4	12	31	47	72
42	7.7	38.2	110	118	6	2.28	7	22	55	84	127
43	8.2	63.5	184	197	8	3.14	9	30	75	115	174
44	8.6	79.2	231	245	9	3.34	10	32	80	123	185
45	8.7	31.1	91	96	3	1.26	4	12	30	46	70
46	8.2	41.4	120	128	5	2.04	6	20	49	75	113
47	7.7	32.1	93	99	5	1.87	6	19	46	71	107
48	8.3	21.7	63	67	3	1.03	3	10	25	38	57
49	8.0	59.9	172	185	8	3.18	10	31	77	117	177

	Data from Andrews (1970a)						Calculated from Andrews' Data				
Site	Length of post-glacial recovery T [10^3 years]	A	Uplift from T to present U_p [m]	Ultimate total uplift from T U_∞ m	Uplift remaining U_r m	Present rate of uplift U_T m/1000 yrs	Uplift in last 2000 years U_{2k} m	Uplift in last 4000 years U_{4k} m	Uplift in last 6000 years U_{6k} m	Uplift in last 7000 years U_{7k} m	Uplift in last 8000 years U_{8k} m
50	9.7	50.0	148	155	4	1.37	4	13	33	50	76
51	9.4	55.3	163	171		1.70					
52	12.4	45.8	137	142	1	.44	1	4	11	16	24
53	8.5	6.9	20	21	1	.30	1	3	7	11	17
54	9.6	40.9	121	127	3	1.17	4	11	28	43	65
55	11.0	25.5	76	79	1	.42	1	4	10	16	23
56	12.5	23.4	70	72	1	.21	1	2	5	8	12
57	10.0	43.3	128	134	3	1.06	3	10	25	39	59
58	9.5	49.2	145	152	4	1.46	4	14	35	54	81

App. Table VII-1 Uplift data from Andrews. Uplift is assumed to be of the form $U(\bar{t}) = U_\infty\, e^{-.390\bar{t}}$, $\bar{t} = T - t$ where t is measured in thousands of years before present so that $\bar{t} = 0$ when $t = T$. The form of uplift may be equivalently expressed $U(\bar{t}) = \dfrac{A}{1 - .677}\, .677^{\bar{t}}$, where $\dfrac{A}{1 - .677} = U_\infty$. We then calculate, as discussed in the text, the uplift remaining, the uplift in the last 6 and 8 thousand years, and the present rate of uplift. Contour maps of the data are given in Figures App. VII-2–5. Locations are given in Figure App. VII-1.

curve (5) with $i = .677$ to the uplift data. The rest of the columns of the table are ours. We calculate the total amount of uplift, u_∞, expected from T by using (6). The amount of uplift remaining is then calculated by using (3). The uplift over the last 2, 4, 6, 7, and 8 thousand years is calculated according to (2). The present rate of uplift is calculated from (4).

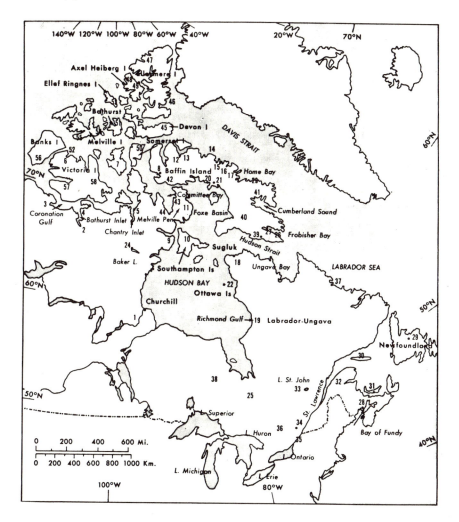

Figure App. VII-1. The locations of sites in Table App. VII-1. The figure, reproduced here with J. T. Andrews' permission and by permission of the National Research Council of Canada, is from the *Canadian Journal of Earth Sciences, 7,* p. 703. The figure also appears in Andrews (1970a, Fig. I-1, p. 2).

The locations of sites in Table App. VII-1 are shown in Figure App. VII-1. Contour plots showing the amount of uplift that has occurred in the last 6 and 8 thousand years, the amount of uplift remaining, and the present rate of uplift (as given in Table App. VII-1) are presented in Figures App. VII-2 through App. VII-5.

Andrews presents similar contour plots for the uplift over the last 6 and 8 thousand years (Andrews, 1970, Figure 2, 3). The only differ-

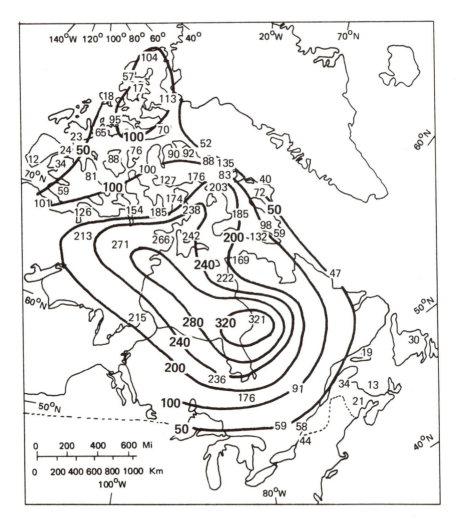

Figure App. VII-2. Uplift in Canada in meters over the last 8000 years as inferred from Andrews' data.

ence is that he presents emergence rather than uplift. Emergence, or
the observed uplift of shorelines, is more significant to the geologist.
For this reason Andrews reconverts his data before contouring them.
For our purpose, the actual uplift of the earth's surface is more important. Uplift can be obtained from emergence if the elevation of sea
level at the time of formation of the ancient shoreline is subtracted
from the emergence data.

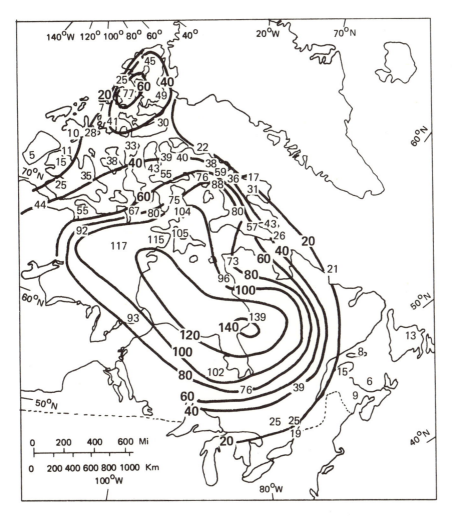

Figure App. VII-3.   Uplift in Canada in meters over the last 6000 years as inferred from
Andrews' data.

Andrews does not use his own data to determine the amount of uplift remaining, even though this follows as a natural consequence of the infinite series approach he uses. Once A (see Table App. VII-1) is determined, all else follows. Instead Andrews uses a complicated approach which works back from an approximate formula proposed by Einarsson through a formula proposed by Gutenberg to Andrews' data. On this basis Andrews estimates that about 160 m of uplift remain in

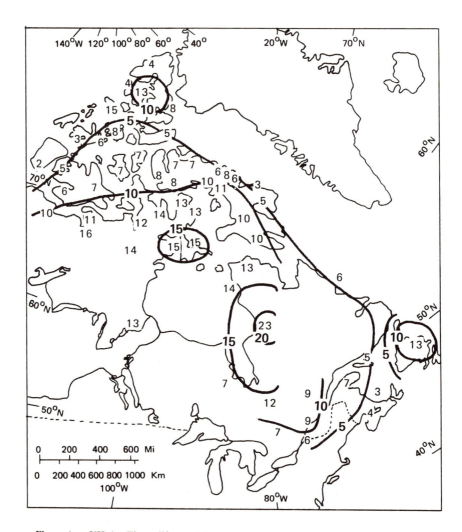

Figure App. VII-4.   The uplift remaining in Canada [m] implied by Andrews' method.

central Canada. We estimate about 20 m remaining uplift (Figure App. VII-4).

Andrews' calculation of the present rate of uplift also differs from ours. Because he apparently felt that the uplift was constrained by equation (1) (as it would be if *all* the uplift that would occur were obliged to occur between $\bar{t} = 0$ and $\bar{t} = T$) he approximated the rate of uplift by

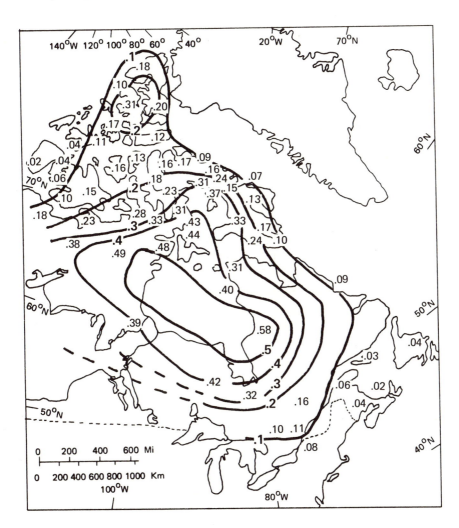

Figure App. VII-5. The present rate of uplift in Canada in meters/100 yrs implied by Andrews' data.

$$\frac{\partial u}{\partial t} = \frac{A}{t}.$$ (Andrews, 1970, p. 114, 118.)

This approximation leads to slightly greater present rates of uplift than those we calculated from Andrews' data (Figure App. VII-5). Andrews calculated the present rate of uplift at the center of uplift to be 1.3 m/100 years. We calculate from his data a maximum central rate of uplift of .5 m/100 years.

We might also point out that the time at which $\bar{t}$ is chosen to equal zero has no real significance[2] as long as it is recognized that the exponential form of the uplift decay cannot be projected back *too far* past the time of deglaciation. The uplift should not change character substantially at the instant the last ice disappears, since substantial thinning of the ice would presumably have occurred before then. Andrews found that this was indeed the case (1970, p. 110). Nevertheless, toward the beginning of deglaciation, the uplift should be controlled by the rate and manner in which the ice load was removed, and the form of the post-glacial uplift cannot be projected to these times.

B. Uplift in Canada Taking Other Analyses into Account

Andrews (1970) made the first regional compilation of uplift in Canada, but he assumed that all locations uplifted exponentially with a single time constant of 2560 years. Recently Walcott (1972) made an extensive compilation of uplift data in Canada. His work is particularly valuable for three reasons:

1) It presents the raw uplift data for 34 areas in Canada and the eastern United States and deduces emergence maps without the bias of an assumed form of uplift.

2) It presents rate of uplift data deduced from water level measurements over the last 50 years in the Great Lakes.

3) It draws attention to a precise leveling survey in Quebec which provides needed rate of uplift data in the inland areas of the province.

---

[2]Andrews associated $\bar{t} = 0$ with the time of deglaciation. It is not strictly necessary to do so, and no changes will result if a later time is chosen. Although the more rapid uplift at early times near the date of deglaciation may appear to give the date of deglaciation significance, this significance is largely fictitious.

Location	Reference	Rate of Uplift [m/100 yrs]	Uplift [m] over Last *n* Thousand Years				
			2*k*	4*k*	6*k*	7*k*	8*k*
Fox Basin	Ives	(<.5)	9	17	78	102	
Greenland	Laska Washburn & Stuiver	.05	.5	2	9	26	56
Ottawa	Andrews	.20	5	24	81	140	
E. Melville	Andrews & Farrand	.66	16	62	138	174	
Elles	Farrand	.33	8	22	44	(64)	
NE Baffin	Loken	.16	4	19	44	56	
NE Fox Basin	Andrews Ives	.42	11	20	72		
James Bay	Farrand	.62	18	54	102	136	
Southampton	Farrand	.60	14	36	60		
N. Bay	Farrand	.50	10	26	48	64	88
Sault	Farrand	.2	4	11	28	38	52
C. Rich	Farrand	.08	2	7	17	22	30

App. Table VII-2 Uplift data from various sources. Present rate of uplift is estimated from the initial slope of the uplift curve.

Walcott also summarizes data on the post-glacial behavior of the east coast of the United States, which we discuss in Section IV.E.3.

Walcott's emergence data are tabulated for particular times in Table App. VII-3 and the present rate of uplift estimated from the emergence curves. Farrand's and others' uplift data are tabulated in the same fashion in Table App. VII-2. The locations of the uplift data for Farrand (and others) and Walcott are shown in Figures App. VII-6 and 7. Figure App. VII-8 shows the present rate of uplift compiled from all sources. Andrews' rate of uplift values are lighter since they are considered less reliable because of the arbitrary uplift form imposed. Figures App. VII-9–13 show the uplift that has occurred over different intervals of time in Canada. Walcott's emergence data is converted to uplift data by adding 5, 10, and 20 m of uplift to the emergence values for 6000, 7000, 8000 years BP, respectively (see Figure IV-15). The uplift in the central areas of Canada at the 8000 year BP is obtained by backward extrapolation of the uplift data in the manner Andrews suggests. These areas were ice-covered in 8000 BP and so provide no direct

| Walcott's Location No. | Name | \multicolumn{5}{c}{Emergence Data [m]} | Estimated from Walcott's Curves Present Rate Uplift [m/100 yrs] |
		$T = 2$	$T = 4$	$T = 6$	$T = 7$	$T = 8$	
1	Churchill	21	54	100	133		.60
2	Cape Henrietta Maria	34	69	115	138		1.14
3	Keewatin	(27)	60	103	140		1.05
4	Ottawa Isl.	7	29	76	124		.25
5	Southampton Isl.	10	38	97	130		.32
6	Ungava Penin.	3	13	51	110		.12
7	Boothia Penin.	9	22	40	52	100	.27
8	Bathurst Inlet	8	27	65	95	145	.25
9	Cape Tanfield	2	6	26	45	80	.14
10	Igloolik Isl.	14	36	80	135		.50
11	Milne Inlet	6	23	52	66	83	.20
12	Inner Fiords	4	12	28	41	60	.075
13	Ipik Bay	8	23	73			.24
14	Outer Fiords	—	—	8	13	21	.20
15	Somerset Isl.	(3)	(9)	21	34	52	.075
16	Cornwalis Isl.	(9)	(21)	35	48	70	.075
17	Bathurst Isl.	(13)	27	50	65	80	.29
18	Cape Storm	6	13.5	30	40	59	.24
19	South Cape Fiord	4.5	13.5	25.5	37		.16
20	Bay Fiord	5.5	17	36	55	88	.15
21	Axel Heilberg Isl.	(11)	24	45	61	93	.15
22	Ottawa					40	
23	Montreal				5	37	
24	Riviere Du Loup					30	
25	Lac St. Jean					80	
26	NW Newfoundland	2.5	6	13	20	30	.09

App. Table VII-3   Emergence data from Walcott's (1972a) compilation.   The emergence data are converted to uplift data for Figures App. VII 9–13 by adding 5, 10, 20 m to the values for emergence for 6000, 7000, and 8000 years BP, respectively.

data.   The uplift 8000 BP is speculative in the central regions and is indicated by dashed contours.

The present rate of uplift in Canada shown in Figure App. VII-8 is in general agreement with both Walcott (1972) and Andrews (1970, p. 123), but some details differ.   Andrews' maximum rate of uplift is just past James Bay.   Walcott places the rate of uplift at Cape Henrietta Maria at 1.7 ± .5 m/100 yrs (average rate over the last 2000 years) and feels the rate of uplift to the north in Richmond Gulf is even higher.

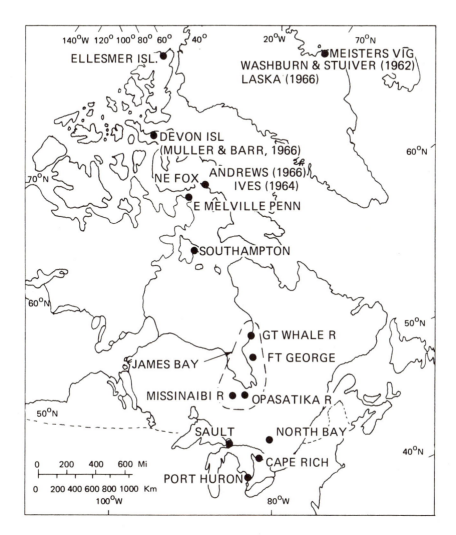

Figure App. VII-6. Locations of points in Figure IV-23 and Table App. VII-2. Unless indicated, data referenced to Farrand, 1962.

Our maximum rate of uplift is centered more over James Bay with a lobe into Quebec and is at the lower limit of Walcott's permissible range. This choice is supported by the low present rate of uplift at the Ottawa Islands and at Churchill. It is also supported to some extent by the tendency in Fox Basin and Keewatin for the present rate of uplift to be greatest where the ice last dissipated (see Figures IV-8 and IV-9 of the text). Thus it is not unreasonable to assume the same condition

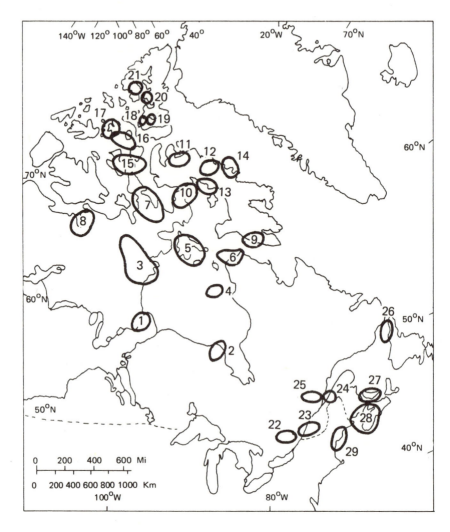

Figure App. VII-7. Areas for which emergence data have been compiled by Walcott (1972a). Data are tabulated in Table App. VII-3.

to hold in Quebec. Nevertheless, the paucity of data points dictates that any choice of uplift contours will be somewhat speculative. This is particularly true in northern Quebec.

We have chosen to take the present rate of uplift at Churchill to be about .75 m/100 years in agreement with the present slope of the uplift curve. We show in Section IV.E.4 that there is reason to believe eustatic sea level has been rising in the recent past (last 50 yrs) at a rate

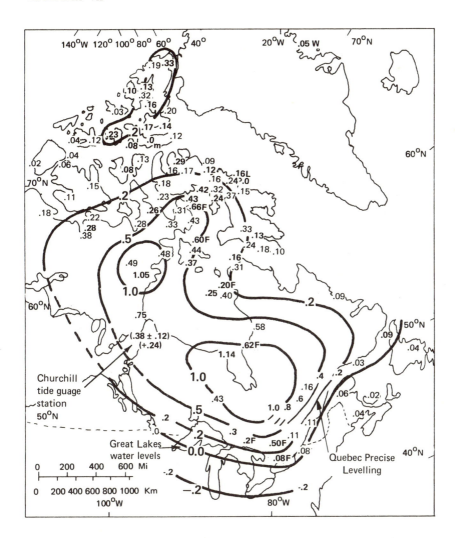

Figure App. VII-8. Present rate of uplift in Canada [m/100 yrs]. Light numbers are Andrews' force fit data. Dark values are estimates from Walcott's data unless indicated by a letter keying them to other sources (see Table App. VII-2). The Great Lakes data are tied in as Walcott suggests. Uplift at the Churchill tide gauge station is in agreement with that inferred from uplift curves if sea level is eustatically rising at present at .24 m/100 yrs. This is also just the value suggested by sea level data along the United States east coast (Chapter IV). Note rate of uplift is greatest in areas most recently deglaciated (see Figures IV-9-10).

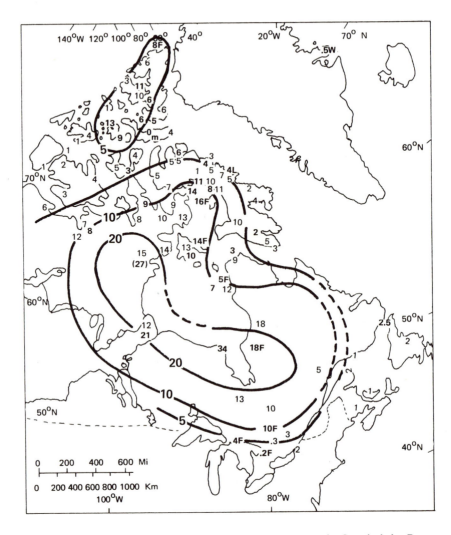

Figure App. VII-9. Uplift over the last 2000 radiocarbon years in Canada [*m*]. Data conventions same as in Figure App. VII-8.

of about .24 m/100 yrs. Such a rate of eustatic sea level rise would bring the tide gauge data at Churchill (Barnett, 1970) into good agreement with the rate inferred from the uplift curve.

The uplift over the last two and four thousand radiocarbon years shown in Figures App. VII-9, 10 differs from Walcott's, primarily in the uplift lobes in Quebec and Keewatin. These lobes are suggested by the present high rate of uplift in those areas, and the lack of data in

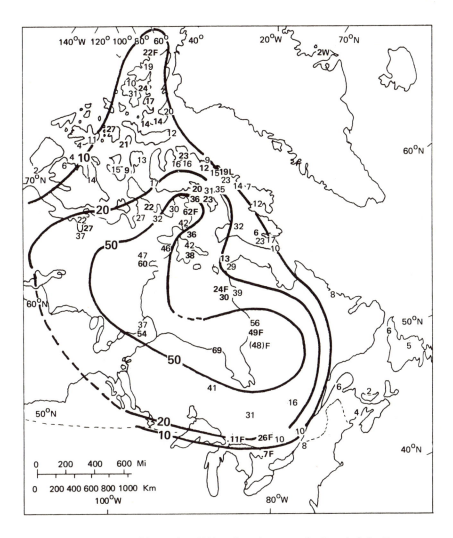

Figure App. VII-10. Uplift over last 4000 radiocarbon years in Canada [*m*]. Data conventions same as in Figure App. VII-8.

those areas certainly permits such interpretation. Our 6000 radiocarbon year uplift map is very close to Walcott's. The values of uplift over the last 7000 and 8000 radiocarbon years must be regarded as less certain than the uplift maps for the shorter intervals.

In addition, a correction we have not made becomes important for uplifts over the last 7000 and 8000 years. We have not accounted for the deformation of the geoid that results from the deformation of the

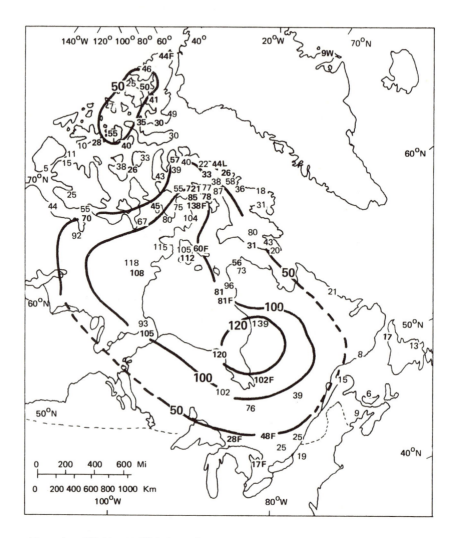

Figure App. VII-11. Uplift [*m*] over last 6000 radiocarbon yrs. 5 m has been added to Walcott's emergence data to convert to uplift data. Conventions same as in Figure App. VII-8.

earth's surface. We show in Section IV.E.7 of the text that the uplift relative to the earth's geoid is about 84% of the uplift relative to the center of the earth. Since sea level is an equipotential and follows the geoid, this means about 50 m should be added to the maximum 300 m of uplift shown in Figure App. VII-13 and about 30 m to the maximum 200 m of Figure App. VII-12. For the uplift over periods shorter than

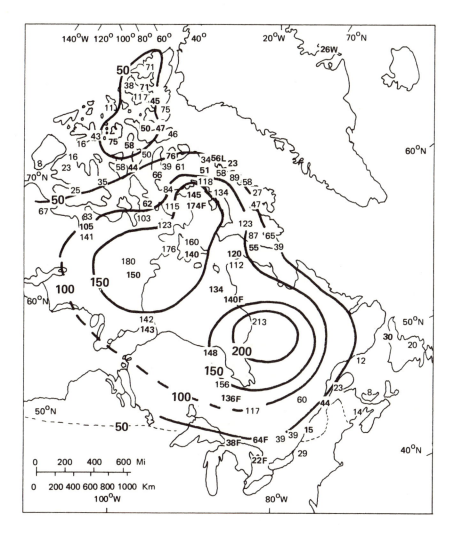

Figure App. VII-12.   Uplift [*m*] over last 7000 radiocarbon yrs. 10 m has been added to Walcott's emergence data to convert to uplift data. Conventions same as Figure App. VII-8. 25–30 m should be added to the central uplift to correct for deformation of the geoid.

7000 years, the correction for the movement of the geoid becomes less unimportant relative to other uncertainties.

From the uplifts over 2, 4, and 6 thousand radiocarbon years, it can be seen that Andrews' values are really in remarkably good agreement with the uplift determined directly from the primary data. Although

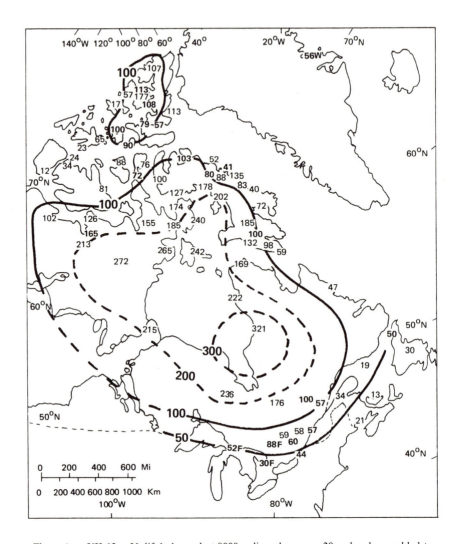

Figure App. VII-13. Uplift [*m*] over last 8000 radiocarbon yrs. 20 m has been added to Walcott's emergence data to convert to uplift data. Conventions same as Figure App. VII-8. Central contours are controlled almost exclusively by Andrews' data. About 50 m should be added to the uplift at the center of uplift to correct for the deformation of the geoid. Contours in this area are speculative since the area was ice covered at this time and Andrews' data are obtained by extrapolation of the uplift curves.

Andrews' method distorts the uplift data somewhat, that distortion is not too great. The success of Andrews' method has some interesting and significant implications, as we discussed in Chapter IV.

BIBLIOGRAPHY

Abramowitz, M. and Stegun, I. A., 1965. *Handbook of Mathematical Functions.* New York: Dover, pp. 331–353.

Ahrens, T. J. and Syono, Yasuhiko, 1967. Calculated Mineral Reactions in the Earth's Mantle. *J. Geophys. Res.* 72:4181–4188.

Akimoto, S. and Fujisawa, H., 1968. Olivine-Spinel Solid Solution Equilibria in the System, $Mg_z SiO_4$–$Fe_z SiO_4$. *J Geophys. Res.* 73:1467–1479.

Alterman, Z., Jarosch, H. and Pekeris, C. L., 1959. Oscillations of the Earth. *Proc. Roy. Soc. Lon.* A, 252:80–95.

———, 1961. Propagation of Rayleigh Waves in the Earth. *Geophys. J.* 4:219–241.

Anderson, B. G., 1960. Sørlandet i sen-og Postglacial tid. *Norgen Geol Undersøkelese.* 210:1–142.

Anderson, D. L., 1967. The Anelasticity of the Mantle. *Geophys. J.* 14:135–164.

Anderson, D. L., and O'Connell, R., 1967. Viscosity of the Earth. *Geophys. J.* 14:287–295.

Andrews, J. T. 1966. Pattern of Coastal Uplift and Deglaciation, West Baffin Island, Northwest Territories. *Geogr. Bull.* 8: No. 2:174–193.

———, 1968a. Postglacial Rebound in Arctic Canada: Similarity and Prediction of Uplift Curves. *Can. J. Earth Sci.* 5:39–47.

———, 1968b. Pattern and Cause of Variability of Postglacial Uplift and Rate of Uplift in Arctic Canada. *J. Geology.* 76:404–425.

———, 1968c. The Pattern and Interpretation of Restrained Postglacial and Residual Rebound in the Area of Hudson Bay. *Earth Science Symposium on Hudson Bay,* P. J. Hood, Editor, Geological Survey Canada, GSC Paper 68–53. pp. 49–62. Ottawa.

———, 1970a. *A Geomorphological Study of Post-Glacial Uplift with Particular Reference to Arctic Canada.* 156p. Oxford: Alden and Mowbray Ltd.

———, 1970b. Present and Postglacial Rates of Uplift for Glaciated Northern and Eastern North America Derived from Postglacial Uplift Curves. *Can. J. Earth Sci.* 7: No. 2:703–715.

Andrews, J. T. and Falconer, G., 1969. Late Glacial and Post-Glacial History and Emergence of the Ottawa Islands, Hudson Bay, Northwest Territories: Evidence on the Deglaciation of Hudson Bay. *Can. J. Earth Sci.* 6:1263–1267.

Archambeau, C. B., Flinn, E. A., and Lambert, D. G., 1969. Fine Structure in the Upper Mantle, *J. Geophys. Res.* 74:5825–5865.

Artyushkov, E. V., 1971. Rheological Properties of the Crust and Upper Mantle According to Data on Isostatic Movements. *J. Geophys. Res.* 76: 1376–1390.

Backus, G. E., 1967. Converting Vector and Tensor Equations to Scalar Equations in Spherical Coordinates. *Geophys. J.* 13:71–101.

Bandy, O. L., Butler, E. A., and Wright, R. C., 1969. Alaskan Upper Miocene Marine Glacial Deposits and the Turborotalia Pachyderma Datum Plane. *Science.* 166:607–608.

Barnett, D. M., 1966. A Re-examination and Re-interpretation of Tide Gauge Data for Churchill, Manitoba. *Can. J. Earth Sci.* 3:77–88.

———, 1970. An Amendment and Extension of Tide Gauge Data Analysis for Churchill, Manitoba. *Can. J. Earth Sci.* 7:626–627.

Bentley, C. R. 1962. Glacial and Subglacial Geography of Antarctica. *Arctic Research, The M. F. Maury Memorial Symposium.* Geophysical Monograph No. 7, NAS-NRC No. 1036, American Geophysics Union. pp. 11–25. Washington, D.C.

Bentley, C. R., Cameron, R. L., Bull, C., Kojima, K., and Gow, A. J., 1964. Physical Characteristics of Antarctic Ice Sheet. *Antarctic Map Folio Series.* Folio 2, American Geographical Society.

Bird, J. B., 1967. *The Physiography of Arctic Canada.* pp. 117–159. Baltimore: The Johns Hopkins Press.

Bloom, A. L., 1967. Pleistocene Shorelines: A New Test of Isostasy. *Bull. Geol. Soc. Am.* 78:1477–1494.

———, 1970a. Paludal Stratigraphy of Truk, Ponape, and Kusaic, Eastern Caroline Islands. *Bull. Geol. Soc. Am.* 81:1895–1904.

———, 1970b. Discussion of Walcott's Paper, Isostatic Response to Loading of the Crust in Canada. *Can. J. Earth Sci.* 7:731–733.

———, 1970c. Holocene Submergence in Micronesia as the Standard for Eustatic Sea-Level Changes. *Quaternia.* XII:145–154. Rome.

———, 1971. Glacial-Eustatic and Isostatic Controls of Sea Level since the Last Glaciation. *The Late Cenozoic Glacial Ages,* K. K. Turekian, Editor. Chapter 13, 606pp. New Haven: Yale University Press.

Bloom, A. L., and Stuiver, M., 1963. Submergence of the Connecticut Coast. *Science.* 139:332–334.

Bowen, F., 1958. *Introduction to Bessel Functions.* 134pp. New York: Dover.

Broecker, W. S., 1966a. Absolute Dating and the Astronomical Theory of Glaciation. *Science.* 151:299–304.

———, 1966b. Glacial Rebound and the Deformation of the Shorelines of Proglacial Lakes. *J. Geophys. Res.* 71:4777–4783.

Broecker, W. S., Ewing, M., and Heezen, B. C., 1960. Evidence for an Abrupt Change in Climate Close to 11,000 Years Ago. *Amer. J. Sci.* 258:429–448.

Broecker, W. S., Thurber, D. L., Goddard, J., Ku Teh-Lunj, Matthews, R. K., and Mesolella, K. G., 1968. Milankovitch Hypothesis Supported by Precise Dating of Coral Reefs and Deep Sea Sediments. *Science.* 159:297–300.

Broecker, W. S., and Van Donk, J., 1970. Insolation Changes, Ice Volumes, and the O^{18} Record in Deep-Sea Cores. *Rev. Geophys.* 8:169–198.

Brotchie, J. F., and Silvester, R., 1969. On Crustal Flexure. *J. Geophys. Res.* 74:5240–5252.

Bryson, R. A., Wendland, W. M., Ives, J. D., and Andrews, J. T., 1969. Radiocarbon Isochrons on the Disintegration of Laurentide Ice Sheet. *Arctic and Alpine Research.* 1:1–14.

Bullen, K. E., 1965. *An Introduction to the Theory of Seismology.* 381pp. Cambridge: Cambridge University Press.

Burgers, J. M., and Collette, 1958. On the Problem of the Postglacial Uplift in Fennoscandia, I and II. *Koninklijke Nederlandse Akademie Van Wettenschappen.* 61:B:221–241.

Carslaw, H. S., and Jaeger, J. C., 1959. *Conduction of Heat in Solids.* 510pp. Oxford: Clarendon Press.

Carslaw, H. S., and Jaeger, J. C., 1963. *Operational Methods in Applied Mathematics.* 359pp. New York: Dover.

Cathles, L. M. 1965. The Physics of Glacial Uplift. 126pp. Unpublished. A.B. thesis, Princeton University.

———, 1971. The Viscosity of the Earth's Mantle. Unpublished Ph.D. Dissertation, Princeton University.

Clark, S. P., 1966. Handbook of Physical Constants, Memoir 97, revised edition. *Bull. Geol. Soc. Am.* 587pp.

Cole, G. H. A., 1962. *Fluid Dynamics.* New York: John Wiley & Son. 238pp.

Crittenden, M. D., 1963. Effective Viscosity of the Earth Derived from Isostatic Loading of Pleistocene Lake Bonneville. *J. Geophys. Res.* 68:1865–1880.

Curray, J. R., 1961. Late Quaternary Sea Level: A Discussion. *Bull. Geol. Soc. Am.* 72:1707–1712.

Curray, J. R., Shepard, F. P., and Veeh, H. H., 1970. Late Quaternary Sea Level Studies in Micronesia: Carmarsel Expedition. *Bull. Geol. Soc. Am.* 81:1865–1880.

Curry, R. R., 1966. Glaciation About 3,000,000 Years Ago in the Sierra Nevada. *Science.* 154:770–771.

Daly, R. A., 1934. *The Changing World of the Ice Age.* 271pp. New Haven: Yale University Press.

Dansgaard, W., Johnsen, S. J., Moller, J., and Langway, G. C., 1969. One Thousand Centuries of Climatic Record from Camp Century on the Greenland Ice Sheet. *Science.* 166:377–381.

Dansgaard, W., and Tauber, H., 1969. Glacier Oxygen-18 Content and Pleistocene Ocean Temperatures. *Science.* 166:499–592.

Dansgaard, W., Johnsen, H. B., and Langway, C. C., 1971. Climatic Record Revealed by the Camp Century Ice Core. *The Late Cenozoic Glacial Ages,* K. K. Turekian, Editor. New Haven: Yale University Press.

DeGeer, E. H., 1954. Skandinaviens Geokronologi. *Geologiska Föreningens I. Stockholm Förhandlingar.* 76:299–329.

Denton, G. H., and Armstrong, R. L., 1968. Glacial Geology and Chronology of the McMurdo Sound Region. *Antarctic Journal.* 3:99–101.

Denton, G. H., and Armstrong, R. L., 1969. Miocene-Pliocene Glaciations in Southern Alaska. *Amer. J. Sci.* 267:1121–1142.

Denton, G. H., Armstrong, R. L., and Stuiver, M. 1969. Histoire Glaciaire et Chronologie de la Region du Detroit de McMurdo, Sude de la Terre Victoria, Antarctide Note Preliminaire. *Revue de Geographie Physique et de Geologie Dynamique.* XI: Fasc. 3:265–278.

Denton, G. H., Armstrong, R. L., and Stuiver, M., 1971. The Late Cenozoic Glacial History of Antarctica. *The Late Cenozoic Glacial Ages,* K. K. Turekian, Editor. New Haven: Yale University Press.

Dicke, R. J., 1966. The Secular Acceleration of the Earth's Rotation. *The Earth-Moon System,* B. G. Marsden and A. G. W. Cameron, Editors. pp. 98–164. New York: Plenum Press.

————, 1969. Average Acceleration of the Earth's Rotation and the Viscosity of the Deep Mantle. *J. Geophys. Res.* 74:5895–5902.

Dohler, G. C., and Ku, L. F., 1970. Presentation and Assessment of Tides and Water Level Records for Geophysical Investigations. *Can. J. Earth Sci.* 7:607–625.

Donn, W. L., Farrand, W. R., and Ewing, M., 1962. Pleistocene Ice Volumes and Sea-Level Lowering. *J. Geology.* 70:206–214.

Donner, J. J., 1969. Land/Sea Level Changes in Southern Finland During the Formation of the Salpausselkä Endmoraines. *Geol. Soc. Finland Bull.* 41: 135–150.

Dorf, E., 1960. Climatic Changes of the Past and Present. *American Scientist.* 48:341–364.

————, 1970. Paleobotanical Evidence of Mesozoic and Cenozoic Climate Changes. *Symposium North American Paleontological Convention, 1969.* pp. 323–346. Lawrence, Kansas: Allen Press, Inc.

Dreimanis, A., 1957. Stratigraphy of the Wisconsin Glacial Stage along the Northwestern Shore of Lake Erie. *Science.* 126:166–168.

Duber, H., and Abate J., 1968. Numerical Inversion of Laplace Transforms by Relating Them to the Finite Fourier Cosine Transform. *J. Assoc. Comp. Math.* 15:115–123.

Durney, B. R., 1964. Deformation of a Purely Elastic Earth Model under Various Tectonic Loads. *Technical Report No. 1,* Prepared under NASA Research Grant NSG-556 to Princeton University by National Aeronautics & Space Administration, April 1.

Ellsaesser, H. W., 1966. Expansion of Hemispheric Meteorological Data in Antisymmetric Surface Spherical Harmonic (Laplace) Series. *J. Applied Meteorology.* 5:263–276.

Elsasser, W. M., 1971a. Sea Floor Spreading on Thermal Convection. *J. Geophys. Res.* 76:1101–1112.

————, 1971b. Existence of an Asthenosphere. *J. Geophys. Res.* 76:7296–7297.

Emery, K. O., and Garrison, L. E., 1967. Sea Levels 7,000 to 20,000 Years Ago. *Science.* 157:684–687.

Emery, K. O., Niino, H. and Sullivan, B., 1971. Post Pleistocene Levels of the

East China Sea. *Late Cenozoic Glacial Ages,* K. K. Turekian, Editor. New Haven: Yale University Press.

Emery, K. O., and Uchupi, E., 1972. *Western North Atlantic Ocean: Topography, Rocks, Structure, Water, Life and Sediments.* Memoir 17. Tulsa: American Association of Petroleum Geologists.

Eringen, A. C., 1962. *Nonlinear Theory of Continuous Media.* New York: McGraw-Hill.

———, 1967. *Mechanics of Continua.* New York: John Wiley & Sons.

Fairbridge, R. W., Editor, 1968. *The Encyclopedia of Geomorphology.* pp. 912–931. New York: Reinhold Book Company.

Farrand, W. R., 1962. Postglacial Uplift in North America. *Amer. J. Sci.* 260:181–199.

Farrand, W. R., and Gajda, R. T., 1962. Isobases on the Wisconsin Marine Limit in Canada. *Geogr. Bull.* 17:5–22.

Farrell, W. E., 1972. Deformation of the Earth by Surface Loads. *Rev. Geophys.* 10:761–797.

Fisher, I. 1959a. A Tentative World Datum from Geoidal Heights Based on the Hough Ellipsoid and the Columbus Geoid. *J. Geophys. Res.* 64:73–84.

———, 1959b. The Impact of the Ice Age on the Present Form of the Geoid. *J. Geophys. Res.* 64:85–87.

Flint, R. F. 1957. *Glacial and Pleistocene Geology.* New York: John Wiley & Sons.

———, 1971. *Glacial and Quaternary Geology.* New York: John Wiley & Sons.

Frye, J. C., 1963. Problems of Interpreting the Bedrock Surface of Illinois. *Ill. Geol. Surv. Reprint Ser.* 11pp.

Frye, J. C., and Leonard, A. B., 1952. Pleistocene Geology of Kansas. *Kansas Geol. Surv. Bull.* 99:230p.

Gantmacher, F. R., 1960. *The Theory of Matrices.* Translated from Russian by K. A. Hirsch. Vols. 1 and 2. New York: Chelsea Publishing Co.

Garland, G. D., 1965. *The Earth's Shape and Gravity.* Oxford: Pergamon Press.

Garrison, L. E., and McMaster, R. L., 1966. Sediments and Geomorphology of the Continental Shelf off Southern New England. *Marine Geology.* 4:273–289.

Gilbert, F., and Backus, G. E., 1966. Propagator Matrices in Elastic Wave and Vibration Problems. *Geophysics.* 31:326–332.

Goetz, C., 1971. High Temperature Rheology of Westerly Granite. *J. Geophys. Res.* 76:1223.

Goldreich, P., and Toomre, A., 1969. Some Remarks on Polar Wandering. *J. Geophys. Res.* 74:2555–2567.

Gordon, R. B., 1965. Diffusion Creep in the Earth's Mantle. *J. Geophys. Res.* 70:2413.

———, 1967. Thermally Activated Processes in the Earth: Creep and Seismic Attenuation. *Geophys. J.* 14:33.

———, 1971. Observation of Crystal Plasticity Under High Pressure with Application to the Earth's Mantle. *J. Geophys. Res.* 76:1248.

Guilcher, A., 1969. Pleistocene and Holocene Sea Level Changes. *Earth Science Reviews.* 5:69–97.

Gutenberg, B., 1941. Changes in Sea Level, Postglacial Uplift, and Mobility of the Earth's Interior. *Bull. Geol. Soc. Am.* 52:721–722.

————, 1954. Postglacial Uplift in the Great Lakes Region. *Archiv Für Meteorologic Geophysik und Bioklimatologic.* 7:243–251.

Haddon, R. A. W., and Bullen, K. E., 1969. An Earth Model Incorporating Free Earth Oscillation Data. *Physics of the Earth and Planetary Interiors.* 2:30–49.

Hales, A. L., Cleary, J. R., Doyle, H. A., Green, R., and Roberts, J., 1968. P-Wave Station Anomalies and the Structure of the Upper Mantle. *J. Geophys. Res.* 73:3885–3895.

Haskell, N. A., 1935. The Motion of a Viscous Fluid under a Surface Load. *Physics.* 6:265–269.

————, 1936. The Motion of a Viscous Fluid under a Surface Load, Part II. *Physics.* 7:56–61.

————, 1937. The Viscosity of the Asthenosphere. *Amer. J. Sci.* 33:22–28.

Hawkins, G. A., 1963. *Multilinear Analysis in Engineering and Science.* 218pp. New York: John Wiley & Sons.

Heiskanen, W. A., and Vening Meinesz, F. A. 1958. *The Earth and Its Gravity Field.* pp. 357–370. New York: McGraw-Hill.

Helmberger, D., and Wiggins, R. A., 1971. Upper Mantle Structure of Midwestern United States. *J. Geophys. Res.* 76:3229–3245.

Herman, Yvonne, 1970. Arctic Paleo-Oceanography in Late Cenozoic Time. *Science.* 169:474–477.

Hicks, S. D. and Shofnos, W., 1965. Yearly Sea Level Variations for the United States. *J. Hydraul. Div. Proc. Amer. Soc. Civil Eng.* 91:23–33.

Hildebrand, H. B., 1962. *Advanced Calculus for Applications.* pp. 102–106. New Jersey: Prentice Hall.

Hollin, J. T., 1962. On Glacial History of Antarctica. *J. Glaciology.* 4:173–195.

————, 1969. Ice Sheet Surges and the Geological Record. *Can. J. Earth Sci.* 6:903–910.

————, 1970. Is the Antarctic Ice Sheet Growing Thicker? *International Symposium on Antarctic Glaciological Exploration* (ISAGE), A. J. Gow *et al.,* editors. pp. 363–374. Gentbrugge: International Association of Scientific Hydrology.

Holmes, A., 1965. *Principles of Physical Geology.* New York: The Ronald Press Company.

Honkasalo, T., 1964. On the Use of Gravity Measurements for Investigation of the Land Upheaval in Fennoscandia. *Fennia.* 89:21–23.

————, 1966. Gravity and Land Upheaval in Fennoscandia. *Acad. Sci. Fennicae.* 90: Ann., Series A III:139–141.

Hoskins, L. M., 1910. The Strain of Gravitating Compressible Elastic Sphere. *Trans. Amer. Math. Soc.* II:203–248.

————, 1920. The Strain of a Gravitating Sphere of Variable Density and Elasticity. *Trans. Amer. Math. Soc.* 21:1–43.

Hoyt, J. H., Henry, V. J., and Weimer, R. J., 1965. Age of Late-Pleistocene Shoreline Deposits, Coastal Georgia. *Means of Correlation of Quaternary Successions*, R. B. Morrison and H. E. Wright, Jr., Editors. *Proceedings of VII Congress International Association for Quaternary Research*. 8:381–393. Salt Lake City: University of Utah Press.

Hyyppä, Esa, 1966. The Late-Quaternary Land Uplift in the Baltic Sphere and the Relation Diagram of the Raised and Tilted Shore Levels. *Acad. Sci. Fennicae*. 90: Ann, Series A III:153–168.

Innes, M. J. S., and Weston, A., 1966. Crustal Uplift of the Canadian Shield and its Relation to the Gravity Field. *Ann. Acad. of Sci. Fennicae*. Series A III:169–176.

Ives, J. D., 1964. Deglaciation and Land Emergence in Northeast Foxe Basin, Northwest Territories. *Geogr. Bull.* 21:54–65.

Jackson, W. D., 1967. *Classical Electrodynamics*. pp. 54–69. New York: John Wiley & Sons.

Jaeger, J. C., 1964. *Elasticity, Fracture and Flow*. 204pp. New York: John Wiley & Sons.

Jahnke, E., and Emde, F., 1945. *Tables of Functions*. pp. 109–124. New York: Dover.

Jeffreys, H., 1959. *The Earth*. 420 pp. Cambridge: Cambridge University Press.

Jeffreys, H., and Jeffreys, B. S., 1956. *Methods of Mathematical Physics*. 714 pp. Cambridge: Cambridge University Press.

Jelgersma, S., 1966. Sea Level Changes during the Last 10,000 Years. *World Climate from 8,000 to 0 B.C., Proceedings of the International Symposium on World Climate 8,000 to 0 B.C.* pp. 54–71. London: Royal Meteorological Society.

Jongsma, D., 1970. Eustatic Sea Level Changes in the Arafura Sea. *Nature*. 228:150–151.

Kääriäinen, Erkki, 1953. On the Recent Uplift of the Earth's Crust in Finland. *Veroffentlichungen des Finnischen Geodatischen Instituten, Helsinki*. 42: 1–106.

———, 1966. Land Uplift in Finland Computed with the Aid of Precise Levellings. *Acad. Sci. Fennicae*. 90: Ann., Series A III:187–190.

Kanamori, H., and Press, Frank. 1970. How Thick Is the Lithosphere? *Nature*. 226:330–331.

Kaula, W. M., 1963. Elastic Models of the Mantle Corresponding to Variations in the External Gravity Field. *J. Geophys. Res.* 68:4967–4978.

———, 1966. Tests and Combinations of Satellite Determinations of the Gravity Field with Gravimetry. *J. Geophys. Res.* 71:5503–5314.

———, 1968. *An Introduction to Planetary Physics, The Terrestrial Planets*. pp. 61–77. New York: John Wiley & Sons.

———, 1970. Earth's Gravity Field: Relation to Global Tectonics. *Science*. 169:982–985.

———, 1972. Global Gravity and Tectonics. *The Nature of the Solid Earth*, E. C. Robertson, Editor. pp. 385–405. New York: McGraw-Hill.

Kaye, C. A., and Barghoorn, E. S., 1964. Late Quaternary Sea Level Changes and Crustal Rise at Boston, Mass., with notes on autocompaction of peat. *Bull. Geol. Soc. Am.* 75: #2:63–80.

King, P. B., 1965. Tectonics of Quaternary Time in Middle North America. *The Quaternary of the United States,* Wright, H. E., and Frey, D. G., Editors. pp. 831–870. Princeton, New Jersey: Princeton University Press.

Knopoff, L., 1964. The Convection Current Hypothesis. *Rev. Geophys.* 2:89–122.

Kupsch, W. O., 1967. Postglacial Uplift, A Review. *Life, Land and Water,* W. J. Mayer-Oakes, Editor. pp. 155–186. Winnipeg: University of Manitoba Press.

Land, Lynton S., Mackenzie, F. T., and Gould, S. J., 1967. Pleistocene History of Bermuda. *Bull. Geol. Soc. Am.* 78:993–1006.

Laska, M. P., 1966. Postglacial Deleveling in Skeldal Northeast Greenland. *Arctic.* 19:349–353.

Lennon, G. W., 1966. An Investigation of Secular Variations of Sea Level in European Waters. *Acad. Sci. Fennicae.* 90: Ann., Series A III:225–263.

Liden, R., 1938. Den Senkvartära Strandförskjutningens Förlopp och Kronolgi i Ångermand. *Geol. Fören. Förhandl.* 60: h 3.

Lliboutry, L. A., 1971. Rheological Properties of the Asthenosphere from Fennoscandian Data. *J. Geophys. Res.* 76:1433–1446.

Løken, O. H., 1965. Postglacial Emergence at the South End of Inugsuin Fiord, Baffin Island, Northwest Territories. *Geogr. Bull.* 7:243–258.

Longman, I. M., 1962. A Green's Function for Determining the Deformation of the Earth under Surface Mass Loads; 1. Theory. *J. Geophys. Res.* 67: 845–850.

———, 1963. A Green's Function for Determining the Deformation of the Earth under Surface Mass Loads; 2. Computations and Numerical Results. *J. Geophys. Res.* 68:485–496.

———, 1966. Computation of Love Numbers and Load Deformation Coefficients for a Model Earth. *Geophys. J.* 11:133–137.

Longwell, C. R., 1960. Interpretation of the Leveling Data. *Comprehensive Survey of Sedimentation in Lake Mead, 1948–1949.* U.S. Geol. Survey Prof. Paper No. 295. pp. 33–38.

Love, A. E. H., 1911. *Some Problems of Geodynamics.* 180 pp. Cambridge: Cambridge University Press.

———, 1944. *The Mathematical Theory of Elasticity.* 643 pp. New York: Dover.

MacDonald, G. J. F., 1960. Tectonic Theories. *Trans. of American Geophysical Union.* 41:168–169.

———, 1963. The Deep Structure of the Continents. *Rev. Geophysics.* 1:587–665.

———, 1966. The Figure and Long-Term Mechanical Properties of the Earth. *Advances in Earth Sciences,* P. M. Hurley, Editor. pp. 199–245. Cambridge, Massachusetts: M.I.T. Press.

MacDonald, G. J. F., and Ness, N. F., 1960. Stability of Phase Transitions Within the Earth. *J. Geophys. Res.* 65:2173–2190.

Matthews, R. K., 1969. Tectonic Implications of Glacio-Eustatic Sea Level Fluctuations. *Earth and Planetary Science Letters.* 5:459–462.

McConnell, R. K., 1965. Isostatic Adjustment in a Layered Earth. *J. Geophys. Res.* 70:5171–5188.

———, 1968a. Viscosity of the Earth's Mantle. *The History of the Earth's Crust,* R. A. Phinney, Editor. pp. 45–57. Princeton, New Jersey: Princeton University Press.

———, 1968b. Viscosity of the Mantle From Relaxation Time Spectra of Isostatic Adjustment. *J. Geophys. Res.* 73: No. 22:7089–7105.

McGinnis, L. D., 1968. Glacial Crustal Bending. *Bull. Geol. Soc. Am.* 79:769–776.

McKenzie, D. P., 1966a. The Shape of the Earth. Unpublished Ph.D. Thesis. Cambridge, Kings College.

———, 1966b. The Viscosity of the Lower Mantle. *J. Geophys. Res.* 71:3995–4010.

———, 1967a. Some Remarks on Heat Flow and Gravity Anomalies. *J. Geophys. Res.* 72:6261–6273.

———, 1967b. The Viscosity of the Mantle. *Geophys. J.* 14:297–305.

———, 1968. The Influence of the Boundary Conditions and Rotation on Convection in the Earth's Mantle. *Geophys. J.* 15:457–500.

Mesolella, K. J., Matthews, R. K., Broecker, W. S., and Thurber, D. L., 1969. The Astronomical Theory of Climatic Change: Barbados Data. *J. Geology.* 77:250–274.

Milliman, J. D., and Emery, K. O., 1968. Sea Levels during the Past 35,000 Years. *Science.* 162:1121–1124.

Moran, J. M., and Bryson, R. A., 1969. The Contribution of Laurentide Ice Wastage to the Eustatic Rise of Sea Level: 10,000 to 6,000 B.P. *Arctic and Alpine Research.* 1:97–104.

Morgan, W. J., 1972. Deep Mantle Convection Plumes and Plate Motions. *Am. Assoc. Pet. Geologists, Bull.* 56: #2:203–213.

Mörner, Nils-Axel, 1969. The Late Quaternary History of the Kattegat Sea and the Swedish West Coast, Deglaciation, Shorelevel Displacement, Chronology, Isostasy and Eustasy. *Sveriger Geologiska Undersökning.* Arsbok 63: Ser. C, No. 640, No. 3:404–453.

Morse, P. M., and Feshbach, J., 1953. *Methods of Theoretical Physics.* Vols. 1 and 2. New York: McGraw-Hill.

Müller, F., and Barr, W., 1966. Postglacial Isostatic Movement in Northeastern Devon Island, Canadian Arctic Archipelago. *Arctic.* 19:263–269.

Munk, W. H., and MacDonald, G. J. F., 1960a. Continentality and the Gravitational Field of the Earth. *J. Geophys. Res.* 65:2169–2172.

Munk, W. H., and MacDonald, G. J. F., 1960b. *The Rotation of the Earth.* 323pp. Cambridge: Cambridge University Press.

Nadai, A., 1950. *The Theory of Flow and Fracture in Solids.* Vol. II. 705 pp. New York: McGraw-Hill.

Newell, H. D., and Bloom, A. L., 1970. The Reef Flan and Two Meter Eustatic Terrace of Some Pacific Atolls. *Bull. Geol. Soc. Am.* 81:1881–1894.

Niskannen, E. 1939. On the Upheaval of Land in Fennoscandia. *Ann. Acad. Sci. Fennicae.* 53: Series A:1–30.

———, 1943. On the Deformation of the Earth's Crust Under the Weight of a Glacial Ice-Load and Related Phenomena. *Ann. Acad. Sci. Fennicae.* 7: Series A: 1–59.

———, 1948. On the Viscosity of the Earth's Interior and Crust. *Ann. Acad. Sci. Fennicae.* 15: Series A:1–28.

———, 1949. On the Elastic Resistance of the Earth's Crust. *Ann. Acad. Sci. Fennicae.* 21: Series A:1–23.

Nye, J. F., 1952. A Method of Calculating the Thickness of the Ice Sheets. *Nature.* 169:529–530.

———, 1959. The Motion of Ice Sheets and Glaciers. *J. Glaciology.* 3:493–507.

———, 1963. The Response of a Glacier to Changes in the Rate of Nourishment and Wastage. *Proc. Roy. Soc., London.* 275: A:87–112.

O'Connell, R. J., 1969. Dynamic Response of Phase Boundaries in the Earth to Surface Loading. Unpublished Ph.D. Thesis, California Institute of Technology.

———, 1971. Pleistocene Glaciation and the Viscosity of the Lower Mantle. *Geophys. J.* 23:299–327.

O'Connell, R. J., and Wasserburg, G. J., 1967. Dynamics of the Motion of Phase Change Boundary to Changes in Pressure. *Rev. Geophys.* 5:329–410.

Parsons, B. D., 1972. Changes in the Earth's Shape. Unpublished Ph.D. Dissertation, Downing College, Cambridge University.

Paterson, W. S. B., 1972. Laurentide Ice Sheet: Estimated Volumes during Late Wisconsin. *Rev. Geophys.* 10: #4:885–917.

Porath, H., 1971. Magnetic Variation Anomalies and Seismic Low Velocity Zone in the Western United States. *J. Geophys. Res.* 76:2643–2648.

Post, R. L., and Griggs, D. T., 1973. The Earth's Mantle: Evidence of Non-Newtonian Flow. *Science.* 181:1242–1244.

Press, F., 1972. The Earth's Interior as Inferred from a Family of Models. *The Nature of the Solid Earth,* E. C. Robertson, Editor. pp. 147–171. New York: McGraw-Hill.

Prest, V. K., 1957. Pleistocene Geology and Surficial Deposits. *Geology & Economic Minerals in Canada.* Canada Geological Survey, Economic Geology Series. #1. C. H. Stockwell, Editor. Chapter VIII, pp. 443–493. Ottawa: Edmond Cloutier.

———, 1969. Retreat of Wisconsin and Recent Ice in North America. *Geological Survey of Canada Map No. 1257A.* Department of Energy, Mines and Resources.

Rayleigh, J. W. S., 1937. *The Theory of Sound.* I: Chapter 10. London: Macmillan & Company.

Redfield, A. C., 1967. Postglacial Changes in Sea Level in the Western North Atlantic Ocean. *Science.* 157:687–692.

Redfield, A. C., and Rubin, M., 1962. The Age of Salt-Marsh Peat and Its

Relation to Recent Changes in Sea Level at Barnstable, Mass. *Proc. Nat. Acad. Sci. U.S.* 48:1728–1735.

Sauramo, M. R., 1939. The Mode of Land Upheaval in Fennoscandia During Late Quaternary Time. *Fennia.* 66:1–28.

———, 1958. Die Geschichte der Ostee. *Acad. Sci. Fennicae.* Ann: A:51.

Scholl, D., Craighead, F. C., and Stuiver, M., 1969. Florida Submergence Curve Revised: Its Relation to Coastal Sedimentation Rates. *Science.* 163: 562–564.

Schytt, V., Hoppe, G., Blake, W., and Grosswald, M.G., 1968. The Extent of the Würm Glaciation in the European Arctic. Internat. Assoc. of Scientific Hydrology, IUGG, General Assembly of Bern, 1967, Commission of Snow and Ice. Publication #79. pp.207–216.

Shepard, F. P. 1960. Rise of Sea Level along Northwest Gulf of Mexico. *Recent Sediments, Northwest Gulf of Mexico,* F. P. Shepard et al., Editors. 394pp. Tulsa: Am. Assoc. Petroleum Geologists.

———, 1963. Thirty-Five Thousand Years of Sea Level. *Essays in Marine Geology in Honor of K. O. Emery,* T. Clements, R. Stevenson, D. Halmos, Editors. 1–10. Los Angeles: University of Southern California Press.

———, 1970. Lagoonal Topography of Caroline and Marshall Islands. *Bull. Geol. Soc. Am.* 81:1905–1914.

Slichter, L. B., and Caputo, Michele, 1960. Deformation of an Earth Model by Surface Pressures. *J. Geophys. Res.* 65:4151–4156.

Stocker, R. L., and Ashby, M. F., 1973. On the Rheology of the Mantle. *Rev. Geophys.* 11:391–426.

Stuiver, M., 1971. Evidence for the Variation of Atmospheric C^{14} Content in the Late Quaternary. *The Late Cenozoic Glacial Ages,* K. K. Turekian, Editor. Chapter 4. New Haven: Yale University Press.

Stuiver, M., and Daddario, J. L., 1963. Submergence of the New Jersey Coast. *Science.* 142:951.

Takeuchi, H., 1963. Time Scales of Isostatic Compensation. *J. Geophys. Res.* 68: No. 8:2357.

Takeuchi, H., and Saito, M., 1962. Statical Deformations and Free Oscillations of a Model Earth. *J. Geophys. Res.* 67:1141–1154.

Takeuchi, H., and Hasegawa, Y., 1964. Viscosity Distribution within the Earth. *Geophys. J.* 9:503–508.

Tanner, W. F., Editor, 1968. Tertiary Sea Level Fluctuations. *Palaeogeography, Palaeoclimatology, Palaeoecology.* 5: Special Issue: No. 1:1–178.

Teng, Ta-Liang, 1968. Attenuation of Body Waves and the Q-Structure of the Mantle. *J. Geophys. Res.* 73:2195–2208.

Thompson, Sir William, and Tait, P. G., 1883. *Treatise on Natural Philosophy.* Vol. II. Cambridge: Cambridge University Press.

Trowbridge, A. C., 1921. Erosional History of the Driftless Ocean. *Univ. Iowa Stud. Natur. Hist.* 9:55–127.

Turcotte, D. L., and Oxburgh, E. R., 1967. Finite Amplitude Convection Cells and Continental Drift. *J. Fluid Mech.* 28:29–42.

————, 1969. Convection in a Mantle with Variable Physical Properties. *J. Geophys. Res.* 74:1458–1474.

Turcotte, D. L., and Schubert, G., 1971. Structure of the Olivine-Spinel Phase Boundary in the Descending Lithosphere. *J. Geophys. Res.* 76:7980–7987.

Turekian, K. K., Editor, 1971. *The Late Cenozoic Glacial Ages.* 606pp. New Haven: Yale University Press.

Van Bremmelen, R. W., and Berlage, H. P., 1935. Versuch Einer Mathmatischen Behandlung Geotektonischer Bewegungen Unter Besonderer Berücksichtegung der Undationstheorie. *Gerlands Beiträge zur Geophysik.* 43:19–55.

Veeh, H. H., and Chappell, J., 1970. Astronomical Theory of Climate Changes: Support from New Guinea. *Science.* 167:862–865.

Vening Meinesz, F. A., 1937. The Determination of the Earth's Plasticity from the Postglacial Uplift of Scandinavia: Isostatic Adjustment. *Koninklijke Akademie Van Wettenschapen Te Amsterdam.* 40: No. 8:654–662.

Verhoogen, J., 1965. Phase Changes and Convection Currents in the Earth's Mantle. *Phil. Trans. Roy. Soc. Lon.* 248: A:276–283.

Walcott, R. J., 1970a. Flexural Rigidity, Thickness, and Viscosity of the Lithosphere. *J. Geophys. Res.* 75:3941–3954.

————, 1970b. Isostatic Response to Loading the Crust in Canada. *Can. J. Earth Sci.* 7:716.

————, 1972a. Late Quaternary Vertical Movements in Eastern North America: Quantitative Evidence of Glacio-Isostatic Rebound. *Rev. Geophysics.* 10: #4:849–884.

————, 1972b. Past Sea Levels, Eustacy and Deformation of the Earth. *Quaternary Research.* 2:1–14.

————, 1973. Structure of the Earth from Glacio-Isostatic Rebound. *Annual Reviews of Earth and Planetary Sciences.* Fred A. Donath, Editor, pp. 15–37. Palo Alto: Annual Reviews Inc.

Wang, Chi-Yuen, 1966. Earth's Zonal Deformations. *J. Geophys. Res.* 71: 1713–1720.

Washburn, A. L., and Stuiver, M., 1962. Radiocarbon-Dated Postglacial Deleveling in Northeast Greenland and Its Implications. *Arctic.* 15:66–73.

Weertman, J., 1963. Rate of Growth and Shrinkage of Non-Equilibrium Ice Sheets. *J. Glaciology.* 5:145–159.

————, 1970. The Creep Strength of the Earth's Mantle. *Rev. Geophys.* 8: 145–168.

Wellman, H. W., 1964. Delayed Isostatic Response and High Sea Levels. *Nature.* 202:1322–1323.

West, R. G., 1968. *Pleistocene Geology and Biology.* pp. 185–213. New York: John Wiley & Sons.

White, R. E., 1971. P-Wave Velocities in the Upper Mantle Beneath the Australian Shield from Earthquake Data. *Geophys. J. R. Astr. Soc.* 24: 109–118.

Wiggins, R., and Helmberger, D. V., 1973. Upper Mantle Structure of the Western United States. *J. Geophys. Res.* 78:1870–1880.

Wilson, A. T., 1964. Origin of Ice Ages: An Ice Shelf Theory of Pleistocene Glaciation. *Nature.* 201: No. 4915:147–149.

Woldstedt, P., 1958. *Das Eiszeitalter.* Vol. I: 8–9. Vol. II: 134–166. Stuttgart: Ferdinand Enke Verlag.

Woollard, G. P., 1962. Crustal Structure in Antarctica. *Antarctic Research, The Matthew Fontain Maury Memorial Symposium.* pp. 55–73. Geophysical Monograph No. 7, NAS-NRC No. 1036, American Geophysical Union, Washington, D.C.

INDEX

For Calculated Isostatic Adjustment Index Table, see pp. 388–389.

373

	Worldwide Uplift					Peripheral Uplift					
	History	Total	Rate	Eustatic Sea Level	Ocean Basin adjustment	General	History	Rate	History	Rate	Abu Dhabi
Deep flow											
Spherical Coordinates (Model #1,1)*	114-115	136	256,258 *	139	136-137		220-230,231	236			260-261
Cartesian Coordinates									185 329-337	189	
Heuristic				140	193,217		219		193 323-326 329-337		
High Viscosity Lower Mantle											
Spherical Coordinates (Model #4)		137	257,259	141	136-137		222-230,235	238			
Cartesian Coordinates											
Increasing viscosity with depth											
Spherical Coordinates (Model #5)		137	257,259		136-137		222-230,234	239			
Cartesian Coordinates											
Non-adiabatic density gradients in upper mantle							237				
Spherical Coordinates (Model #2)		136	256,258		136-137		222-230,232				261
Heuristic											
Asthenosphere 155-173											
Spherical Coordinates (Model #3)		136	257,259		136-137		222-230,233	237			261
Cartesian Coordinates					169-170	159			185 323-337		
Heuristic											
Channel flow 156-160							209-211,213				
Cartesian Coordinates							168-170,186		186,336	178,189	
Heuristic	158-159					158	219				
Lithosphere 144-155											
Cartesian Coordinates							169		185,186 329-337		
Heuristic		147,149									
Elastic											
Spherical Coordinates											
Cartesian Coordinates						331-333	209-211		185,186 331-337		
Heuristic						148-150					

* For model definitions see Table IV-1, pg. 110.

Canada Uplift				Bonneville	Greenland	Fennoscandian Uplift					Harmonic Spectra
History	Central History	Rate	Central Rate			History	Central History	Rate	Central Rate	Strand-line Data	
241-248	181,240	251-253	250				194,195 197		194	181	105,181
						185	187	189		181	59,181
217-219						329-337					75,86,98
						193,321					
						323-326					
						327,329-337					
	240	254	250				194		194		106
	240		250				194		194		107
											44,45
	240		250				194		194		108
											88,90,100
	240		250				194		194		106
				161-163,171	163-165	185,329-337				181	181
											91
209-211	212										
165,168 219						186,336 321	187	187		183	183
				151,161,181	163-5,181	151,185-186				181,183	181,183
					154	329-337					53
											80,304-305
209-211				150		185,186 331-337					
217				150							44,45,59,80

Library of Congress Cataloging in Publication Data

Cathles, Lawrence M   1943–
  The viscosity of the earth's mantle.

  Bibliography: p.
  1. Earth—Mantle. 2. Isostasy. 3. Glacial epoch.
I. Title.
QE511.C36 1975   551.1′2   74-16162
ISBN 0-691-08140-9